AF816328

Progress in Botany

Volume 70

Ulrich Lüttge • Wolfram Beyschlag
Burkhard Büdel • Dennis Francis
Editors

Progress in Botany 70

Springer

Editors
Professor Dr. U. Lüttge
TU Darmstadt, Institut für Botanik
FB Biologie (10)
Schnittspahnstrasse 3–5
64287 Darmstadt
Germany
luettge@bio.tu-darmstadt.de

Professor Dr. W. Beyschlag
Fakultät für Biologie
Lehrstuhl für Experimentelle Ökologie
und Ökosystembiologie
Universität Bielefeld
Universitätsstrasse 25
33615 Bielefeld
Germany
w.beyschlag@uni-bielefeld.de

Professor Dr. B. Büdel
TU Kaiserslautern
FB Biologie
Abt. Allgemeine Botanik
Erwin-Schrödinger-Str.
Gebäude 13/2
67663 Kaiserslautern
Germany
buedel@rhrk.uni-kl.de

Dr. D. Francis
University of Cardiff
Cardiff School of Biosciences
Cardiff
United Kingdom CF10 3TL
francisd@cardiff.ac.uk

ISBN 978-3-642-08799-8 e-ISBN 978-3-540-68421-3

Progress in Botany ISSN 0340–4773

Softcover reprint of the hardcover 1st edition 2009

Cover design: WMXDesign GmbH, Heidelberg

Printed on acid-free paper

9 8 7 6 5 4 3 2 1

springer.com

Contents

Contributors

Michael F. Allen
Center for Conservation Biology, Department of Plant Pathology,
University of California, Riverside, CA 92521-0334, USA
michael.allen@ucr.edu

Edna Antony
Water Technology Centre for Eastern Region, Chandrasehkarpur,
Bhubangewar 751023, India

Steven R. Archer
School of Natural Resources, University of Arizona, Tucson, AZ 85721, USA

Maaike Y. Bader
Functional Ecology of Plants, Institute of Biology and Environmental Sciences,
University of Oldenburg, P.O. Box 2503, 26111 Oldenburg, Germany

Anne M. Borland
Institute for Research on Environment and Sustainability, Newcastle University,
Newcastle-upon-Tyne NE1 7RU, UK
a.m.borland@ncl.ac.uk

Dennis Francis
School of Biosciences, Cardiff University, Cardiff CF10 3TL, UK
francisd@cardiff.ac.uk

Burkhard Frenzel
Institut für Botanik und Botanischer Garten 210, Universität Hohenheim,
D-70593 Stuttgart, Germany
bfrenzel@uni-hohenheim.de

Sebastian Gayler
Institute of Soil Ecology, Helmholtz Center Munchen, German Research Center
for Environmental Health

Hans-Walter Heldt
Albrecht-von Haller-Institut für Pflanzenwissenschaften,
Justus-von-Liebig-Weg 11, D37077 Göttingen, Priv., Ludwig-Beck-Str.5,
D37075 Göttingen, Germany
HansWalterHeldt@aol.com

Marian Kazda
Institute of Systematic Botany and Ecology, Albert-Einstein-Allee 11,
D-89081 Ulm, Ulm University, Germany
marian.kazda@uni-ulm.de

David Lloyd
Microbiology (BIOSI1), Main Building, Cardiff University, P.O. Box,
Cardiff CF10 3TL, Wales, UK
lloydd@cardiff.ac.uk

Miroslav Nikolic
Center for Multidisciplinary Studies, University of Belgrade,
SRB-11030 Belgrade, Serbia

Vic Norris
Laboratory AMMIS, University of Rouen, F-76821 Mont-Saint-Aignan Cedex, France

Eckart Priesack
Institute of Soil Ecology, Helmholtz Center München, German Research Center
for Environmental Health
priesack@gsf.de

Camille Ripoll
Laboratory AMMIS, University of Rouen, F-76821 Mont-Saint-Aignan Cedex,
France

Volker Römheld
Institute for Plant Nutrition (330), University of Hohenheim,
D-70593 Stuttgart, Germany

Iris Schmid
Institute of Systematic Botany and Ecology, Albert-Einstein-Allee 11,
D-89081 Ulm, Ulm University, Germany

Thorsten Seidel
Department of Biochemistry and Physiology of Plants, W5,
University of Bielefeld, 33501 Bielefeld, Germany
thorsten.seidel@uni-bielefeld.de

Michel Thellier
Laboratory AMMIS, University of Rouen,
F-76821 Mont-Saint-Aignan Cedex, France
Michel.Thellier@univ-rouen.fr

Heather L. Throop
Department of Biology, New Mexico State University, Las Cruces,
NM 88003, USA
throop@nmsu.edu

Gerhard Zotz
Functional Ecology of Plants, Institute of Biology and Environmental Sciences,
University of Oldenburg, P.O. Box 2503, 26111 Oldenburg, Germany
gerhard.zotz@uni-oldenburg.de

Review

Curriculum Vitae

Professor Emeritus at the Department of Plant Sciences, Albrecht-von-Haller-Institute for Plant Sciences, Georg-August-University of Göttingen, was born 3 January 1934 in Berlin

1954–1955	Study of chemistry at the University Innsbruck (Austria),
1955–1960	Continuation of study at the University of Marburg (Germany)
1957–1958	Stipend from the Adolf Todt Foundation for half a year study at the Institute of Chemistry, University of Edinburgh, Scotland
1961	Graduation with the degree of a Diplom Chemiker
1960	Married to Fiona Stewart, M.A., three sons
1962	Promotion Dr. Phil. at the University of Marburg

Thesis: Phosphate containing metabolites in tissues and in isolated mitochondria from rat and pigeon. (Institute of Physiological Chemistry) Prof. Theodor Bücher, Prof. Martin Klingenberg

1962–1968	Research Assistant at the Institute of Physiological Chemistry, University Marburg
1968–1976	Senior Researcher at the Institute for Physiological Chemistry and Physical Biochemistry, University of Munich
1976–1978	Professor at the Institute for Physiological Chemistry and Physical Biochemistry, University of Munich
1978	Professor of Biochemistry and Director of the Division of Plant Biochemistry, University of Göttingen
2002	Professor Emeritus of Biochemistry, University of Göttingen

Honors, Awards

1980	Miller Professor University Urbana Ill.
1982	Research Fellow of the Royal Society, University of Sheffield, ARC
1990	Elected Member of the Akademie der Wissenschaften zu Goettingen
1993	Max-Planck-Research Prize of the Alexander-von Humboldt Stiftung and the Max-Planck-Gesellschaft
1996	Corresponding Membership Award of the American Society of Plant Physiologists also: Corresponding Membership Award of the Australian Society of Plant Physiologists
2002	Invited Guest Professor Universities of Delhi and Hyderabad (India)
2002	Highly Cited Researcher, Institute for Scientific Information (ISI)

Textbook

Pflanzenbiochemie (Spektrum-Verlag 1. Auflage 1996, 4. Auflage, 2008) Plant Biochemistry (Elsevier Academic Press, USA 2004), also Japanese, Chinese and Indian editions, Russian edition in preparation.

Activities

2000–2006	Representative of the Union of German Academies of Science in the InterAcademic Panel (IAP), Coordinator of the IAP Initiative on Genetically Modified Plants.

From Liver to Leaves: Memories of a Plant Biochemist

H.-W. Heldt

Contents

Abstract This is a report of more than 40 years of my own work and that with my group, beginning in Marburg in the group of Martin Klingenberg in the institute of Theodor Bücher, continuing in Munich in the institute of Martin Klingenberg and finally in Goettingen.

A wide spectrum of sophisticated methods developed in our group was used to study the metabolism of liver as well as leaves, ranging from the discovery of the mitochondrial ATP/ADP translocator to the finding of a number of chloroplast translocators. It was followed by studies of the regulation of the Calvin Cycle, of starch and sucrose synthesis, the role of mitochondrial oxidative phosphorylation in photosynthesis, the redox transfer within a leaf cell, assay of metabolite gradients in C_4 plants and the relationship between subcellular metabolite concentrations in intact leaves and in the phloem sap. These results were made possible by a very fruitful collaboration of our team and many colleagues from Germany and abroad.

1 Liver

At school, chemistry was my favourite subject, and it had always been my intention to study chemistry in order to become a chemist working in industry. I studied pure chemistry at the universities of Innsbruck (Austria), Edinburgh (Scotland) and Marburg. During my preparation for the diploma exam I came across a section on biochemistry in Klages's textbook on organic chemistry. At the time, biochemistry

H.-W. Heldt
Albrecht-von Haller-Institut für Pflanzenwissenschaften, Justus-von-Liebig-Weg 11, D37077 Goettingen, Priv., Ludwig-Beck-Str.5, D37075 Goettingen, Germany
e-mail: HansWalterHeldt@aol.com

U. Lüttge et al. (eds.), *Progress in Botany 70*,
DOI:10.1007/978-3-540-68421-3, © Springer-Verlag Berlin Heidelberg 2009

was not taught to chemistry students. What I read about ATP, the glycolytic and citric acid cycle fascinated me and I made up my mind to write my diploma thesis on a biochemical topic which in Marburg was then only possible in the Institute of Physiological Chemistry at the Medical Faculty. I needed the consent of Prof. Karl Dimroth, the head of the department of organic chemistry, which he granted, but asking me: "What do you want to do with this later?"

It was an extremely fortunate choice and when I started my work in 1959 at the Institute of Physiological Chemistry in Marburg, it was probably one of the most modern in the world. Its director, Prof. Theodor Bücher, had been working in Otto Warburg's laboratory of the Kaiser-Wilhelm-Institut in Berlin when he discovered the glycolytic enzyme phosphoglycerate kinase. This was only published after the war (Bücher 1947). He then had a laboratory at the Eppendorf University clinic in Hamburg, where, together with Dr. Netheler, he designed the legendary Eppendorf photometer to be used as a tool to carry out enzymatic assays according to Otto Warburg, where reactions are coupled to pyridine nucleotide dehydrogenases. He established an enzymic assay of ethanol based on alcohol dehydrogenase, which became very important for identifying drunk drivers. He also worked out a protocol to isolate five glycolytic enzymes in one process from rabbit muscle. This procedure was applied by the Boehringer Company, marking its start as a supplier of biochemicals, and opened up the prospect of using enzymatic assays as a routine method.

In 1953, Theodor Bücher was appointed professor in Marburg and transformed the erstwhile small division into the large Institute of Physiological Chemistry with completely modernised laboratories. He designed the institute to facilitate the study of metabolism from many different angles by a number of research groups led by senior researchers. The group of Rudolf Czok isolated enzymes, that of Dirk Pette (1965) assayed the proportions of the activities of a large number of metabolic enzymes in different tissues, that of Hans Jürgen Hohorst assayed metabolite levels in freeze-stopped tissues (Hohorst et al. 1959), Hans Schimassek (1963) performed metabolic studies with perfused rat liver, Roland Kirsten and his wife performed amino acid analyses in tissues with a novel automated microscale amino acid analyser designed in the institute (Kirsten and Kirsten 1962), Klaus Papenberg assayed the nucleotide content of human liver biopsy samples by ion exchange chromatography (Schnitger et al. 1959) and last but not least Dr. Klingenberg studied the respiratory metabolism of isolated mitochondria (Klingenberg and Slenczka 1959). Later, many of these persons became full professors at various universities. One very important person was Hans Schnitger, an ingenious inventor of scientific instruments. As the ingredients of enzymatic analyses were valuable, the reaction volumes had to be reduced, which led to the invention in the institute of the so-called microliter system. Hans Schnitger designed the Eppendorf pipette, which was built in our workshop and which we used long before it became commercially available. The plastic Eppendorf vessel, now used in almost every laboratory in the world, was developed. In a way, Marburg was then a birthplace of modern biochemistry.

It was the aim of Theodor Bücher to utilise the knowledge of biochemical studies in medicine. Many prominent medical professors worked in his institute to learn biochemical methods or cooperated with him. He cooperated closely with the married couple Dr. Schmidt and Dr. Schmidt in Kassel, pioneers in the introduction of the measurement of enzyme activities in blood samples, nowadays a medical routine. I explain all this in so much detail, since for me as a chemistry student with no knowledge whatsoever of biochemistry or physiology, the scientific environment of the Marburg institute was of crucial importance for my later work.

Prof. Bücher assigned me to Klaus Papenberg, who used a microscale ion exchange chromatography apparatus, constructed by Hans Schnitger and built in the workshop of the institute, to analyse nucleotide patterns in human liver biopsy samples of about 40 mg (Schnitger et al. 1959). The ion exchange column was a 200-cm long plastic tube with an inner diameter of 0.5 mm and filled with Dowex anion exchanger. It was a forerunner of high pressure liquid chromatography. The fluid was initially pumped by a peristaltic pump and later by a high pressure piston pump designed once more by Hans Schnitger. A chromatogram took about 7 days, during which about 500 samples of 0.12 ml each were collected in Teflon racks. Each sample had to be transferred to a microcuvette for the measurement of UV-absorption by a Beckmann DU photometer, and transferred back into the rack, where the samples were dried and digested by acid for phosphate determination by the molybdate method, where each sample had to be assayed again photometrically. For this reason, the whole method was very time consuming and required the assistance of a technician. Later, the ultraviolet measurements were carried out with a micro-flow cuvette. It was my job to run the chromatographies and to identify the peaks of the ultraviolet and phosphorous measurements. The whole method was designed to analyse the nucleotide pattern of extracts from liver biopsies as a potential diagnostic tool for liver diseases. I sometimes accompanied Klaus Papenberg to a hospital in Kassel to hold the vessel with liquid nitrogen for fixing a liver biopsy from patients in an operation theatre. These biopsies were taken in addition to those taken for normal diagnostic purposes. Later, however, it was realised that the analytical approach was not viable, because the 5-s between the withdrawal of the biopsy and the fixing were enough to distort the nucleotide pattern. Unfortunately, one day on his way from Kassel to Marburg Klaus, Papenberg had a serious car accident and never returned to the laboratory. I, an inexperienced diploma student, was left with a laboratory and a technician. As nobody told me what to do, I decided, using the available technique, to analyse the content of nucleotides and other metabolites containing phosphate in rat liver and presented these data in my diploma thesis (Heldt 1963). I presume Prof. Bücher was satisfied with my work, since I had described the entire nucleotide outfit of liver tissue.

I was able to retain my laboratory, consisting of two rooms, a technician and with a great view of the Elisabeth Church. As I received no further instruction, but had access to all the resources, it was more or less up to me how to continue these studies for my Ph.D. thesis. The work often kept me occupied late into the night. That was when I became acquainted with Martin Klingenberg, who had his laboratory at the other end of the building. Martin Klingenberg had been a post-doctoral student with Britton Chance at the Johnson Foundation in Philadelphia, USA, where he used a dual-wavelength spectrophotometer to study cytochromes and by which he discovered cytochrome P450. When coming to Marburg, Martin Klingenberg built a dual-wavelength spectrophotometer for himself and used it for his fundamental studies of the function of the mitochondrial respiratory chain. One of his achievements was the unveiling of the relationship between the redox state of the mitochondrial respiratory carriers and the phosphorylation potential of the generated ATP as assayed with isolated mitochondria (Klingenberg and Schollmeyer 1960). Theodor Bücher and he published the legendary report "Wege des Wasserstoffs in der lebendigen Organisation" in which the different redox potentials of NADH and NADPH in the mitochondria and the cytosolic compartment were characterised and their function defined (Bücher and Klingenberg 1958).

During these long nocturnal discussions, Martin Klingenberg became my mentor and directed my interest towards mitochondria. With my chromatography, I assayed nucleotides and other phosphorylated metabolites contained in whole tissues, isolated mitochondria from rat liver, rat heart and pigeon breast and determined how much of the total metabolites in these tissues were located in the mitochondria. In order to learn about the turnover of the mitochondrial metabolites, I briefly incubated isolated mitochondria with ^{32}P-phosphate and assayed the incorporation by attaching to the chromatography apparatus a flow detector for ^{32}P- radioactivity which I had built for myself (Heldt and Klingenberg 1965). I incorporated all these results in my Ph.D. thesis. Officially, Theodor Bücher was my "Doctor Father", but in actual fact it was Martin Klingenberg.

Upon finishing my Ph.D. it would have been normal for me to go into industry, where the prospects for a career were then very good. To my great surprise, Martin Klingenberg asked me if I would not like to stay and I very gladly agreed. So I became a member of the Klingenberg group, but with my laboratory at the opposite side of the building, and with my own technician. Hans Jacobs, a medic seeking experience in biochemistry, joined me for a longer period. Chromatographic results had shown that creatine phosphate was radioactively labelled in isolated mitochondria. This was surprising since at the time creatine kinase was known as a cytosolic enzyme. From these studies, we discovered a mitochondrial creatine kinase isoenzyme different from the cytosolic one (Jacobs et al. 1964).

My investigations had shown that mitochondria contain endogenous nucleotides, which cannot be washed out. The question arose: how are these mitochondrial adenine nucleotides able to communicate with those outside? To study the kinetics of the incorporation of ^{32}P into the endogenous and external ATP of isolated mitochondria, I constructed a reaction vessel with an outflow at the bottom, which was closed by a programmable electric valve, and attached to a manually driven sample collector containing reaction vessels with perchloric acid for a metabolic quench. The mitochondria were kept in a deenergised state by incubating them in an anaerobic medium, and the reaction was started by blowing oxygen together with ^{32}P-phosphate into the stirred mitochondrial suspension, of which samples were taken at intervals of seconds. These studies showed that the internal ADP was more rapidly phosphorylated than the external ADP, especially at low temperatures (Heldt et al. 1965; Heldt and Klingenberg 1968). The inhibitory effect of atractyloside was particularly interesting in these experiments. It has been known for over 2,000 years that the thistle *Atractylis gumnifera* growing in Mediterranean areas contains a deadly poison, and in the nineteenth century atractyloside was isolated as the poisonous agent. Biochemical investigations carried out much later indicated that atractyloside inhibited mitochondrial respiration at the site of mitochondrial ATP synthesis (Bruni et al. 1962). To our surprise, our experiments showed clearly that atractyloside did not affect the phosphorylation of the mitochondrial internal ADP, but inhibited only the phosphorylation of external ADP (Heldt et al. 1965). Our results demonstrated that in mitochondrial oxidative phosphorylation a translocation step is involved, catalysing the transfer of ATP out of the mitochondria. These results were in line with experiments carried out by Erich Pfaff, who had established in the Klingenberg laboratory the method of silicon layer filtering centrifugation where a mitochondrial suspension is placed in a centrifuge tube above a layer of silicon oil. Upon centrifugation the organelles migrate out of the suspension medium through the silicon layer into a layer of perchloric acid for quenching. Erich Pfaff observed in isolated mitochondria an exchange of endogenous and external adenine nucleotides which was inhibited by atractyloside (Pfaff et al. 1965, 1969; Heldt and Pfaff 1969). For me, a young post-doctoral student, to be involved in the discovery of the ATP/ADP translocator was a great boost.

I am very grateful to Martin Klingenberg for promoting my participation at international conferences and for giving me the opportunity to discuss our results with leading scientists. I attended the International Biochemistry Congress in New York 1964. This was an exciting time for research on mitochondria. At that time, it was believed that there would be a phosphorylated intermediate in mitochondrial oxidative phosphorylation. It was a sensation when Webster and Green announced that they had found it, but at the New York Congress David Green recalled these findings as they had been forged. With great clamour, Paul Boyer claimed phosphohistidine as an intermediate of oxidative phosphorylation (Peter et al. 1963). I checked this by measuring phosphorylation kinetics and found that Boyer was mistaken; phosphohistidine was in fact an intermediate of substrate level phosphorylation by succinic thiokinase (Heldt 1967). A satellite meeting named "Compostium" on Britton Chance's farm, where the top people of mitochondria were discussing research, was very thrilling. Pallade presented beautiful pictures of the ultrastructure of mitochondria and there were discussions about the small particles which Sjostrand had found inside the mitochondrial inner membrane, later to be identified as F_1 particles of the ATP synthase.

One year later, I was invited as a speaker at the first Bari Symposium. This meeting was very generously hosted by Ernesto Quagliariello. Along with the world experts on mitochondrial metabolism some young scientists had been invited as speakers to this 6-day meeting. I was able to present our results on the function of the mitochondrial ATP/ADP translocator (Heldt 1967), and had lively disputes with prominent scientists opposing this view, where in the end I was proved right. This was a great experience for me, a relatively young post-doctoral student. At this Bari meeting, I first heard Peter Mitchell speaking about the chemiosmotic theory, followed by very animated discussions with the opponents of his theory, for example

Lars Ernster, Britton Chance and Bill Slater. I also witnessed these discussions at the Round Table Conference in Poligiano el Mare in 1966, hosted by Ernesto Quagliariello, where I was also an invited speaker. The problem with the presentations of Peter Mitchell was that he proposed very detailed mechanisms on how a proton gradient is formed by mitochondrial electron transport and how this gradient is utilised to synthesise ATP. All these proposed mechanisms turned out to be more or less wrong, but the general concept of a proton gradient as an intermediate of mitochondrial phosphorylation remained. In later years, I was generously invited a number of times as a speaker to so-called Bari meetings taking place in various towns in Apulia. I am very grateful for the great hospitality of Ernesto Quaglioriello, Fernando Palmieri and other Italian colleagues involved.

I have already mentioned that we had ruled out phosphohistidine as an intermediate of mitochondrial oxidative phosphorylation, as proposed by Paul Boyer; instead we found it to be an intermediate of substrate level phosphorylation via succinic thiokinase. Kinetic measurements carried out with Hans Jacobs showed that endogenous mitochondrial GDP was phosphorylated to GTP by this substrate level phosphorylation (Heldt et al. 1964). We had also found that endogenous guanine nucleotides are confined to the mitochondrial matrix; they are not transported by the ATP/ADP translocator. Experiments by Klaus Schwalbach revealed that the newly formed GTP transferred its terminal phosphate residue not to ADP, as one might have expected, but to AMP via a mitochondrial GTP-AMP-phosphotransferase which we were able to identify (Heldt and Schwalbach 1967).

Herrick Baltscheffsky from Stockholm had observed that chromatophores of the photosynthetic bacterium *Rhodospirillum rubrum* incorporated radioactively labelled phosphate upon illumination. He visited us and asked me to identify which endogenous nucleotide is the acceptor of the labelled phosphate. My chromatographic analysis revealed that the labelled product was not a nucleotide but inorganic pyrophosphate. This showed for the first time that in photosynthetic bacteria besides ATP inorganic pyrophosphate can also be a product of photophosphorylation (Baltscheffsky et al. 1966), and opened up a new field of research which occupied Baltscheffsky's group for many years.

In the meantime, Theodor Bücher had left Marburg to become director of the Institute of Physiological Chemistry at the Medical Faculty of the University of Munich. His successor in Marburg was Peter Karlson. I got a tenured position with administrative duties. Martin Klingenberg was promoted to a professorship and attempts were made to induce him to stay in Marburg. He was, however, offered chairs at the universities of Munich, Constance and a directorship at a Max-Planck Institute in Göttingen. It was great experience for me when he asked me to assist him in his negotiations by compiling lists of the necessary equipment and summing up the required outlay. I accompanied him on his negotiations in Göttingen and Constance. In both places, the sites where the institutes were to be built were still green fields. At that time, the University of Constance was situated in the ballroom of the Insel Hotel, which screens divided into small compartments for professors' offices. From the ballroom balcony you could have a good view of the whole university with the professors at work.

My assistance in these negotiations taught me how to tackle such matters and this stood me in good stead later when I was offered a chair in Göttingen and was confronted with similar negotiations. Martin Klingenberg finally decided to accept the offer of a chair of Physical Biochemistry at the University of Munich next to the chair of Physiological Chemistry of Theodor Bücher. My job was to order the equipment required and I was very much involved in organising the removal of the laboratory from Marburg to Munich.

When we finally moved into the new building I was provided with a spacious laboratory for my group. Martin Klingenberg gave me full independence in my scientific work; it was up to me to collaborate with him on equal terms. I found this very generous and had many stimulating discussions with him resulting in several fundamental publications. Of course my administrative duties still remained. As I was responsible for financial matters, I had to check all the bills before they were transferred to the accounting office. Other administrative duties were carried out by Achim Kröger and Gebhard von Jagow. We were, as good friends, an excellent team. This was all good training for later in Göttingen. In Munich, I received my "habilitation" – venia legendi – at the medical faculty with the title of "Privatdozent" and was busy giving practical laboratory courses, lecturing and examining medical students on their "Physikum" after five semesters. These exams provided me with the opportunity to select some particularly able medical students and to offer them the chance to do their research for their medical doctor with me. This is how I came to have some really excellent students working in my laboratory.

When continuing the studies of the mitochondrial ATP/ADP translocator, I found that in isolated mitochondria in the course of oxidative phosphorylation the ATP/ADP ratio in the supernatant was very much higher than in the matrix. These studies led to the conclusion that in oxidative phosphorylation the energy of the proton gradient is not only required for the ATP-synthesis within the mitochondria, but also for the export of ATP into the cytosol against a concentration gradient (Heldt et al. 1972a). At first we met with great opposition since our findings questioned the existing models of the proton stoichiometry of photophosphorylation, and it took quite a while until it was generally accepted. The question remained, however, whether the differences in ATP ratios, which we observed with isolated mitochondria, actually existed in living tissues.

At some meeting or other I heard about the method of nonaqueous fractionation of frozen and freeze dried tissues, developed by Behrens (1938) to isolate nuclei and later employed by Heber (1957) to separate chloroplasts from leaf tissue. I decided to try this method for the analysis of the adenine nucleotide levels in the mitochondrial and extramitochondrial compartment of perfused liver and won over the medical student Peter Schmucker for this rather risky project. We visited Ulrich Heber in Düsseldorf to learn how he used the nonaqueous method to isolate chloroplasts. He purified the chloroplasts by several differential centrifugations. Back in Munich we decided to do it a bit differently. The haemoglobin-free perfused liver (the perfusion was done by Roland Scholz; Scholz et al. 1969) was freeze stopped by a pair of tongs cooled in liquid nitrogen, the frozen material freeze dried and thoroughly homogenised to fine particles of subcellular matter. In contrast to the

Heber method, we separated the homogenate by a single centrifugation with a density gradient of heptane and tetrachlorocarbon or trichloroethylene. In this gradient the mitochondrial material was mainly in the top region. The gradient was divided into six fractions which were collected. Extracts were prepared from each fraction for measurement of metabolite contents and of the activity of mitochondrial and cytosolic marker enzymes. The metabolite contents and marker enzyme activities in the different fractions enabled the extrapolation of mitochondrial and cytosolic metabolite levels. Not surprisingly, this experimental approach turned out to be extremely difficult. It required the work of two other collaborators, Sybille Soboll and Rembert Elbers, before this method was sufficiently reliable to present for the first time data on subcellular metabolite contents in living liver tissue. We also found that in living liver tissue the ATP/ADP ratio in the cytosol was much higher than in the mitochondria, confirming our concept that the export of ATP from the mitochondria required an active transport driven by the proton gradient (Elbers et al. 1974; Soboll et al. 1978). The same method was used to investigate the subcellular distribution of di- and tricarboxylates in living liver tissue (Soboll et al. 1980).

I also had a close cooperation with Hans Rafael from Heidelberg, who visited us several times to study the function of guanine nucleotides in the regulation of the thermogenetic activity of brown fat mitochondria (Rafael et al. 1972; Rafael and Heldt 1976).

Since our silicon filtering centrifugation was a suitable tool for studying metabolite transport, we tried to use it for studying transport across the plasma membrane of isolated animal cells. I went to the laboratory of Hans Schimassek in Heidelberg to learn how to prepare isolated liver cells, and with these cells Hartmut Baur characterised for the first time the transport of hexoses across the liver plasma membrane (Baur and Heldt 1977).

2 Leaves

Heinrich Strotmann investigated in my laboratory the role of phosphate containing metabolites in photosynthetic reactions of the alga *Chlorella pyrenoidosa* (Strotmann and Heldt 1969). Discussions with him made me curious as to how metabolites are transported into chloroplasts. Although I was then ignorant of the metabolism involved in photosynthesis, I realised that chloroplasts share similarities with mitochondria. In order to learn what was known about chloroplast transport I attended the first International Congress on Photosynthesis in Freudenstadt in 1968. A lecture by David Walker (1969) on the permeability of the chloroplast envelope was for me the most valuable information of this meeting. From oxygen electrode traces of isolated chloroplasts performing CO_2 fixation in the presence of various metabolites he concluded that the chloroplast envelope is readily permeable to phosphate, 3-phosphoglycerate and dihydroxyacetone phosphate. Similar conclusions had been reached by Ulrich Heber (Heber et al. 1967) and James Bassham (Bassham et al. 1968), but nothing was known about the mechanism of this permeability. I realized that here was some work for us. Three years later at the Second International Congress on Photosynthesis in Stresa (Italy) I was already an invited symposium speaker on the topic of chloroplast transport (Heldt et al. 1972b).

At the Freudenstadt congress I asked Dr. Cockburn, a collaborator of David Walker, for the protocol to isolate intact chloroplasts from spinach leaves. Back in Munich, I bought spinach at the market and made my first chloroplast isolations. It turned out that, as chloroplasts are larger than mitochondria, they were particularly suitable for transport measurements by silicon layer centrifugation. Naturally I first looked at ATP/ADP transport. The experiments showed that chloroplasts had indeed a specific ATP/ADP translocator, but unlike that of mitochondria, it was not inhibited by actractyloside, and its nucleotide specificity was different. It became quite obvious that the chloroplast transporter was not suited for the export of ATP generated by photosynthesis, but rather for the supply of chloroplasts with ATP from the cytosol during the dark period (Heldt 1969). Much later we also identified this translocator in pea root plastids (Schünemann et al. 1993). By now the primary structure of the plastidic ATP/ADP translocator had been determined by the group of Eckehard Neuhaus (Kampfenkel et al. 1995).

As the chloroplast envelope consists of two membranes, the question arose: in which one is the ATP/ADP translocator localised? To answer this question I sought the cooperation of Frieder Sauer, an electron microscopist in the Bücher group. Electron micrographs were made with isolated chloroplasts at different osmolarities of the suspension medium and, in parallel experiments, the permeability to radioactively labelled solutes such as GTP, sucrose, dextrane and H_2O was measured by silicon layer filtration. A comparison of the relative spaces obtained from the permeability measurements, and the planimetric evaluation of the spaces between the two membranes and of the stroma in the corresponding electromicrographs, revealed that the inner envelope membrane was the osmotic barrier of the chloroplasts and also the site of specific metabolite transport, whereas the outer membrane appeared to be unspecifically permeable to metabolites of low molecular weight, e.g. nucleotides and sugar phosphates (Heldt and Sauer 1971). Later, it was shown that the permeability of the outer envelope membrane was due to a pore-forming protein, a putative porine (Flügge and Benz 1984).

Next, I studied with the help of my technician Lynn Rapley the transport of radioactively labelled phosphate and phosphorylated metabolites into spinach chloroplasts (Heldt and Rapley 1970a, b). We observed a rapid uptake of phosphate and 3-phosphoglycerate, whereas the uptake of hexose-and pentose phosphate was low. The uptake of phosphate was competitively inhibited by triose phosphate and 3-phosphoglycerate, but not by 2-phosphoglycerate. We also investigated the uptake of carboxylates. These studies revealed that certain dicarboxylates, such as L-malate, oxaloacetate, 2-oxoglutarate, fumarate, succinate, aspartate and glutamate, are transported into the chloroplasts by specific counter exchange. Malonate, maleinate, citrate or monocarboxylates were not transported (Heldt and Rapley 1970b). These relatively simple experiments led to the discovery of the chloroplast triose phosphate–phosphate translocator and a specific dicarboxylate transport, which, as was realised later, is catalysed by several translocators with overlapping specificities. By chance, I had struck a vein of gold which almost immediately admitted me to the higher ranks of photosynthesis researchers. From that time, my laboratory in Munich worked in parallel on two different subjects, the study of liver metabolism as already described and the study of photosynthesis metabolism. It is a remarkable sign of the freedom I enjoyed for my research that it was possible to work on photosynthesis in a medical faculty with students of medicine.

The specifity of the triose phosphate translocator was further characterised by Karl Werdan (Werdan and Heldt 1972) and Rainer Fliege (Fliege et al. 1978). Ingo Flügge, who had studied biochemistry in Tübingen, joined my group to work on his Ph.D. and participated in these investigations. He was particularly interested in the protein nature of the triosephosphate–phosphate translocator and thus complemented our studies very effectively. He identified the translocator protein and characterised its substrate binding site in very detailed studies (Flügge and Heldt 1976, 1977, 1978, 1981, 1986). Later, with his own group in Cologne, he determined the primary structure of this and many other translocators (Weber et al. 2005). The dicarboxylate transport was characterised in detail by Karl Lehner (Lehner and Heldt 1978). Chong Ja Chon, a Korean lady, who had studied nutritional sciences in Germany, also participated in our studies. In the evenings, she wrote a book on Chinese cuisine and was helped by the crew with the difficult German language. This book, distributed by a prominent German publisher, was a bestseller for very many years. Thus, our laboratory did not only produce scientific information.

Together with Ulrich Heber we investigated the transport of monosaccharides into spinach chloroplasts. We discovered a specific translocator transporting several hexoses and pentoses, such as D-glucose and D-ribose, but essentially no L-glucose (Schäfer et al. 1977). The specificity of this transport turned out to be strikingly similar to the specificity of the glucose transporter of liver cells, which we had also identified (Baur and Heldt 1977).

The question arose: how does the inorganic carbon required for CO_2 fixation reach the chloroplasts? It had been discussed at that time that chloroplasts may have a bicarbonate translocator. Karl Werdan measured the uptake of radioactively labelled bicarbonate into isolated chloroplasts by means of silicon layer filtering centrifugation and showed that the uptake of bicarbonate was due to a diffusion of CO_2 across the envelope into the chloroplast stroma, where it equilibrated with bicarbonate as catalysed by carbonic anhydrase (Werdan et al. 1972).

Our results indicated that the accumulation of bicarbonate in the chloroplast stroma followed a pH gradient. It was known from isolated thylakoid preparations that photosynthetic electron transport results in a transport of protons into the thylakoid space. The question arose: to what extent the pH in the chloroplast stroma is affected by this proton transport? Karl Werdan developed a technique to determine the pH in the stroma and the thylakoid space of intact chloroplasts by measuring the uptake of dimethyloxazolidine dione (DMO, an anion, accumulating in alkaline environment) and of methylamine (a cation accumulating in acidic environment) by means of silicon layer filtering centrifugation, which allowed the evaluation of the pH in the stroma and thylakoid space of intact chloroplasts. Using this method, we showed for the first time that the change from darkness to illumination resulted in intact chloroplasts in the generation of a pH gradient across the thylakoid membrane of about 2.5 pH units and an alkalisation of the stroma by about 0.6 pH units (Heldt et al. 1973). The method found recognition. Gernot Falkner from Salzburg working with this method in our laboratory obtained very similar results with the blue green alga *Anacystis nidulans* (Falkner et al. 1974). Kurt Jungermann from Freiburg came to us to measure the intracellular pH in the bacterium *Clostridium pasteurianum* (Riebeling et al. 1973).

Further studies by Archie Portis from the USA revealed that the light dependent alkalisation of the stroma is accompanied by an increase of Mg^{2+} concentration by about 3 mM (Portis and Heldt 1976). We asked ourselves whether such changes of pH and Mg^{2+} concentration in the stroma might have a regulatory function on the enzymes of the Calvin cycle. In order to check this we altered the stromal pH by adding acetate, and found that the CO_2 fixation by illuminated chloroplasts did indeed depend to a great extent on the pH. The change of stromal pH which we had observed during the transition from darkness to illumination was sufficient to change the rate of CO_2 fixation from almost zero to maximal activity (Werdan et al. 1975). This result demonstrated that the alkalisation of the stroma occurring upon illumination has a regulatory function in switching on the Calvin cycle in the light.

In order to localise the site of these switches in the Calvin cycle, we needed a method to analyse metabolite levels in intact chloroplasts in the steady state of photosynthetic CO_2 fixation. Ross Lilley from Australia developed a suitable method. He illuminated chloroplasts in the presence of ^{32}P-phosphate and CO_2, and rapidly separated the chloroplasts from the suspension medium by silicon layer filtering centrifugation followed by quenching with acid. The resultant extract was subjected to ion exchange chromatography. To detect the metabolites ^{32}P, radioactivity was monitored in the effluent of the column (Lilley et al. 1977). This method was used to measure the levels of Calvin cycle intermediates in chloroplasts when the pH and the Mg^{2+} concentration in the stroma were altered. To lower the pH in the stroma we utilised a method devised by the group of Ulrich Heber (Purczeld et al. 1978). Our studies revealed that changes in the pH and the stromal Mg^{2+} concentration mainly affected the activities of the fructose- and sedoheptulose bisphosphatases, demonstrating that these two enzymes were important regulatory steps for turning on the Calvin cycle upon illumination (Flügge et al. 1980).

It turned out that these were not the only factors regulating the Calvin cycle. Daniel Arnon and Bob Buchanan (Buchanan et al. 1967) had found an activation of fructose-bisphosphatase by reduced ferredoxin. Later Bob Buchanan's group demonstrated that the ferredoxin dependent activation of fructose- and sedoheptulose bisphosphatase was mediated by a protein factor (Schürmann and Buchanan 1975), which was subsequently identified as thioredoxin. It had been a matter of debate whether this reductive activation observed from experiments with extracted enzymes represented a mechanism of light activation or was just a protective mechanism to circumvent the inactivation of the enzymes by oxidising agents. In order to check whether such light dependent reductive activation of enzymes actually occurred in intact chloroplasts, William Laing from New Zealand devised a method for measuring the kinetics of the activation and inactivation of Calvin cycle enzymes of chloroplasts. Samples were withdrawn from a chloroplast suspension subjected to a dark–light and light–dark transient and were injected into a medium where the chloroplasts were immediately lysed and subjected to an enzymic assay of 15–30 s duration. Using this method he was able to demonstrate that, after the onset of illumination, the activities of fructose- and sedoheptuolose bisphosphatases rose 20- and 30-fold, respectively, and rapidly decreased again when the light was turned off (Laing et al. 1981). The catalytic activities of the activated and inactivated enzymes depended strongly on the pH and Mg^{2+} concentration in the medium. Also, a large increase in the activity of phosphoribulokinase was found upon illumination. As there had been discussions about the validity of these results obtained with isolated chloroplasts, Wolfgang Wirtz in Göttingen subsequently devised a method to assay the activation state of Calvin cycle enzymes in whole leaves, by which the enzymes were extracted from pea leaves within 5 s followed by an immediate activity assay (Wirtz et al. 1982). He found that fructose- and sedoheptulose bisphosphatases and also phosphoribulokinase were very rapidly activated upon illumination. Later, Andreas Gardemann and Dieter Schimkat (Gardemann et al. 1982, 1986) found that the light activated fructose- and sedoheptulose bisphosphatase were under a feedback control of their products fructose-6-phosphate and sedoheptulose-7-phosphate, respectively, and both enzymes were inhibited by glycerate, an intermediate of the photorespiratory cycle. Moreover, phosphoribulokinase was found to be inhibited by 3-phosphoglycerate. The latter control appeared to be a device to shut down the formation of ribulose bisphosphate when its carboxylation yielded sufficient 3-phosphoglycerate. Together with George Lorimer, we found a phosphate requirement for the light activation of ribulose-1, 5-bisphosphate carboxylase in spinach chloroplasts (Heldt et al. 1978). Altogether these long-term studies of our group showed that the regulation of the Calvin cycle is very complex, involving light activation by reductive interconversion through thioredoxin, by changes of the Mg^{2+} concentration and pH in the stroma, and feedback control of the light activated enzymes. Juan Sanchez from Spain (Sanchez and Heldt 1990) and Burgi Riens (Riens and Heldt 1992) observed that nitrate reductase, the key enzyme of nitrate assimilation was also light-activated.

During the day, a large part of the photosynthate in a leaf is temporarily stored within the chloroplasts as starch, to be released in the following night. Experimenting with the isolated enzyme, Jack Preiss (Preiss et al. 1967) suggested that the rate of starch synthesis is controlled by the enzyme ADP glucose pyrophosphorylase, which he found to be activated by 3-phosphoglycerate (signalling a high level of photosynthate within the stroma) and inhibited by inorganic phosphate (P_i) (signalling deficiency). It remained to be checked whether these conclusions drawn from *in vitro* experiments were relevant to the regulation in intact chloroplasts. We tackled this problem in cooperation with the groups of David Walker in Sheffield and Ulrich Heber in Würzburg. Through additions to the medium we altered the ratio of PGA/P_i in the stroma and measured the rate of starch synthesis. We found a very good relationship between the PGA/P_i levels assayed in the chloroplasts and the rate of starch synthesis, which clearly supported the view of Jack Preiss about the regulation of starch synthesis (Heldt et al. 1977).

At the International Photosynthesis Congress in Reading (1977) I met Mark Stitt, who had just completed his Ph.D. with Tom ap Rees in Cambridge. He accepted a post-doctoral position in my group and this was the beginning of a long and very fruitful collaboration.

I was offered the chair in Biochemistry at the Biological Faculty of the University of Göttingen, which I accepted in 1978. It was my responsibility to give lectures to biology students on the entire field of biochemistry every week during both semesters, which I did with enjoyment until my retirement in 2002. My institution was located in the Botany building and the research topic designated "Biochemistry of Plants". I discontinued the studies on animal metabolism when I left Munich and this work was continued by my former collaborator, Sibylle Soboll, in Düsseldorf. The laboratory in Göttingen required extensive refurbishing which took 1½ years. I am grateful to Martin Klingenberg for allowing me to remain in my laboratory in his institute in Munich during this long transition period. In order to carry out my teaching responsibilities I commuted between Munich and Göttingen during this time. When we finally moved to Göttingen in 1980 everything was already in place. Due to the organisational talents and hard work of the crew, particularly of Ingo Flügge and Mark Stitt, the laboratory was fully operating one week after our arrival. Hans Rurainski and Klaus Peter Heise, collaborators of my predecessor Prof. Jacobi, were already present in Göttingen and continued their research independently. The whole staff contributed with great commitment to the complete reform of the three different laboratory courses which we had to offer to our undergraduate students. For this reason, all members of our group, including diploma and Ph.D. students, always had heavy teaching duties which they carried out enthusiastically. Through our teaching efforts we won a large number of very able graduate students who were the main scientific working force in our group.

Some of the activities in Göttingen have already been mentioned. One major line of our research was still the elucidation of the regulation of photosynthesis metabolism, and therefore we continued our studies of starch metabolism. At that time, starch breakdown was an enigma. Mark Stitt measured the starch breakdown in isolated intact starch-loaded spinach chloroplasts upon darkening and observed similar rates as in intact leaves. He found that during illumination a synthesis and degradation of starch occurred simultaneously (Stitt and Heldt 1981a). Upon darkening, phosphorolytic degradation delivering phosphorylated end-products, and a hydrolytic degradation delivering maltose, occurred in parallel to a similar extent, where the ratio between these two processes was found to be governed by the phosphate level (Stitt and Heldt 1981b). These were crucial results for later studies on starch degradation.

In Göttingen, we had a very lively joint biochemistry seminar organised by Kurt Jungermann and Walter Neupert of the Biochemistry Department of the Medical Faculty, Hans-Dieter Söling of the Department of Clinical Biochemistry and myself from Plant Biochemistry. This seminar was attended regularly by all our scientific members of the different laboratories. It was the legendary time when people from different branches of biochemistry could still communicate. Almost every week, we organisers would sit together with the invited speaker for dinner. On one of these occasions Hans-Dieter Söling told me that fructose-2.6-bisphosphate had been found in liver metabolism as a signal substance regulating fructose-1.6-bisphosphatase and fructose-6-phosphate kinase. As he was working with this substance, he suggested checking whether fructose 2.6 bisphosphate also had a regulatory function in plants. A subsequent experiment which Mark Stitt carried out with the Söling group was highly successful. They found that fructose 2.6 bisphosphate was indeed contained in spinach leaves and that it strongly inhibited the cytosolic fructose-1.6-bisphosphatase, whereas its effect on the chloroplastic enzyme was low (Stitt et al. 1982a). This result opened up the field for the detailed studies by Mark Stitt and others from our group to elucidate the function of fructose-2.6-bisphosphate in the regulation of sucrose biosynthesis in plants (Stitt et al. 1983, 1984a, b; Herzog et al. 1984). It turned out that Bob Buchanan in the USA had also discovered, through contact with investigators of liver metabolism, the role of fructose-2.6-bisphosphate in the regulation of cytosolic leaf metabolism. Instead of running a competition we exchanged our results with Bob, resulting in a joint publication of our two groups on this topic (Cseke et al. 1984).

So far, in order to study photosynthesis metabolism, we had determined metabolite levels in isolated chloroplasts. Under these conditions, interactions with the cytosol and other organelles are lost. For this reason, we attempted to assay metabolite levels of chloroplasts contained in plant leaf protoplasts. Wolfgang Wirtz devised a method where the protoplasts incubated within a centrifuge tube were disrupted by centrifugation through a nylon net, releasing intact chloroplasts which passed through a layer of silicon oil into perchloric acid while the remaining cytoplasmic components were retained above the silicon and simultaneously quenched by acid. Cross contamination was corrected for by assay of marker enzymes (Wirtz et al. 1980). This method has been employed to assay the metabolite levels in the chloroplast and extrachloroplast compartment during the induction phase of photosynthesis (Stitt et al. 1980). The understanding of the ATP metabolism of whole leaf cells required the assay of metabolite levels in the chloroplastic, mitochondrial and cytosolic compartment of leaf protoplasts. Ross Lilley, on another visit from Australia, devised a method where protoplasts were disrupted by forcing them through a net followed by membrane filtration through different filters by which fractions enriched in the subcellular components were obtained and quenched in acid within 0.1 s (Lilley et al. 1982). Using this method, the adenine nucleotide levels in the chloroplast stroma, the mitochondria and the cytosol have been assayed in wheat leaf protoplasts under various conditions. The ATP/ADP was found to be much higher in the cytosol than in the mitochondria, demonstrating that an ATP gradient is maintained by the mitochondrial ATP/ADP translocator in plants, as we had earlier observed with liver. The cytosolic ATP/ADP ratio was found to decrease in the light, which contradicted the widespread assumption that export of photosynthetically produced ATP from the chloroplasts leads to an increase in the cytosolic ATP/ADP ratio, which then inhibits oxidative phosphorylation by the mitochondria in the light (Stitt et al. 1982b). This conclusion was later confirmed when Silke Krömer and Mark Stitt observed that a selective inhibition of mitochondrial ATP synthesis inhibited sucrose biosynthesis of illuminated protoplasts, which refuted the existing dogma that the function of mitochondria in photosynthesising cells is mainly to supply ATP during the dark period, whereas during the light the ATP supply would be met by photophosphorylation (Krömer et al. 1988; Krömer and Heldt 1991a). Our results clearly demonstrated that during the light period the mitochondria and not the chloroplasts are mainly responsible for the supply of the cytosol with ATP, which has by now been fully accepted.

In order to study the metabolism of plant mitochondria, I visited Roland Douce in Grenoble to learn how to prepare mitochondria from plant leaves. Our subsequent investigations showed that plant leaf mitochondria, besides delivering ATP to the cytosol, have other important functions in plant metabolism. Holger Ebbighausen discovered that plant mitochondria, in contrast to the liver mitochondria I had earlier worked with, contained a specific oxaloacetate translocator enabling a transfer of redox equivalents from the mitochondria to the cytosol via a malate–oxaloacetate shuttle (Ebbighausen et al. 1985; Krömer and Heldt 1991b). Earlier, Hal Hatch from Australia on a visit to our laboratory found that chloroplasts also had a specific oxaloacetate translocator facilitating a malate–oxaloacetate shuttle (Hatch et al. 1984). Studies by Dieter Heinecke, Burgi Riens and others, employing subcellular metabolite analyses, as will be described later, evaluated the redox states of the various subcellular compartments (Heineke et al. 1991) and showed that the transfer of redox power via malate oxaloacetate shuttles, from the chloroplasts as well as from the mitochondria, was able to serve nitrate reduction in the cytosol and also hydroxypyruvate reduction-yielding glycerate, a partial step of the photorespiratory cycle in the peroxisomes. Later, Agepati Raghavendra from India demonstrated with a reconstituted system consisting of isolated mitochondria and peroxisomes from spinach leaves that a redox transfer between the mitochondria and the peroxisomes does indeed operate as part of the photorespiratory cycle. The addition of glycine, resulting in the generation of serine and of reducing equivalents by the mitochondria, led to the formation of glycerate by the peroxisomes (Raghavendra et al. 1998). Moreover, Iris Hanning showed that mitochondria provide the carbon skeletons for amino acid synthesis in the course of nitrate assimilation (Hanning and Heldt 1993).

We studied the function of peroxisomes as they play a role in the photorespiratory cycle. Holger Ebbighausen improved the known method for the preparation of intact peroxisomes from spinach leaves. Ralf Heupel, in cooperation with David Robinson of the Cell Biology Division in the Institute of Plant Physiology in Göttingen, found that the strict compartmentation of peroxisomal metabolism was not due to a boundary function of the surrounding membrane as in mitochondria and chloroplasts, but was caused by metabolite channelling in a multienzyme complex (Heupel et al. 1991). The transfer of low molecular weight metabolites such as malate and oxaloacetate was found to occur through porin-like channels, which Sigrun Reumann characterised in cooperation with Roland Benz from Würzburg by a lipid bilayer technique (Reumann et al. 1995, 1998). The same kind of porin-like channel was also found in glyoxisomes (Reumann et al. 1997).

Our research on metabolite transport continued. Siglinde Borchert found that amyloplasts from roots, in contrast to chloroplasts, contain a specific translocator transporting glucose-6-phosphate in counterexchange with phosphate (Borchert et al. 1989). This translocator turned out to be important for the starch metabolism in potato tuber amyloplasts (Schott et al. 1995). The extensive work which Ingo Flügge did in Göttingen in characterising the protein of the triosephosphate-phosphate translocator has already been mentioned. Together with Woo from Australia he studied the characteristics of glutamate and oxoglutarate transport (Woo et al. 1987a, b), and with Mark Stitt he found a light dependent transport of pyruvate into chloroplasts (Flügge et al. 1985). In cooperation with Enrico Martinoia and Ulrich Heber from Würzburg, he characterised an energy dependent transport of malate into isolated leaf vacuoles (Martinoia et al. 1985). With Juichi Ohnishi from Japan he studied the transport of metabolites into chloroplasts from bundle sheath and mesophyll cells of a C_4-plant (Ohnishi et al. 1989, 1990).

During mutual laboratory visits, our interest in C_4-metabolism was awoken through the increasing cooperation with the group of Hal Hatch in Australia. The concept of the NADP-type of C_4 metabolism implied a diffusion of malate through plasmodesmata from the mesophyll cells into bundle sheath cells, driven by a concentration gradient between these cells. This convincing assumption remained to be verified. Mark Stitt devised a method for the evaluation of metabolite concentration gradients between the two cell types of a C_4 plant. He achieved a partial separation of mesophyll and bundle sheath cells by homogenising freeze-stopped leaf material in liquid nitrogen followed by filtration through nylon nets with different aperture sizes. The assay of metabolite contents and marker enzyme activities in these fractions allowed an extrapolation of the concentrations in pure mesophyll and bundle sheath cells (Stitt and Heldt 1985). This method revealed that there was indeed a difference in the malate concentrations between the mesophyll and bundle sheath cells, amounting to 18 mM. Subsequently, together with Hal's group in Australia, the aperture of the plasmodesmata of bundle sheath cells was studied by Hendrik Weiner (Weiner et al. 1988) and Estela Valle from Argentina (Valle et al. 1989). Jointly, these results allowed an evaluation of the diffusive fluxes by Hal Hatch and his group which fully supported the concept of NADP-C_4-photosynthesis initially brought forward.

The photosynthesis metabolism in a plant leaf is the result of the interplay of the different organelles within a leaf cell. To understand this interplay it is necessary to know the metabolite concentrations in the different compartments. As dealt with before, we had assayed subcellular metabolite concentrations in plant protoplasts, but protoplasts have the disadvantage that they are "congested", as the products of photosynthesis cannot be exported from the cells. I described earlier how in Munich we developed a method based on nonaqueous fractionation of frozen and freeze-dried liver tissue in nonaqueous solvents to analyse subcellular metabolite contents, and how we learned about this method from Ulrich Heber's technique to isolate chloroplasts. In Göttingen, Richard Gerhard adapted our method for liver to spinach leaves which had been frozen in liquid nitrogen. After overcoming many difficulties, he established a reliable method to analyse subcellular metabolite levels in the chloroplastic, cytosolic and vacuolar compartments from intact leaves performing photosynthesis (Gerhardt and Heldt 1984). This method was very labour intensive, because in the 6–8 fractions obtained from the density gradient centrifugation of the homogenised dry cellular matter, not only the metabolites but also the activities of at least three marker enzymes had to be measured by enzymic assays. Since the samples were very small, these assays were performed with a highly sensitive dual wavelength photometer.

Each fractionation was repeated several times in order to obtain reliable data on metabolite contents. Because of the different volumes of the subcellular compartments the information of contents can be misleading. In order to express our data as concentrations we needed to know the volumes of the different subcellular compartments in a leaf. For this we started a very fruitful collaboration with David Robinson of the Division of Cell Biology. In his laboratory, Heike Winter prepared a very large number of electron micrographs from barley and spinach leaves from which she evaluated, by a stereological method, the volumes of the various subcellular compartments within an intact leaf (Winter et al. 1993, 1994). Later, Kirsten Leidreiter also performed these measurements with potato leaves (Leidreiter et al. 1995a). These studies were very labour intensive, but we were rewarded by gaining for the first time an insight into metabolite concentrations in the various subcellular compartments of a living plant. One interesting result of these studies was that, in the mesophyll of the three investigated plants from distant families, the relative sizes of the metabolic subcompartments and the distribution of metabolites between these compartments were quite similar (Leidreiter et al. 1995b). Obviously, photosynthesis metabolism in these diverse plants follows a basic concept. The measurements yielded a new insight into the distribution of metabolites within a photosynthesising cell. In contrast to expectations, the sucrose concentration in the vacuole was found to be much lower than in the cytosol, and amino acids turned out to be almost excluded from the vacuole. On the other hand, hexoses were found to be highly accumulated in the vacuole, keeping the cytosolic concentration extremely low. In many subsequent studies we used the subcellular metabolite analyses as a tool to investigate the regulation of sucrose biosynthesis. Moreover, Dieter Heinecke and many others in our laboratory employed this method, in cooperation with the group of Lothar Willmitzer, to characterise the metabolism in a large number of genetically modified cultivars in order to identify the effects of genetic engineering on the metabolism (Heineke et al. 1992, 1994a, b; Riesmeier et al. 1993; Büssis et al. 1997).

We became interested in the relationship between the metabolism in the photosynthesizing leaf cells studied so far and the export of the products of carbon and nitrate assimilation via the phloem system to the sink tissues of the plant. In the laboratory of Walter Eschrich of the Institute of Forest Botany in Göttingen, Hendrik Weiner and Gertrud Lohaus learned a technique for phloem sap analysis, where the stylet of an aphid feeding on a leaf is severed by a laser beam and phloem sap is collected from the stump of the stylet in a microcapillary (Weiner et al. 1991). Although the phloem sap samples were only a few microlitres, our analytical techniques allowed the assay of the concentration of a large number of metabolites in the phloem sap samples. Moreover, Gertrud Lohaus devised a technique to determine the metabolite concentration in the extracellular compartment of leaves. Together with subcellular metabolite analyses by Heike Winter and Burgi Riens, Gertrud Lohaus succeeded in characterising the process of apoplastic phloem loading in a leaf. Whereas sucrose, according to the expectations of earlier studies from other groups, was found to be highly accumulated in the phloem, to our surprise this did not hold for amino acids, the products of nitrate assimilation. In comparing different plants we found that the amino acid concentrations in the phloem sap were similar to those in the cytosol of the source cells (Riens et al. 1991; Winter et al. 1992; Lohaus et al. 1995). Apparently, no special transport form existed for the assimilated nitrogen, as assumed earlier, but the pattern of amino acids transported, being different in different plants, just reflected the corresponding pattern in the source cell.

Our analytical methods enabled us to study the productivity of crop plants. By comparing the contents of sucrose and amino acids in the leaves, the phloem sap and the tap roots of high and low sugar-producing sugar beet, Gertrud Lohaus found that the decisive factor in a high sucrose accumulation in sugar beet tap roots is a very efficient phloem loading in the leaves (Lohaus et al. 1994). The investigation of the factors governing the protein content of maize showed that a high-protein strain differed from a low-protein strain in delivering asparagine as a product of nitrate assimilation in the root in addition to the amino acids supplied by nitrate assimilation in the leaves (Lohaus et al. 1998). In cooperation with Olga Voitsekhovskaja and Yuri Gamalai from St. Petersburg, our analytical methods were extended to the study of symplastic phloem loader plants where the phloem is loaded via plasmodesmata (Voitsekhovskaja et al. 2006).

In Munich, I made the acquaintance of Barry Osmond. He invited me to visit him in 1981 in Canberra at the Research School of Biological Sciences of the Australian University, and on this occasion I also visited Hal Hatch at the division of Plant Industry at the CSIRO. This trip marked the beginning of a lasting friendship with Barry and Hal and an intense scientific cooperation between our groups. Barry Osmond worked in our laboratory and Hal Hatch and I were fortunate to receive jointly the Max-Planck-Forschungspreis which enabled us to visit each other. Hal worked twice for longer periods in our laboratory, Hendrik Weiner worked in Hal's and Gertrud Lohaus in Barry's laboratories in Canberra. I spent my three sabbaticals in Canberra as a visiting fellow of the Australian University, sponsored by Barry Osmond, and as a visitor in the laboratory of Hal Hatch in the CSIRO. These visits gave me the opportunity to work at the laboratory bench with my own hands as in former times. On my last sabbatical visit in Hal's laboratory shortly before I became an emeritus, I decided to try to identify and isolate from leaves the sucrose phosphate phosphatase, which was at that time the last unknown enzyme of the reaction chain of plant sucrose biosynthesis. Previously, our group in Göttingen and other groups had failed to identify this enzyme. With the help of Hal Hatch and his co-worker John Lunn, I managed to identify the enzyme protein and it was investigated further and sequenced after my departure (Lunn et al. 2000). This gave me the satisfaction that I could still work successfully at the laboratory bench.

Through these years I developed a very strong affiliation to the science and people of Australia. I felt very proud when the Australian Society of Plant Biologists awarded me the honour of a corresponding membership, and I was asked to accompany the late President Johannes Rau on behalf of the DFG and the Humboldt Foundation on his state visit to Australia in 2001.

When I became a professor emeritus in 2002, the division of plant biochemistry, now under the directorship of Ivo Feußner, moved into larger facilities in a new building. I then decided not to continue active research which, because of my status as emeritus, would have had to be on a smaller scale, but to utilize my experience in teaching and research and the resources of the institute in writing and updating a textbook on plant biochemistry for undergraduate students. This textbook has now been published not only in Germany but also in the USA, in Japan (in the Japanese language) and in affordable editions in India and China. The preparation of the fourth German edition is in progress.

When I, a chemist, entered the field of biochemistry in Marburg, my only knowledge of biology was what I had learned at school. In the institute led by Theodor Bücher, I learned how to investigate biological processes by performing quantitative biochemical analyses. I also learned that to obtain novel results, one should at first develop a method others did not have. From that time, my work and that of my group was driven by the curiosity of how a living cell, may it be from liver or leaves, operates. I never had a master plan for this, but one result would raise another question which I was then curious to solve. Thus, by chance I entered the field of metabolite transport from results obtained with liver tissue, and again by chance finally ended up in the field of photosynthesis metabolism. Our results were only possible because I always had excellent co-workers who did not wait for instructions but followed their own ideas. While I was in charge of plant biochemistry in Göttingen, my crew, besides their scientific work, fully participated in the extensive teaching and administrative duties. Birgit Piechulla and Katharina Pawlowski, who both had independent research groups in our institution, were also members of the team. My secretary, the late Anne Brandeck, managed all business matters most admirably and gave me such good advice that I was hardly bothered by administrative matters. Thus, I was able to concentrate fully on my duties in teaching and research. We profited greatly from the numerous foreign researchers who were funded by the Humboldt Foundation and the DAAD. Most of them became my personal friends, and most of them after returning to their home country retained strong ties to Germany. Our work would not have been possible without the continuous and very generous support by the Deutsche Forschungsgemeinschaft through the "Normalverfahren", by which we received all the financial resources we needed. I greatly appreciated the simplicity of writing grant proposals, which took not much more than a long weekend, and the freedom of the usage of the funds. With part of the funds I often did other work than that I had actually proposed, but it did not matter as long as we were successful. I felt rather relieved when in Munich I had to leave the SFB 51 since my work on photosynthesis did not fit in. From that time I rejected all chances of joining an SFB in order to keep my freedom. I feel that networks comprise a danger in stimulating one to carry out investigations which one might otherwise not have done and thus distract from one's own ideas and in the end hinder creativity. I do not believe that networks are a prerequisite for cooperation. As described in this report, I have cooperated with very many researchers from different institutions resulting in joint publications without networks being involved.

References

Baltscheffsky H, Stedingk LV, Heldt HW, Klingenberg M (1966) Inorganic pyrophosphate: formation in bacterial photophosphorylation. Science 153: 1120–1122

Bassham JA, Kirk M, Jensen RG (1968) Photosynthesis by isolated chloroplasts. Biochim Biophys Acta 153: 211–218

Baur H, Heldt HW (1977) Transport of hexoses across the liver cell membrane. Eur J Biochem 74: 397–403

Behrens M (1938) Zell-und Gewebetrennung. In: Abderhalden (ed.) Handbuch biologischer Arbeitsmethoden, pp. 1363–1392

Borchert S, Groβe H, Heldt HW (1989) Specific transport of inorganic phosphate, glucose 6-phosphate, dihydroxyacetone phosphate and 3-phosphoglycerate into amyloplasts from pea roots. FEBS Lett 253: 183–186

Bruni A, Contessa AR, Luciani S (1962) Atractyloside as an inhibitor of energy-transfer reactions in liver mitochondria. Biochim Biophys Acta 60: 301–311

Buchanan BB, Kalberer P, Arnon DI (1967) Ferredoxin-activated fructose diphosphatase in isolated chloroplasts. Biochem Biophys Res Commun 29: 74–79

Bücher T (1947) Über ein phosphatübertragendes Gärungsferment. Biochim Biophys Acta 1: 292–314

Bücher T, Klingenberg M (1958) Wege des Wasserstoffs in der lebendigen Organisation. Angew Chem 70: 552–570

Büssis D, Heineke D, Sonnewald U, Willmitzer L, Raschke K, Heldt HW (1997) Solute accumulation and decreased photosynthesis in leaves of potato plants expressing yeast-derived invertase either in the apoplast, vacuole or cytosol. Planta 202: 126–136

Cseke C, Balogh A, Wong, LH, Buchanan BB, Stitt M, Herzog B, Heldt HW (1984) Fructose 2,6-bisphosphate: a regulator of carbon processing in leaves. Trends Biochem Sci 9: 533–535

Ebbighausen H, Chen J, Heldt HW (1985) Oxaloacetate translocator in plant mitochondria. Biochim Biophys Acta 810: 184–199

Elbers R, Heldt HW, Schmucker P, Soboll S, Wiese H (1974) Measurement of the ATP/ADP ratio in mitochondria and in the extramitochondrial compartment by fractionation of freeze stopped liver tissue in none aqueous media. Hoppe-Seyler's Z Physiol Chem 355: 378–393

Falkner G, Horner F, Werdan K, Heldt HW (1974) Energieabhängige Phosphataufnahme der Blaualge *Anacystis nidulans*. Ber Deut Bot Ges 87: 263–266

Fliege R, Flügge UI, Werdan K, Heldt HW (1978) Specific transport of inorganic phosphate, 3-phosphoglycerate and triosephosphates across the inner membrane of the envelope in spinach chloroplasts. Biochim Biophys Acta 502: 232–247

Flügge UI, Benz R (1984) Pore-forming activity in the outer membrane of the chloroplast envelope. FEBS Lett 169: 85–89

Flügge UI, Heldt HW (1976) Identification of a protein involved in phosphate transport in chloroplasts. FEBS Lett 68: 259–262

Flügge UI, Heldt HW (1977) Specific labelling of a protein involved in phosphate transport of chloroplasts by pyridoxal-5′-phosphate. FEBS Lett 82: 29–33

Flügge UI, Heldt HW (1978) Specific labelling of the active site of the phosphate translocator in spinach chloroplasts by 2,4,6-trinitrobenzene sulfonate. Biochim Biophys Res Commun 84: 37–44

Flügge UI, Heldt HW (1981) The phosphate translocator of the chloroplast envelope. Isolation of the carrier protein and reconstitution of transport. Biochim Biophys Acta 638: 296–304

Flügge UI, Heldt HW (1986) Chloroplast phosphate-triose phosphate-phosphoglycerate translocator: its identification, isolation, and reconstitution. Methods Enzymol 125: 716–730

Flügge UI, Freisl M, Heldt HW (1980) The mechanism of the control of carbon fixation by the pH in the chloroplast stroma. Planta 149: 48–51

Flügge UI, Stitt M, Heldt HW (1985) Light-driven uptake of pyruvate into mesophyll chloroplasts from maize. FEBS Lett 183: 335–339

Gardemann A, Stitt M, Heldt HW (1982) Control of CO_2 fixation. Regulation of spinach ribulose-5-phosphate kinase by stromal metabolite levels. Biochim Biophys Acta 722: 51–60

Gardemann A, Schimkat D, Heldt HW (1986) Control of CO_2 fixation. Regulation of stromal fructose-1,6-bisphosphatase in spinach by pH and Mg^{++} concentration. Planta 168: 530–545

Gerhardt R, Heldt HW (1984) Measurement of subcellular metabolite levels in leaves by fractionation of freeze-stopped material in nonaqueous media. Plant Physiol 75: 542–547

Hanning I, Heldt HW (1993) On the function of mitochondrial metabolism during photosynthesis in spinach leaves (*Spinacia oleracea* L.). Partitioning between respiration and export of redox equivalents and precursors for nitrate assimilation products. Plant Physiol 103: 1147–1154

Hatch MD, Dröscher L, Flügge UI, Heldt HW (1984) A specific translocator for oxaloacetate transport in chloroplasts. FEBS Lett 178: 15–19

Heber U (1957) Zur Frage der Lokalisation von löslichen Zuckern in der Pflanzenzelle. Ber Deut Bot Ges 70: 371–376

Heber U, Santarius KA, Hudson MA, Hallier UW (1967) Intrazellulärer Transport von Zwischenprodukten der Photosynthese im Photosynthese Gleichgewicht und im Dunkel-Licht Wechsel. Z Naturforsch 22b: 1189–1199

Heineke D, Riens B, Grosse H, Hoferichter P, Peter U, Flügge UI, Heldt HW (1991) Redox transfer across the inner chloroplast envelope membrane. Plant Physiol 95: 1131–1137

Heineke D, Sonnewald U, Büssis D, Günter G, Leidreiter K, Wilke I, Raschke K, Willmitzer L, Heldt HW (1992) Apoplastic expression of yeast-derived invertase in potato: effects on photosynthesis, leaf solute composition, water relations and tuber composition. Plant Physiol 100: 301–308

Heineke D, Kruse A, Flügge UI, Frommer WB, Riesmeier JW, Willmitzer L, Heldt HW (1994a) Effect of antisense repression of the chloroplast triose phosphate translocator on photosynthesis metabolism in transgenic potato plants. Planta 193: 174–180

Heineke D, Wildenberger K, Sonnewald U, Willmitzer L, Heldt HW (1994b) Accumulation of hexoses in leaf vacuoles: studies with transgenic tobacco plants expressing yeast derived invertase in the cytosol, vacuole or apoplast. Planta 194: 29–33

Heldt HW (1963) Phosphathaltige Metabolite in Anionenaustauschchromatogrammen säurelöslicher Extrakte aus Rattenleber im Nano-Bereich. Biochem Z 337: 397–413

Heldt HW (1967) The effect of low temperature and of atractyloside on the reaction of endogenous and exogenous nucleotides. In: E. Quagliariello, E.C. Slater, S. Papa, J.M. Tager (eds.) Mitochondrial Structure and Compartmentation, Editrice Adriatica, Bari, pp. 260–267

Heldt HW (1969) Adenine nucleotide translocation in spinach chloroplasts. FEBS Lett 5: 11–14

Heldt HW, Klingenberg M (1965) Endogenous nucleotides of mitochondria participating in phosphate transfer reactions as studied with ^{32}P labelled orthophosphate and ultramicroscale ion exchange chromatography. Biochem Z 343: 433–451

Heldt HW, Klingenberg M (1968) Differences between the reactivity of endogenous and exogenous ADP of mitochondria as studied at low temperatures. Eur J Biochem 4: 1–8

Heldt HW, Pfaff E (1969) Quantitative evaluation of the correlation between the phosphorylation of endogenous and exogenous ADP in mitochondria. Eur J Biochem 10: 494–500

Heldt HW, Rapley L (1970a) Unspecific permeation and specific uptake of substances in spinach chloroplasts. FEBS Lett 7: 139–142

Heldt HW, Rapley L (1970b) Specific transport of inorganic phosphate, 3-phosphoglyceryate and dihydroxyacetonephosphate, and of dicarboxylates across the inner membrane of spinach chloroplasts. FEBS Lett 10: 143–148

Heldt HW, Sauer F (1971) The inner membrane of the chloroplast envelope as the site of specific metabolite transport. Biochim Biophys Acta 234: 83–91

Heldt HW, Schwalbach K (1967) The participation of GTP-AMP-P transferase in substrate level phosphate transfer of rat liver mitochondria. Eur J Biochem 1: 199–206

Heldt HW, Jacobs H, Klingenberg M (1964) Evidence for the participation of endogenous guanosine triphosphate in substrate level phosphate transfer in intact mitochondria. Biochem Biophys Res Commun 17: 130–135

Heldt HW, Jacobs H, Klingenberg M (1965) Endogenous ADP of mitochondria, an early phosphate acceptor of oxidative phosphorylation as disclosed by kinetic studies with ^{14}C labeled ADP and ATP and with atractyloside. Biochem Biophys Res Commun 18: 174–179

Heldt HW, Klingenberg M, Milovancev M (1972a) Differences between the ATP/ADP ratios in the mitochondrial matrix and in the extramitochondrial space. Eur J Biochem 30: 434–440

Heldt HW, Sauer F, Rapley L (1972b) Differentiation of the permeability properties of the two membranes of the chloroplast envelope. Proceedings of the Second International Congress on Photosynthesis, Stresa. W. Junk NV, Den Haag, pp. 1345–1355

Heldt HW, Werdan K, Milovancev M, Geller G (1973) Alkalization of the chloroplast stroma caused by light dependent proton flux into the thylakoid space. Biochim Biophys Acta 314: 224–241

Heldt HW, Chon JC, Maronde D, Herold A, Stankovic ZS, Walker DA, Kraminer A, Kirk MR, Heber U (1977) The role of orthophosphate and other factors in the regulation of starch formation in leaves and isolated chloroplasts. Plant Physiol 59: 1146–1155

Heldt HW, Chon CJ, Lorimer GH (1978) Phosphate requirement for the light activation of ribulose-1,5-bisphosphate carboxylase in intact spinach chloroplasts. FEBS Lett 92: 234–240

Herzog B, Stitt M, Heldt HW (1984) Control of photosynthetic sucrose synthesis by fructose 2,6 bisphosphate. III. Properties of the cytosolic fructose 1,6-bisphosphate. Plant Physiol 75: 561–565

Heupel R, Markgraf T, Robinson DG, Heldt HW (1991) Compartmentation studies on spinach leaf peroxisomes. Evidence for channeling of photorespiratory metabolites in peroxisomes devoid of intact boundary membrane. Plant Physiol 96: 971–979

Hohorst HJ, Kreutz FH, Bücher T (1959) Über Metabolitgehalte und Metabolitkonzentrationen in der Leber der Ratte. Biochem Z 332: 18–46

Jacobs H, Heldt HW, Klingenberg M (1964) High activity of creatine kinase in mitochondria from muscle and brain and evidence for a separate mitochondrial isoenzyme of creatine kinase. Biochem Biophys Res Commun 16: 516–521

Kampfenkel K, Moehlmann T, Batz O, van Montagu M, Inze D, Neuhaus HE (1995) Molecular characterization of an *Arabidopsis thaliana* cDNA encoding a novel putative adenylate translocator of higher plants. FEBS Lett 74: 351–355

Kirsten E, Kirsten R (1962) A nanomole adaptation of the automatic amino acid analysis according to Spackmann, Stein, and Moore. Biochem Biophys Res Comm 7: 76–80

Klingenberg M, Schollmeyer P (1960) Zur Reversiblität der oxidativen Phosphorylierung. Adenosintriphosphat-abhängige Atmungskontrolle und Reduktion von Diphosphopyridinnukl eotid in Mitochondrien. Biochem Z 333: 335–350

Klingenberg M, Slenczka W (1959) Pyridinnukleotide in Mitochondrien. Eine Analyse ihrer Redoxbeziehungen. Biochem Z 331: 480–517

Krömer S, Heldt HW (1991a) On the role of mitochondrial phosphorylation in photosynthesis metabolism as studied by the effect of oligomycin on photosynthesis in protoplasts and leaves of barley (*Hordeum vulgare*). Plant Physiol 95: 1270–1276

Krömer S, Heldt HW (1991b) Respiration of pea leaf mitochondria and redox transfer between the mitochondrial and extramitochondrial compartment. Biochim Biophys Acta 1057: 42–50

Krömer S, Stitt M, Heldt HW (1988) Mitochondrial oxidative phosphorylation participating in photosynthetic metabolism of a leaf cells. FEBS Lett 226: 352–356

Laing WA, Stitt M, Heldt HW (1981) Control of CO_2 fixation. Changes in the activity of ribulosephosphate kinase and fructose- and sedoheptulose-bisphosphatase in chloroplasts. Biochim Biophys Acta 637: 348–559

Lehner K, Heldt HW (1978) Dicarboxylate transport across the inner membrane of the chloroplast envelope. Biochim Biophys Acta 501: 531–544

Leidreiter K, Kruse A, Heineke D, Robinson DG, Heldt HW (1995a) Subcellular volumes and metabolite concentrations in potato (*Solanum tuberosum* cv. Désirée) leaves. Bot Acta 108: 439–444

Leidreiter K, Kruse A, Riens B, Winter H, Lohaus G, Robinson DG, Heineke D, Heldt HW (1995b) Subcellular compartmentation of metabolites in plant cells. In: P. Mathis (ed.) Photosynthesis: from Light to Biosphere, Vol. V, pp. 483–486. Kluwer, Netherlands

Lilley R, Chon CJ, Mosbach A, Heldt HW (1977) The distribution of metabolites between spinach chloroplasts and medium during photosynthesis in vitro. Biochim Biophys Acta 460: 259–272

Lilley RMcC, Stitt M, Mader G, Heldt HW (1982) Rapid fractionation of wheat leaf protoplasts using membrane filtration. Plant Physiol 70: 965–970

Lohaus G, Burba M, Heldt HW (1994) Comparison of the contents of sucrose and amino acids in the leaves, the phloem sap and the roots of high and low sugar producing hybrids of sugar beet (*Beta vulgaris L.*). J Expt Bot 45: 1097–1101

Lohaus G, Winter H, Riens B, Heldt HW (1995) Further studies of the phloem loading process in leaves of barley and spinach: comparison of metabolite concentrations in the apoplastic compartment with those in the cytosolic compartment and in the sieve tubes. Bot Acta 108: 270–275

Lohaus G, Büker M, Huβmann M, Soave C, Heldt HW (1998) Transport of amino acids with special emphasis on the synthesis and transport of asparagine in the Illinois low protein and Illinois high protein strains of maize. Planta 205: 181–188

Lunn JE, Ashton AR, Hatch MD, Heldt HW (2000) Purification, molecular cloning, and sequence analysis of sucrose-6-phosphate phosphohydrolase from plants. Proc Natl Acad Sci USA 97: 12914–12919

Martinoia E, Flügge UI, Kaiser G, Heber U, Heldt HW (1985) Energy-dependent uptake of malate into vacuoles isolated from barley mesophyll protoplasts. Biochim Biophys Acta 806: 311–319

Ohnishi JI, Flügge UI, Heldt HW (1989) Phosphate translocator of mesophyll and bundle sheath chloroplasts of a C_4 plant, *Panicum miliaceum* L. Identification and kinetic characterization. Plant Physiol 91: 1507–1511

Ohnishi JI, Flügge UI, Heldt HW, Kanai R (1990) Involvement of Na^+ in active uptake of pyruvate in mesophyll chloroplasts of some C_4 plants: Na^+/pyruvate cotransport. Plant Physiol 94: 950–955

Peter JB, Hultquist DE, DeLuca M, Kreil G, Boyer PD (1963) Bound phosphohistidine as an intermediate in a phosphorylation reaction of oxidative phosphorylation catalyzed by mitochondrial extracts. J Biol Chem 238: 1182–1184

Pette D (1965) Plan und Muster im zellulären Stoffwechsel. Naturwissenschaften 22: 597–616

Pfaff E, Klingenberg M, Heldt HW (1965) Unspecific permeation and specific exchange of adenine nucleotides in liver mitochondria. Biochim Biophys Acta 104: 312–315

Pfaff E, Heldt HW, Klingenberg M (1969) Kinetics of the adenine nucleotide exchange. Eur J Biochem 10: 484–493

Portis AR, Heldt HW (1976) Light dependent changes of the Mg^{++} concentration in the stroma in relation to the Mg^{++} dependency of CO_2 fixation in intact chloroplasts. Biochim Biophys Acta 449: 434–446

Preiss J, Ghosh MP, Wittkop J (1967) In: TW Goodwin (ed.) Biochemistry of Chloroplasts: Regulation of the biosynthesis of starch in spinach leaf chloroplasts, Vol. 2, pp. 131–152. Academic Press, New York

Purczeld P, Chon CJ, Portis AR, Heldt HW, Heber U (1978) The mechanism of the control of carbon fixation by the pH in the chloroplast stroma. Studies with nitrite mediated proton transfer across the envelope. Biochim Biophys Acta 501: 488–498

Rafael J, Heldt HW (1976) Binding of guanine nucleotides to the outer surface of the inner membrane of guinea pig brown fat mitochondria in correlation with the thermogenetic activity of the tissue. FEBS Lett 63: 304–308

Rafael J, Heldt HW, Hohorst HJ (1972) Guanosine-diphosphate causing changes in the phosphorylation pattern of adenine nucleotides in mitochondria from brown adipose tissue. FEBS Lett 28: 125–128

Raghavendra AS, Reumann S, Heldt HW (1998) Participation of mitochondrial metabolism in photorespiration. Plant Physiol 116: 1333–1337

Reumann S, Maier E, Benz R, Heldt HW (1995) The membrane of leaf peroxisomes contains a porin-like channel. J Biol Chem 270: 17559–17565

Reumann S, Bettermann M, Benz R, Heldt HW (1997) Evidence for the presence of a porin in the membrane of glyoxysomes of Castor bean. Plant Physiol 115: 891–899

Reumann S, Maier E, Heldt HW, Benz R (1998) Permeability properties of the porin of spinach leaf peroxisomes. Eur J Biochem 251: 359–366

Riebeling V, Jungermann K, Werdan K, Heldt HW, Thauer RK (1973) Der intrazelluläre pH in *Clostridium pasteurianum* während des Wachstums auf Glucose. Hoppe-Seyler's Z Physiol Chem 254: 1234–1235

Riens B, Heldt HW (1992) Decrease of nitrate reductase activity in spinach leaves during a light-dark transition. Plant Physiol 98: 573–577

Riens B, Lohaus G, Heineke D, Heldt HW (1991) Amino acid and sucrose determined in the cytosolic, chloroplastic, and vacuolar compartments and in the phloem sap of spinach leaves. Plant Physiol 97: 227–233

Riesmeier JW, Flügge UI, Schulz B, Heineke D, Heldt HW, Willmitzer L, Frommer WB (1993) Antisense repression of the chloroplast triose-phosphate translocator affects carbon partitioning in transgenic potato plants. Proc Natl Acad Sci USA 90: 6160–6164

Sanchez J, Heldt HW (1990) On the regulation of spinach nitrate reductase. Plant Physiol 92: 884–889

Schäfer G, Heber U, Heldt HW (1977) Glucose transport into spinach chloroplasts. Plant Physiol 60: 286–289

Schimassek H (1963) Metabolite des Kohlenhydratstoffwechsels der isoliert perfundierten Rattenleber. Biochem Z 336: 460–467

Schnitger H, Papenberg K, Ganse E, Czok R, Bücher T, Adam H (1959) Chromatographie phosphathaltiger Metabolite eines Leberpunktats. Biochem Z 332: 167–185

Scholz R, Thurman RG, Williamson JR, Chance B, Bücher T (1969) Flavin- and pyridine nucleotide oxidation–reduction changes in perfused rat liver. J Biol Chem 244: 2317–2324

Schott K, Borchert S, Müller-Röber B, Heldt HW (1995) Transport of inorganic phosphate and C_3- and C_6-sugar phosphates across the envelope membranes of potato tuber amyloplasts. Planta 197: 647–652

Schünemann D, Borchert S, Flügge UI, Heldt HW (1993) ATP/ADP translocator from pea root plastids: Comparison with translocators from spinach chloroplasts and pea leaf mitochondria. Plant Physiol 103: 131–137

Schürmann P, Buchanan BB (1975) Role of ferredoxin in the activation of sedoheptulose diphosphatase in isolated chloroplasts. Biochim Biophys Acta 376: 189–192

Soboll S, Scholz R, Heldt HW (1978) Subcellular metabolite concentrations: Dependence of mitochondrial and cytosolic ATP systems on the metabolic state of perfused rat liver. Eur J Biochem 87: 377–390

Soboll S, Elbers R, Scholz R, Heldt HW (1980) Subcellular distribution of di- and tricarboxylates and pH gradients in perfused rat liver. Hoppe-Seyler's Z Physiol Chem 361: 69–76

Stitt M, Heldt HW (1981a) Simultaneous synthesis and degradation of starch in spinach chloroplasts in the light. Biochim Biophys Acta 638: 1–11

Stitt M, Heldt HW (1981b) Physiological rates of starch breakdown in isolated intact spinach chloroplasts. Plant Physiol 68: 755–761

Stitt M, Heldt HW (1985) Generation and maintenance of concentration gradients between the mesophyll and bundle sheath in maize. Biochim Biophys Acta 808: 400–414

Stitt M, Wirtz W, Heldt HW (1980) Metabolite levels during induction in the chloroplast and extrachloroplast compartments of spinach protoplasts. Biochim Biophys Acta 593: 85–102

Stitt M, Mieskes G, Söling HD, Heldt HW (1982a) On a possible role of fructose-2,6-bisphosphate in regulating photosynthetic metabolism in leaves. FEBS Lett 145: 217–222

Stitt M, Lilley RMcC, Heldt HW (1982b) Adenine nucleotide levels in the cytosol, chloroplasts, and mitochondria of wheat leaf protoplasts. Plant Physiol 70: 971–977

Stitt M, Gerhardt R, Kürzel B, Heldt HW (1983) A role of fructose-2,6-bisphosphate in the regulation of sucrose synthesis in spinach leaves. Plant Physiol 72: 1139–1141

Stitt M, Herzog B, Heldt HW (1984a) Control of photosynthetic sucrose synthesis by fructose 2,6 bisphosphate. I. Coordination of CO_2 fixation and sucrose synthesis. Plant Physiol 75: 548–553

Stitt M, Kürzel B, Heldt HW (1984b) Control of photosynthetic sucrose synthesis by fructose 2,6-bisphosphate. II. Partitioning between sucrose and starch. Plant Physiol 75: 554–560

Strotmann H, Heldt HW (1969) Phosphate containing metabolites participating in photosynthetic reactions of *Chlorella pyrenoidosa*. Proceedings of the First International Congress on Photosynthesis, Freudenstadt, pp. 1131–1140

Valle EM, Craig S, Hatch MD, Heldt HW (1989) Permeability and ultrastructure of bundle sheath cells isolated from C_4 plants: structure-function studies and the role of plasmodesmata. Bot Acta 102: 276–282

Voitsekhovskaja OV, Koroleva OA, Bantashev DR, Knop C, Tomos AD, Gamalei YV, Heldt HW, Lohaus G (2006) Phloem loading in two Scrophulariaceae species: what can drive symplastic flow via plasmodesmata. Plant Physiol 140: 383–395

Walker DA (1969) Permeability of the chloroplast envelope. Proceedings of the First International Congress on Photosynthesis, Freudenstadt, pp. 250–257

Weber APM, Schwacke R, Flügge UI (2005) Solute transporters of the plastid envelope membrane. Annu Rev Plant Biol 56: 133–164

Weiner H, Burnell JN, Woodrow IE, Heldt HW, Hatch MD (1988) Metabolite diffusion into bundle sheath cells from C_4 plants: Relation to C_4 photosynthesis and plasmodesmatal function. Plant Physiol 88: 815–822

Weiner H, Blechschmidt-Schneider S, Mohme H, Eschrich W, Heldt HW (1991) Phloem transport of amino acids: Comparison of amino acid contents of maize leaves and of the sieve tube exudate. Plant Physiol Biochem 29: 19–23

Werdan K, Heldt HW (1972) The phosphate translocator of spinach chloroplasts. Proceedings of the Second International Congress on Photosynthesis, Stresa. W. Junk NV Publishers, Den Haag, pp. 1337–1344

Werdan K, Heldt HW, Geller G (1972) Accumulation of bicarbonate in intact chloroplasts following a pH gradient. Biochim Biophys Acta 283: 430–441

Werdan K, Heldt HW, Milovancev M (1975) The role of pH in the regulation of carbon fixation in the chloroplast stroma. Studies on CO_2 fixation in the light and in the dark. Biochim Biophys Acta 369: 276–292

Winter H, Lohaus G, Heldt HW (1992) Phloem transport of amino acids in relation to their cytosolic levels in barley leaves. Plant Physiol 99: 996–1004

Winter H, Robinson DG, Heldt HW (1993) Subcellular volumes and metabolite concentrations in barley leaves. Planta 191: 180–190.

Winter H, Robinson DG, Heldt HW (1994) Subcellular volumes and metabolite concentrations in spinach leaves. Planta 193: 532–555

Wirtz W, Stitt M, Heldt HW (1980) Enzymic determination of metabolites in the subcellular compartments of spinach protoplasts. Plant Physiol 66: 187–193

Wirtz W, Stitt M, Heldt HW (1982) Light activation of Calvin cycle enzymes as measured in pea leaves. FEBS Lett 142: 223–226

Woo KC, Boyle FA, Flügge UI, Heldt HW (1987a) ^{15}N-ammonia assimilation, 2-oxoglutarate transport, and glutamate export in spinach chloroplasts in the presence of dicarboxylates in the light. Plant Physiol 85: 621–625

Woo KC, Flügge UI, Heldt HW (1987b) A two-translocator model for the transport of 2-oxoglutarate and glutamate in chloroplasts during ammonia assimilation in the light. Plant Physiol 84: 624–632

Genetics

What's New in the Plant Cell Cycle?

D. Francis

Contents

Abstract The cell cycle comprises mitosis, post-mitotic interphase (G1), DNA synthetic phase (S-phase), post-synthetic phase (G2) and mitosis (M). In this review, new papers on the plant cell cycle (2007–2008), where the authors have focused on the critical transitions of the plant cell cycle, G1/S and G2/M, have been highlighted. Interestingly, disruption of one subset of cyclin-dependent protein kinases (a CDK B-type) disrupts the normal functioning of shoot but not root apical meristems in Arabidopsis. XAL1 is an agamous-like protein that has a differential role in roots compare with shoots; *xal1* mutant alleles have reduced root meristem size and lengthened cell cycles. E2F proteins are transcription factors that drive cells into S-phase. E2Fc may be targeted for degradation by the molecular machinery leading to the 26S proteasome. The Retinoblastoma (Rb) protein is a negative regulator of E2F. New evidence is also presented whereby starving plant cells of sucrose did not affect expression of Rb which would be consistent with an arrest of those cells in G1 in these conditions. Dynamin-related proteins are

D. Francis
School of Biosciences, Cardiff University, Cardiff CF10 3TL, UK
e-mail: francisd@cardiff.ac.uk

U. Lüttge et al. (eds.), *Progress in Botany 70*,
DOI:10.1007/978-3-540-68421-3, © Springer-Verlag Berlin Heidelberg 2009

important at cytokinesis, and recent evidence suggests that one subset plays a crucial role in cell wall synthesis. These new papers are reviewed against a background of what is currently known about regulation of the plant cell cycle.

1 Introduction

The cell cycle is a series of events that a proliferative cell must traverse before it can divide. Classically, the cell cycle is separated into four successive phases: mitosis (M), G1 (post-mitotic interphase), S phase (DNA Synthetic phase) and G2, post-synthetic phase (Fig. 1).

The major transitions are G2 to M, when proliferative cells achieve mitotic competence, and G1 to S-phase, when cells gear-up for nuclear DNA replication. However, in reality, the cell cycle ensures correct timing of both DNA replication and mitosis. If timing begins to malfunction in cells that are deficient in core cell cycle genes, catastrophic down-spiralling of cell size or unscheduled mitoses could follow. There is some published literature on the plant cell cycle that assigns a cell cycle transition as a checkpoint. In my view, checkpoints are stress-induced blocks on the cell cycle that a cell must clear before undergoing the transition. Also, and as clearly noted by O'Connell et al. (1997), a checkpoint gene is only expressed in response to a stress (e.g. UVB). Mostly, in this review I consider what is new in the "normal" cell cycle, if such a condition actually exists.

Perhaps not surprisingly, given the universality of DNA, the basic events of the cell cycle are common to all eukaryotes. All proliferative cells must replicate their nuclear DNA before they are competent for mitosis. It therefore follows that many of the genes that regulate the G2/M and G1/S transitions are similar among all eukaryotes. However, there are certain aspects of the plant cell cycle that are unique. For example, cytokinesis is very different between plants and animals. During plant cytokinesis, a new cell wall is formed between two recently divided

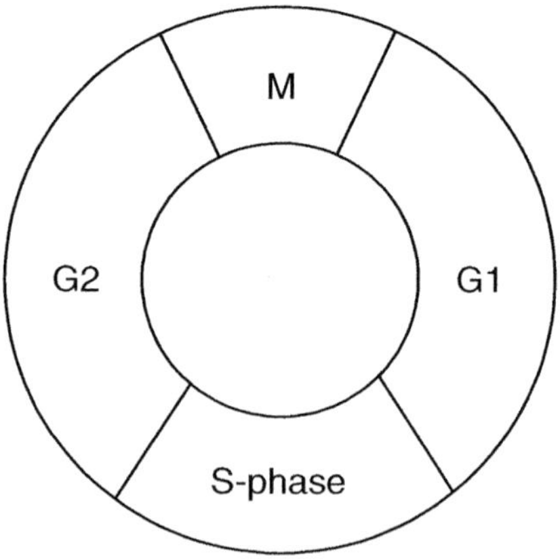

Fig. 1 An idealised cell cycle comprising G1, S-phase, G2 and M

cells. For this to occur, unique plant genes are expressed and structures necessary for cytokinesis are formed that do not have a counter player in animal cell cycles. Another difference is that either G1 or G2 are natural arrest points in the plant cell cycle whereas in animals, G1 is the natural arrest point.

In this short review, the aim will be to paint a general picture of the plant cell cycle and to highlight, in the author's opinion, what is relatively new. The reader can also be updated via two very recent publications on the plant cell cycle (Inze 2007; Bryant and Francis 2008).

2 Cyclin-Dependent Kinases and Cyclins

The major phase transitions of the cell cycle, G1/S and G2/M, are driven by cyclin dependent protein kinases (CDKs). Currently 152 CDKS have been identified from 41 plant species (Dudits et al. 2007). In animals, CDKs are numbered whilst plant CDKs are suffixed by Arabic letters in seven classes (A to G) along with an eighth rather more anomalous class, the CDK-like kinases (CKLs). CDKs phosphorylate substrates to drive a cell through a particular phase transition. CDKA was first cloned in *Arabidopsis*, *Arath;CDKA;1,* but originally was known as *Atcdc2a* (Ferreira et al. 1991). The gene could complement a *cdc2*$^{-}$ mutant of fission yeast, and its encoded protein exhibited a highly conserved sequence of amino acids, the so-called PSTAIRE motif in its α1 helix. A second *Arabidopsis* CDK (cdc2b), now known as one of the *Arath;CDKB* types was identified, but it cannot complement the yeast *cdc2* mutants and PSTAIRE is replaced by PPTALRE reflecting evolutionary divergence between the B and A-types (Segers et al. 1996). It was subsequently discovered that the B-type was a member of quite a large family of CDKs that are unique to plants (Dudits et al. 2007). However, despite the ever-growing number of plant CDKs rather little is known about their substrate specificity, and the picture is not much clearer in animal cell cycles in this regard.

CDKs, C and G, may or may not operate in the cell cycle and much more needs to be learnt about them. Arguably, therefore most is known about the A- and B-type CDKs. *Arath; CDKA;1* is expressed more or less constitutively both with respect to the cell cycle and different regions of the plant, and its encoded kinase activity is highest at G1/S and G2/M in *Arabidopsis*, whilst in the tobacco BY-2 cells, CKDA activity is relatively constant from S-phase to G2/M (Joubes et al. 2000). In contrast, B-type CDKs are transcriptionally regulated during the cell cycle and their expression tends to be limited to the meristems of the plant (Fereira et al. 1994). In BY-2 cells, CDK kinase activity peaks in G2 (Sorrell et al. 2001).

For some time now, the consensus view has been that a functional link between cell cycle genes and development of the plant is tenuous. For example, a dominant negative *Arath;CDKA;1* allele when expressed in tobacco resulted in fewer but larger cells, but the development of those plants was not unduly affected (Hemerly et al. 1995). Plants expressing a DN allele of *Arath;CDKB1;1* also displayed a normal shape, but again with fewer and larger cells with long G1 phases (Porceddu et al. 2001). Thus, the data were consistent with a null effect of cell cycle manipulation on development. Indeed, the data support the organismal theory of development where cell division is merely a functional consequence of inherent developmental programmes (Kaplan and Hagemann 1991). However, recently, a disruption of CDKB2 led to impaired cell cycle progression and induced severe defects in the shoot apical meristem of *Arabidopsis*. Also, evidence was provided that *CDKB2* is a necessary part of a coordinated programme of gene expression including *WUSCHEL* and *SHOOTMERISTEMLESS* that is necessary for correct functioning of the vegetative shoot apical meristem (Andersen et al. 2008). Indeed, these data are consistent with a requirement for meristem organisation and cell division to be carefully coordinated for normal shoot apical meristem function. As I have stated several times before, the organismal and cellular theories of development are unwisely portrayed as mutually exclusive hypotheses. I agree with the recent conclusion of Wang et al. (2008), from their studies of inhibitors of CDK activity (see Sect. 6), that cell division, growth and development are intrinsically intertwined.

3 Cyclins

CDKs only have catalytic activity when bound to a partner cyclin and cease to be catalytic following an almost instantaneous destruction of the cyclin. This was first discovered in sea clams as a protein that increased in concentration as meiocytes prepared for meiosis and then suddenly disappeared when these meiotic cells began to divide (Evans et al. 1983). This well-timed destruction is best characterised for cyclins that degrade in anaphase of mitosis. Plant cells cannot complete mitosis normally if CDK activity persists beyond anaphase (Weingartner et al. 2004). The timing of G1/S cyclin degradation is less instantaneous than that of the G2/M cyclin at anaphase (Morgan 2007) although, to my knowledge, there are few data on the kinetics of cyclin degradation in plants.

In lower eukaryotes, such as budding and fission yeast, there is one CDK (Cdc2 in fission yeast and CDC28 in budding yeast) that binds with different cyclins at different stages of the cell cycle (Morgan 2007). However, in animals and plants, different cyclins and different CDKs function at different points in the cell cycle (Table 1).

Table 1 A generic listing of cyclin-dependent kinases (CDKs) in higher plants (for a comprehensive listing see Dudits et al. 2007) together with likely plant cyclin partners (for an up-to-date listing see Nieuwland et al. 2007)

	G1/S	"Likely cyclin partner	G2/M	
CDKA	✓	D/A (P?)	✓	D/A
CDKB		B (P)	✓	B
CDKC		C/T[1]		Phosphorylation of C-terminal of large sub-unit of RNA polymeraseII
CDKD		H		CAK kinase
CDKE		?		?
CDKF		?		CAKAK kinase
CDKG		?		?

[1]John Doonan, personal communication

In plants, over 100 cyclins have been identified and are grouped according to their homology with animal cyclins which themselves have been apportioned into 13 classes (A–L and T). In *Arabidopsis thaliana,* the A-, B- and D-type cyclins are important during the cell cycle but there are a host of others identified, including C-, H-, L-, P- and T-types which may or may not be integral to the cell cycle (Table 1).

4 G0/G1/S-Phase

Like animal cells, plant cells will arrest in G1 with the 2C amount of nuclear DNA. They do this if deprived of nutrient (Van't Hof 1966) or if stressed in other ways such as low temperature (Francis and Barlow 1988). Such non-cycling cells are said to be in G0. A stunning example of G0 cells in the plant is the quiescent centre in root apical meristems (RAMs). First predicted and then discovered by Clowes, this is a tiny population of cells; for example, 300 out of 250,000 in broad bean RAMs (Clowes 1967) and as few as four in *Arabidopsis* (van den Berg et al. 1997). These cells show low rates of DNA, RNA and protein synthesis (Clowes 1956). In higher plants, it was hypothesised that G0 cells are located in a central position within the QC, whereas on the margins of the QC there are slowly cycling cells with very long G1 phases. Indeed, these very slowly proliferating cells have been identified in RAMs of *Zea mays* (Clowes 1970). When the root cap is damaged, cells of the QC spring to life and rapidly repopulate the cap before returning to their quiescent or slow cycling status (Barlow 1974). Such a population of cells conforms to what animal cell biologists would call stem cells. To have stem cells in RAMs has always seemed to be botanically bizarre (at least to this author!) and maybe we should rename them as founder cells.

Some time ago, Feldman elegantly showed that the QC could be excised and cultured, and from which an entire plant could be regenerated (Feldman and Torey 1975). These observations were consistent with the idea that cells of the quiescent centre are pluripotent. However, and at the time somewhat surprisingly, when all cells of the QC were laser ablated, subtending cells of the cortex were recruited into a rapidly reconstituted QC. Hence the QC and contacting apical initials proved to be surprisingly plastic. Moreover, if single QC cells were laser ablated, cells in direct contact with the now defunct QC cell accumulated starch, leading to the conclusion that cells of the quiescent centre inhibit the differentiation of surface-contacting initial cells (van den Berg et al. 1997). How exactly quiescence is imposed on QC cells is unknown, although the recent discovery of a homeodomain *WUSCHEL*-like gene, *QHB*, would suggest feed-back loops that involve plant growth regular-induced signal transduction chains that stabilise the population of QC cells (Kamiya et al. 2003).

Recently, a MADS-box gene, *XAL1*, was shown to be involved in regulating root growth. *xal1* mutant alleles had reduced meristem length, increased cell cycle duration and reduced rates of cell production compared with wild type. The results were surprising given that *XAL1* is very closely related to the

AGAMOUS MADS-box clade that specifies carpel morphogenesis. The authors proposed that *XAL1* may have a dual role in RAMs and SAMs through a differential coupling with different types of hormonal signals that specify root or shoot morphogenesis (Tapia-López et al. 2008). In my view, this is a very nice example of a homeodomain gene interfacing between developmental, cell cycle and growth controls and a further reminder of distinct differences in the functioning RAMs and SAMs.

4.1 G1/S Cyclins

As in animal cells, D-type cyclins are to the fore in activating cells from G0 to G1 (Menges and Muray 2008). Interestingly, the latter authors took one step back and considered the issue of arrest in G0 induced by stress factors. It transpires that expression of at least one member of the cyclin classes: A-, B-, C-, D- P and H are down-regulated in response to heat stress. We now begin to get a clue about why that there are so many cyclins in plants compared with animals. It seems that this array of regulatory proteins form part of the plasticity of responses that plants must have in order to cope with unpredictable changes to the environment. Indeed, the sudden stops and starts to plant growth in the wild must be at least partly governed by very fine tuning of plant cyclin gene expression.

4.2 E2F

The E2F-DP family of transcription factors plays a big part in activating cells into S-phase. As is often the case, the abbreviations of genes can have a complicated history that bares only a passing relationship to the function the protein in question. E2F is no exception. It was first identified when a promoter of a human adenovirus, E2, was activated by a host cytoplasmic binding factor (protein) (Dyson 1998). DP is a dimerisation protein, a comparatively simpler abbreviation. Currently, three plant E2Fs are well characterised (a, b and c) and there are at least three more varying in nomenclature from E2F d, e through to DEL1 2 and 3, and they may not require a DP to function normally (Ramirez-Parra et al. 2003, 2007).

Many genes have E2F recognition sites in their promoters, and the general picture is that E2F is necessary to activate S-phase genes that drive cells into nuclear DNA replication (Ramirez-Parra et al. 2003). Recently, in *Arabidopsis*, E2Fc, which is a putative negative regulator of other proteins, was shown to be targeted for degradation by SKP2A (Jurado 2008). The latter protein is part of a rather larger complex, SKP/CULLIN/F box complex that targets and steers proteins to the 26S proteasome for degradation (del Pozo et al. 2002a, b, 2006). *SKP2A* over-expression can stimulate cell division in meristems, but this might be because it targets E2FC for degradation (Jurado et al. 2008).

4.3 Retinoblastoma Protein

The E2F transcription factors are regulated negatively by the so-called retinoblastoma protein; there is one RB in *Arabidopsis* and several related (RBR) ones. At the G1/S transition, RB is hyperphosphorylated by CDKA that binds to several D-type cyclins (Bonnioti and Gutierrez 2001); this high level of phosphorylation destabilises binding between RB and E2F and the latter is then released to activate S-phase genes. Whether RB can regulate the number of founder cells in the quiescent centre of RAMs (see above) was tested recently. A local reduction in RBR expression resulted in an increase in the number of founder cells whilst RBR over-expression had the converse effect before ultimately arresting mitotic cells. They measured the frequency of mitotic figures in *Arabidopsis* roots and concluded that manipulation of RBR did not affect cell cycle times (Wildwater et al., 2005). However, it must be noted that the mitotic index is only a sensitive measure of rates of division if M-phase remains constant, and that any changes in the cell cycle time are due to changes in other component phases of the cell cycle.

A very well-known phenomenon in the plant cell cycle is that depleting cells of sucrose results in their natural arrest in either G1 or G2 (Van't Hof 1966). One feature of sucrose starvation in *Arabidopsis* cell cultures is that high expression of *Arath;RBR1* persisted despite the depletion of sucrose from the medium (Hirano and Sekine 2008). What this implies is that *Arath;RBR1* cannot be hyperphosphorylated and, as such, would remain bound to E2F. This would fit with down-regulation of D-type cyclins and as a consequence a down-regulation in CDK activity in cells that have been starved of sucrose (Murray et al. 1998).

5 S-Phase

During S-phase, DNA is replicated semi-conservatively so that the DNA molecule of a one-armed chromosome is unwound and each unwound strand acts as a template for replication of nascent DNA. S-phase terminates when two DNA molecules are fully replicated and housed in double-armed chromosomes. If only it was as simple as that!

DNA replication along the chromosome is achieved by firing of initiation points spaced at regular intervals along the chromosome. We calculated that 30,000 initiation points function to replicate all 14 chromosomes in a diploid rye genome, and this increased to over a million in the 42 chromosomes of the allohexaploid genome of *Tricitum aestivum* (AD Kidd, D Francis, MD Bennett, unpublished data). An initiation site and its two termini span a length of DNA known as a replicon (Vant'Hof 1985). However, there is no evidence to support a temporal and controlled activation of successive replicons along the chromosome. The situation is much more complicated with different clusters of replicons firing at different times during S-phase. So, we have a perplexing problem that, although DNA replication is continuous from one initiation site (at least in the leading strand!), DNA replication is most likely to be discontinuous on any given chromosome at any specific point during S-phase.

Eukaryote DNA contains genic DNA interspersed at random between vast domains of non-genic DNA. However, although the amount of DNA to be replicated is typically vast, S-phase can last for a staggeringly short period of 10 min in *Drosophila* embryos, to several hours in species with very large genome sizes (Francis et al. 2008). The plasticity of S-phase can be explained partly by the activation of many replication origins for fast S-phases and by the quiescence of weak replication origins between strong ones as S-phase gets longer. Unlike budding yeast, there appear to be no specific sequences of DNA that configure a replication origin in higher eukaryotes, although an origin recognition complex (ORC) of six polypeptides typically binds to and is characteristic of a DNA replication origin. In budding yeast, a detailed but not necessarily complete map of proteins that interact to replicate nuclear DNA has been characterized, but space does not permit a detailed description of them (Diffley and Cocker 1992; Diffley and Labib 2002; Diffley 2004). Recently, some of the components of this network have been cloned in plants but currently we lack functional evidence for their role *in planta* (see Table 2).

Table 2 An array of well-characterised yeast and animal genes/proteins known to participate in the initiation of DNA replication together with known plant orthologues whose role *in planta* remains to be confirmed

Budding yeast gene	Type of protein	Function or probable function	Plant orthologue
ORCs[a]	Each ORC comprises six non-catalytic (structural (?)) polypeptides	Guard the replication origin	✓
CDC6p[b]	AAA Atpase	Attaches to ORC and "facilitates" binding of MCM proteins to ORC	✓ but max. expressed in early S-phase in BY-2 cells
CDT1[c]	Binding protein	Attaches to ORC to assist initiation DNA replication but must be degraded to prevent re-replication genome	✓
MCM10	An essential S-phase protein	Required for initiation of DNA replication but *not* part of pre-replication complex (PRC)	
MCM2–7[d–f]	Bind either side of ORC	"Have helicase-like activity facilitating" unwinding of DNA at the replication origin (part of PRC) although not when purified	✓
CDC7	Protein kinase	Phosphorylate ORC and MCMs enabling proper function of MCMs and binding of additional proteins of the replication complex (RC)	

(continued)

Table 2 (Continued)

Budding yeast gene	Type of protein	Function or probable function	Plant orthologue
DBF4	Regulatory subunit		
CDC45[g]	Binding protein	Functions alongside *CDC7/DBF4* to trigger initiation of DNA replication after the PRC has formed	✓
GINS1–4	Binding proteins factor	Most likely stabilise and promote helicase activity of MCMs in the RC	
SLD	Binding proteins	Part of RC aiding DNA unwinding and replication	

[a]Gavin et al. (1995)
[b]Ramos et al. (2001)
[c]Lin et al. (1999)
[d]Bryant et al. (2001)
[e]Springer et al. (1995)
[f]Sabelli et al. (1996)
[g]Stevens et al. (2004)

6 G2/M

During mitosis, an exact partitioning of chromosomes occurs so that two identical sets of chromosomes become housed in new nuclei in smaller cells. Once complete, the growth cycle begins once more as the new cells prepare for the next mitosis. Hence, G2/M is a critical transition in the life of the cell. Unerringly, most cells of a healthy organism undergo mitosis correctly at each and every mitosis. This is despite environmental insults to DNA of the staggering order of $2\times10^5-4\times10^5$ single strand breaks cell^{-1} day^{-1} (Bray et al. 2008). The clear picture in animals and the emerging one in plants is that cells have multiple mechanisms to stop themselves from dividing inappropriately, but have only a fleeting opportunity to divide providing every aspect of mitotic competence is fulfilled. The inhibitory pathways will be explored here.

6.1 Negative Regulation by Lack of Cyclins

At the biochemical level, a mitotic CDK can be switched off quite simply through lack of the partner cyclin. Mitotic cyclins peak in concentration at G2/M and are degraded in anaphase of mitosis. The dependency on the correct timing of cyclin synthesis and degradation was aptly demonstrated when *Arabidopsis* was transformed with a mitotic cyclin lacking a destruction box. This created a transgenic line with non-degradable cyclin. The G2/M, prophase and metaphase stages were completed normally, but both anaphase and telophase became chaotic as the mitotic cells lurched into aberrant divisions (Weingartner et al. 2004). Hence, not only is the carefully timed activation of CDK activity critical for G2/M but its inactivation by a well-timed destruction of the partner cyclin is also equally important for a successful mitosis.

The activation of binding between a cyclin and its CDK is regulated by phosphorylation of the CDK at a threonine residue (160/167) towards the carboxy terminus of the protein. This phosphorylation is now known to be catalysed by a cyclin-dependent activating kinase (CAK). In plants, this function seems to be performed by D-type CDKs which themselves are activated by an H-type cyclin (Table 1). In turn, the binding between CDKD and cyclinH is catalysed by CDKF which in effect is CAK activating kinase (reviewed by Dudits et al. 2007). One could imagine that this phosphorylation cascade operates indefinitely as there might be a previous activation between a CAK and it partner cyclin that is phosphorylated by another CAK, and so on. If so, the mother of all CAK kinases is unknown (in both plants and animals). However, it is probably the case that the CDK-cyclin complex, once activated, is capable of phosphorylating and activating an up-stream CAK kinase in a positive feed-back loop, although evidence of this in plants still awaits us.

6.2 Negative Regulation by WEE1

In most eukaryotes, another form of repression of CDKs at G2/M is exercised by WEE1 kinase. This kinase phosphorylates the CDK at its tyrosine15 residue close to the amino terminal of the CDK. In fission yeast, there is functional redundancy between WEE1 and MIK1 kinase (Lundgren et al. 1991). In humans, WEE1 kinase activity is high during interphase but drops in mitosis (McGowan and Russell 1995). Furthermore, CDK-cyclinB1 is partitioned away from the nucleus in interphase where its kinase activity is repressed by another Wee1-like kinase, MYT1 (Mueller et al. 1995), which phosphorylates both Thr14 and Tyr15 of CDK1 (the latter is an orthologous to *Arath;CDKA;1*). Hence, there is a double check on repressing CDK, keeping it out of the nucleus in interphase whilst WEE1 remains nuclear-located, capable of preventing or suppressing any promiscuous CDK activity during interphase (Baldin and Docommin 1995). At G2/M, CDK1-B1 enters the nucleus, WEE1 is partitioned into the outer reaches of the cytoplasm and, seemingly, any residual nuclear WEE1 is hyperphosphorylated by the CDK prior to proteolytic destruction of WEE1 (Watanabe et al. 2005). During interphase, nuclear WEE1 is tethered to a 14-3-3 protein which shields the WEE1 kinase from hyperphosphorylation and might also be important for the partitioning and movement of WEE1 from nucleus to cytoplasm at prophase of mitosis (Morgan 2007). And so, for the proliferative cell, we have an almost perfect inverse spatial and molecular relationship between WEE1 and the CDK, which prevents unscheduled CDK activity from initiating mitosis in unfavourable conditions.

The current status of WEE1 in plants is far from clear. *WEE1* orthologues have been cloned in maize, *Arabidopsis* and rice (Sun et al. 1999; Sorrell et al. 2002; Guo et al. 2007). In *Arabidopsis*, *WEE1* expression is limited to proliferative regions of the plant whilst its over-expression in fission yeast induces a long cell phenotype, a classic screen for *WEE1* orthologues (Sorrell et al. 2002). Moreover, the plant WEE1 kinase is active in interphase but is very low during mitosis, and it can bind to Non-Epsilon plant 14-3-3 proteins in a cell cycle dependent manner (D. Francis et al., unpublished data). However, *Arath;WEE1* could not complement a *wee1/mik1* double mutant of fission yeast (Sorrell et al. 2002). Also, a WEE1 T DNA insertion line (*wee1*-1) develops normally and WEE1 itself was only highly expressed in *Arabidopsis* plants treated with hydroxyurea (De Schutter et al. 2007). The latter perturbs DNA replication by competing with endogenous nucleotide pools within the cell. The conclusion drawn from this work is that *WEE1*'s role in the plant is restricted to stress responses so that an over-expressed WEE1 will prevent cells from dividing until DNA damage is repaired or DNA replication is normalised. We have unpublished data (manuscript submitted for publication) indicating that *wee1-1* does exhibit a subtle root phenotype, so that perhaps WEE1 is an integral component of the processes that governs root elongation, but that there might be functional redundancy for WEE1 at the G2/M transition. Alternatively, if it is accepted that all plants in the wild are under stress, or able to respond rapidly to stress, then reconciling these different observations about *WEE1* may be easier than we think; time will tell.

Equally, if not more puzzling, is the lack of a bona fide CDC25 gene in higher plants that is orthologous to yeast, animal and lower plant CDC25s. In most eukaryotes, CDC25 phosphatase is the lone positive activator that drives a cell into mitosis at the slender G2/M window of opportunity. A small CDC25 in *Arabidopsis* has been cloned that can exhibit phosphatase activity (Landrieu et al. 2004) and is able to induce a small cell size phenotype in fission yeast (Sorrell et al. 2005). Also, dephosphorylation of plant CDK at G2/M was demonstrated through cytokinin treatment in cultured tobacco (*Nicotiana plumbaginofolia*) cells (Zhang et al. 1996, 2005). However, *Arath;CDC25* only encodes a catalytic domain and is expressed weakly in most plant tissues (Sorrell et al. 2005). These data, and the knowledge that *Arath;CDC25* protein can also exhibit arsenate reductase activity (Bleeker et al. 2006) led Boudolf et al. (2006) to hypothesise that the G2/M transition is regulated firstly through CDKB that phosphorylates an inhibitor of CDKA activity (see below). It is proposed that the phosphorylated inhibitor then releases CDKA to drive cells into mitosis (Fig. 2).

6.3 Negative Regulation of CDKS by the ICKS

A third level of CDK repression in interphase is through interactors (or sometimes called inhibitors) of CDK activity (ICKs). The first ICK was identified in *Arabidopsis* (Wang et al. 1997). There are seven ICKs in *Arabidopsis* although

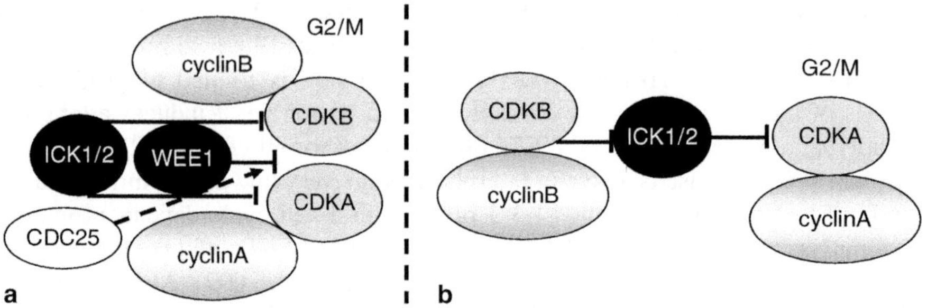

Fig. 2 (**a**) An orthodox model for G2/M where WEE1 and ICK1/2 repress whilst cdc25 activates the CDKs, (**b**) a model based on the lack of a bone fide plant CDC25 where CDKB inactivates ICK1/2 that releases CDKA to drive cells into mitosis

2–7 are more commonly referred to as KRP- (Kip-related-) -proteins. *ICK1* can be induced by abscisic acid (Wang et al. 1998) and salt stress (Ruggiero et al. 2004; Wang et al. 2007 for a review of ICKs). Yeast two hybrid protein–protein interaction data show that whilst ICK1 interacts with CDKA;1 it does not interact with B-type CDKs (Wang et al. 2007). This might have repercussions for the Boudolf et al. (2006) model of a CDKB repressing ICK1 repressing CDKA model of releasing cells into mitosis (Fig. 2).

7 Cytokinesis

At telophase of mitosis, two newly partitioned sets of chromosomes are each housed in two new nuclei spaced apart in the mother cell. Cytokinesis in plants occurs when a new cell wall is placed between the newly constituted nuclei. It is a topic that has been extremely well investigated over the past 10 years, and the published literature on plant cytokinesis really merits a PiB chapter of its own. My inclusion of cytokinesis in this chapter will do little justice to the enormous effort that has and is being expended to research it. Nevertheless, not mentioning cyto kinesis would be unforgivable.

At cytokinesis, residual microtubules from the mitotic tubular machinery are left at the division plane. This is the phragmoplast which acts as the foundation for the new cell wall. As one might expect, there are cytokinesis-specific genes, one of which is *KNOLLE* that encodes a syntaxin protein (Lauber et al. 1997; Völker et al. 2001); *knolle* mutant embryos have irregular cell patterns and are deficient in the cytokinesis process (elegantly reviewed by Jürgens 2005). An interacting gene is KEULE that binds to KNOLLE (Asaad et al. 1996). Moreover, there is a complex of proteins comprising SNARE, SYP31 and CDC48 that interact with KNOLLE to regulate the process of vesicle fusion at the cell plate. (Heese et al. 2001; Feiler et al. 1995). The so called SNARE-complex also interacts with the AAA Atpase, NSF which all form another part of the protein complex that is essential for cytokinesis to take place (Rancour et al. 2002). Cell plate formation may be considered as a modified form of vesicle secretion per se (reviewed by Sanderfoot 2007). This is turn seems to rely on the Golgi complex as the provider of vesicles that contribute to cell wall plate formation. The trafficking of these vesicles to the cell plate is under the control of specific motor proteins as mentioned above.

The finishing touch to cytokinins is obviously the new cell wall. In effect, islands of cellulose synthase complex are transported to the outer surface of the plasma membrane, and there they lay down cellulose microfibrils in paths resembling snails criss-crossing an open surface (Burgess 1985). These walls ultimately comprise an interwoven network cellulose and the non-fibrillar pectins and hemicelluloses, together with (hydroxy)proline-rich glycoproteins. In *Arabidopsis*, AtEXT3, a structural glycoprotein, is considered to be part of the scaffold that enables the construction of the primary cell wall (Cannon et al. 2008).

Dynamin-related proteins are large GTPases that are involved in cytokinesis (Praefke and McMahon 2004). Of the 16 DRPs in *Arabidopsis*, three accumulate in the cell plate (Lauber et al. 1997; Otegui et al. 2001). Very recently a *rsw* null mutant of *Arabidopsis* was characterised with a reduced content of cellulose in its cell walls. The mutant phenotype of short and swollen roots is believed to be because of changes in plasma membrane dynamics that inhibit cellulose synthesis (Collings et al. 2008).

8 Conclusions

Clearly, bit by bit, the plant cell cycle is being accounted for on a gene by gene basis, but it does have gaps and riddles. Why so many cyclins, why no CDC25? How exactly does the cell cycle interface with growth, development and extra cellular signalling? The proteomics era offers the possibility of understanding plant cell cycle genes in functional networks of the sort that are beginning to be resolved regulating cytokinesis. The uniqueness of plants with their sessile growth habit will no doubt continue to pose unusual answers to what might seem relatively straightforward questions about the plant cell cycle. In this short review, I have tried to focus on what is new in the plant cell cycle in the past two years (at least to me) but, undoubtedly, there is much more to come.

References

Andersen SU, Buechel S, Zhao Z, Ljung K, Novák O, Busch W, Schuster C, Lohman JU (2008) Requirement of B2-type cyclin dependent kinases for meristem integrity in *Arabidopsis thaliana*. Plant Cell 20: 88–100

Asaad FF, Mayer U, Wanner G, Jürgens G (1996) The KUELE gene is involved in cytokinesis in Arabidopsis. Mol Gen Genet 253: 267–277

Baldin B, Ducommin B (1995) Subcellular localisation of human wee1 kinase during the cell cycle. J Cell Sci 108: 2425–2432

Barlow PW (1974) Regeneration of the cap of primary roots of Zea mays. New Phytol 73: 937–954

van den Berg C, Willemsen V, Hendriks G, Weisbeek P, Scheres B (1997) Short-range control of cell differentiation in the *Arabidopsis* root meristem. Nature 390: 287–289

Bleeker PM, Hakvoort HW, Bliek M, Souer E, Schat H (2006) Enhanced arsenate reduction by a CDC25-like tyrosine phosphatase explains increased phytochelatin accumulation in arsenate-tolerant *Holcus lanatus*. Plant J 45: 917–929

Boniotti MB, Gutiierrez C (2001) A cell cycle regulated kinase activity phosphorylates plant retinoblastoma protein and contains, in *Arabidopsis*, a CDKA/cyclin D complex. Plant J 28: 341–350

Boudolf V, Inze D, De Veylder L (2006) What if higher plants lack a CDC25 phosphatase? Trends Plant Sci 11: 474–479

Bray C, Sunderland PA, Waterworth WM, West CE (2008) DNA ligase – a means to and end joining. In: Bryant JA, Francis D (eds) The eukaryotic cell cycle. Taylor and Francis, Abingdon, UK, pp. 203–218

Bryant JA, Francis D (2008) The eukaryotic cell cycle. Taylor and Francis, Abingdon, UK

Bryant JA, Moore K, Aves SJ (2001) Origins and complexes: the initiation of DNA replication. J Exp Bot 52: 193–202

Burgress J (1985) An introduction to plant cell development. Cambridge University Press, Cambridge, UK

Cannon MC, Terneus K, Hall Q, Tan L, Wang Y, Wegenhart BL, Chen L, Lamport DTA, Chen Y, Kieliszewski MJ (2008) Self-assembly of the plant cell wall requires an extensin scaffold. Proc Natl Acad Sci USA 105: 2226–2231

Clowes FAL (1956) Localisation of nucleic acid synthesis in root meristems. J Exp Bot 7: 307–312

Clowes FAL (1958) Development of quiescent centres in root meristems. New Phytol 57: 85–88

Clowes FAL (1967) The quiescent centre. Phytomorphology 17: 132–140

Clowes FAL (1970) The immediate response of the quiescent centre to X-rays. New Phytolgist 69: 1–18

Collings DA, Gebbie LK, Howles PA, Hurley PA, Birch RJ, Cork AH, Hocart CH, Arioli T, Williamson RE (2008) *Arabidopsis* dynamin-like protein DRP1A: a null mutant with widespread defects in endocytotosis, cellulose synthesis, cytokinesis, and cell expansion. J Exp Bot, doi:10.1093/jxb/erm324

De Schutter K, Joubes J, Cools T, Verekest A, Corellou F, Babiychuk E, Can Der Scheren E, Beeckman T, Kushnir S, Inze D, De Veylder L (2007) *Arabidopsis* WEE1 kinase controls cell cycle arrest in response to activation of the DNA integrity checkpoint. Plant Cell 19: 211–225

del Pozo JC, Boniotti MB, Gutierrez C (2002a) *Arabidopsis* E2Fc functions in cell division and is degraded by the ubiquitin-SCF^{AtSKP2} pathway in response to light. Plant Cell 14: 3057–3071

del Pozo JC, Dharmasiri S, Helman H, Walker L, Gray WM, Estelle M (2002b) AXR1-ECR1-dependent conjugation of RUB1 to the *Arabidopsis* Cullin AtCUL is required for auxin response. Plant Cell 14: 421–433

del Pozo JC, Diaz-Trivino S, Cisneros N, Gutierrez C (2006) The balance between cell division and endoreduplication depends on E2Fc-DPb, transcription factors regulated by the ubiquitin –SCF^{SKP2A} pathway. Plant Cell 18: 2224–2235

Diffley JF (2004) Regulation of early events in chromosome replication. Curr Biol 14: R778–R786

Diffley JFX, Cocker JH (1992) Protein–DNA interactions at a yeast replication origin. Nature 357: 169–172

Diffley JF, Labib K (2002) The chromosome replication cycle. J Cell Sci 115: 869–872

Dudits D, Cserháti M, Miskolczi P, Horváth (2007) The growing family of plant cyclin-dependent kinases with multiple functions in cellular and developmental regulation. In: Inze D (ed) Cell cycle control and plant development. Blackwell, Oxford, UK, pp. 1–30

Dyson N (1998) The regulation of E2F by pRB-family proteins. Genes Dev 12: 2245–2262

Evans T, Rosenthall E, Youngbloom J, Distel D, Hunt T (1983) Cyclin: a protein specified by maternal RNA in sea urchin eggs that is destroyed at each cleavage division. Cell 33: 389–396

Feiler HS, Desprez T, Santoni V, Kronenberger J, Caboche M, TraasJ (1995) The higher plant *Arabidopsis thaliana* encodes a functional CDC48 homologue which is highly expressed in dividing and expanding cells. EMBO J 14: 5626–5637

Feldman LJ, Torey JG (1975) The quiescent centre and primary vascular pattern formation in cultured roots of Zea. Can J Bot 53: 2796–2803

Ferreira PC, Hemerly AS, Villarroel R, Van Montagu M, Inze D (1991) The *Arabidopsis* functional homolog of the p34cdc2 protein kinase. Plant Cell 3: 531–540

Ferreira P, Hemerly A, de Almeida Engler J, Bergounioux C, Burssens S, Van Montagu M, Engler G, Inze D (1994) Three discrete classes of *Arabidopsis* cyclins are expressed during different intervals of the cell cycle. Proc Natl Acad Sci USA 91: 11313–11317

Francis D, Barlow PW (1988) Temperature and the cell cycle. Symp Soc Exp Biol 42: 181–202

Francis D, Davies MS, Barlow PW (2008) A strong nucleotypic effect on the cell cycle regardless of ploidy level. Ann Bot 101(6): 747–757

Gavin KA, Hidaka M, Stillman B (1995) Conserved initiator proteins in eukaryotes. Science 270: 1667–1771

Guo J, Song J, Wang F, Zhang XS (2007) Genome-wide identification and expression analysis of rice cell cycle genes. Plant Mol Biol 64: 349–360

Heese M, Gansel X, Sticher L, Wick, P, Grebe M, Granier F, Jürgens, G (2001) Functional characterization of the KNOLLE-interacting t-SNARE AtSNAP33 and its role in plant cytokinesis. J Cell Biol 155: 239–249

Hemerly A, de Almeida Engler J, Bergounioux C et al. (1995) Dominant negative mutants of Cdc2 kinase uncouple cell division from iterative plant development. EMBO J 14: 3925–3936

Hirano H, H, A Sekine M (2008) *Arabidopsis* retinoblastoma-related protein 1 is involved in G1 phase cell cycle arrest caused by sucrose starvation. Plant Mol Biol 66: 259–275

Inze D (2007) Cell cycle control and lant development. Blackwell, Oxford, UK

Joubes J, Chevalier C, Dudits D, Heberle-Bors E, Inze D, Umeda M, Renaudi JP (2000) CDK-related protein kinases in plants. Plant Mol Biol 43: 607–620

Jurado S, Díaz-Triviño S, Abraham Z, Manzano C, Gutierrez C, del Pozo C (2008) SKP2A, an F-box protein that regulates cell division, is degraded via the ubiquitin pathway. Plant J 53: 828–841

Jürgens G (2005) Plant cytokinesis: fission by fusion. Trends Cell Biol 5: 277–283

Kamiya N, Nagasaki H, Morikami A, Sato Y, Matsuoka M (2003) Isolation and characterisation of rice WUCHSEL-type homeobox gene that is specifically expressed in the central cells of a quiescent center in the root apical meristem. Plant J 35: 429–441

Kaplan DR, Hagemann W (1991) The relationship of cell and organism in vascular plants – Are cells the building blocks of plant form? Bioscience 41: 693–703

Landrieu I, da Costa M, De Veylder L, Dewitte F, Vandepoele K, Hassan S, Wieruszeski J-M, Faure J-D, Van Montague M, Inze D, Lippens G (2004) A small CDC25 dual-specificity tyrosine-phosphatase isoform in *Arabidopsis thaliana*. Proc Natl Acad Sci USA 101: 13380–13385

Lauber MH, Waizenegger I, Steinmann T, Schwarz H, Mayer U, Hwang I, Lukowitz W, Jurgens G (1997) The *Arabidopsis* KNOLLE protein is a cytokinesis-specific syntaxin. J Cell Biol 139: 1485–1493

Lin XY, Kaul SS, Roundsley SD et al. (1999) Sequence and analysis of chromosome 2 of the plant *Arabidopsis thaliana*. Nature 402: 761–773

Lukowitz W, Mayer U, Jurgens G (1996) Cytokinesis in the Arabidopsis embryo involves the syntaxin-related KNOLLE gene product. Cell 84: 61–71

Lundgren K, Walworth N, Booher R, Dembski M, Kirschner M, Beach D (1991) mik1 and wee1 cooperate in the inhibitory tyrosine phosphorylation of cdc2. Cell 64: 1111–1122

McGowan CH, Russell P (1995) Cell cycle regulation of human WEE1. EMBO J 14: 2166–2175
Menges M, Muray JAH (2008) Plant D-type cyclins: structure, roles and functions. In:.Bryant JA, Francis D (eds) The eukaryotic cell cycle. Taylor and Francis, Abingdon, UK, pp. 1–28
Morgan DO (2007) The cell cycle. Principles of control. New Science Press, Oxford, UK
Mueller PR, Coleman TR, Kumagai A, Dunphy WG (1995) Myt1: a membrane-associated inhibitory kinase that phosphorylates Cdc2 on both threonine-14 and tyrosine-15. Science 270: 86–90
Murray JAH, Freeman D, Greenwood J, Huntley R, Makerh J, Riou-Khamlichi C, Sorrell DA, Cockcroft C, Carmichael JP, Soni R, Shaah ZH (1998) Plant D cyclins and retinoblastoma protein homologues. In: Francis D, Dudits D, Inze D (eds) Plant cell division. Portland Press, Colchester, UK, pp. 99–127
Nieuwland J, Menges M, Murray JAH (2007) The plant cyclins. In: Inze D (ed) Cell cycle control and plant development. Blackwell, Oxford, UK, pp. 31–61
O'Connell MJ, Raleigh JM, Verkade HM, Nurse P (1997) Chk1 is a WEE1 kinase in the G2 DNA damage checkpoint inhibiting cdc2 by Y15 phosphorylation. EMBO J 16: 545–554
Otegui MS, Mastronarde DN, Kang BH, Bednarek SY, Staehelin LA (2001) Three-dimensional analysis of syncytial type cell plates during endosperm cellularisation visualised by high resolution electron tomography. Plant Cell 13: 2033–2051
Porceddu A, Stals H, Reichheld JP, Segers G, De Veylder L, Barroco RP, Casteels P, Van Montagu M, Inze D, Mironov V (2001) A plant-specific cyclin-dependent kinase is involved in the control of G2/M progression in plants. J Biol Chem 276: 36354–36360
Praefka GJ, McMahon HT (2004) The dynamin superfamily: universal membrane tubulation and fission molecules? Nature Rev Mol Cell Biol 5: 133–147
Ramirez-Parra E, Frundt C, Gutierrez C (2003) A genome-wide identification of E2F-regulated genes in *Arabidopsis*. Plant J 33: 801–811
Ramirez-Parra E, del Pozo JC, Desvoyes B, de la Paz Sanchez M, Gutierrez C (2007) E2F-DP transcription factors. In: Inze D (ed) Cell cycle control and plant development. Blackwell, Oxford, UK, pp. 138–163
Ramos GB, de Almeida Engler J, Ferreira PC, Hemerly AS (2001) DNA replication in plants: characterization of a cdc6 homologue from *Arabidopsis thaliana*. J Exp Bot 52: 2239–2240
Rancour DM, Dickey CE, Park S, Bednarek SY (2002) Characterization of AtCDC48. Evidence for multiple membrane fusion mechanisms at the plane of cell division in plants. Plant Physiol 130: 1241–1253
Ruggiero B, Kowa H, Manabe Y et al. (2004) Uncoupling the effects of abscisic acid on plant growth and water relations. Analysis of sto1/nced3, an abscisic acid-deficient but salt stress tolerant mutant in *Arabidopsis*. Plant Physiol 136: 3134–3147
Sabelli PA, Burgess SR, Kush AK, Young MR, Shewry PR (1996) cDNA cloning and characterisation of a maize homologue of the MCM proteins required for the initiation of DNA replication. Mol Gen Genet 252: 125–136
Sanderfoot A (2007) Vesicle traffic at cytokinesis. In: Verma DPS, Hong Z (eds). Cell division control in plants. Springer, Berlin, pp. 289–302
Segers H, Gadisseur I, Bergounoux C et al. (1996) The *Arabidopsis* cyclin-dependent kinase gene cdc2b is preferentially expressed during S-phase and G2 phase of the cell cycle. Plant J 10: 601–612
Sorrell DA, Menges M, Healy JMS, Deveaux Y, Amano C, Kagami H, Shinmo A, Doonan JH, Sekine M, Murray JAH (2001) Cell cycle regulation of cyclin-dependent kinases in tobacco cultivar Bright Yellow-2 cells. Plant Physiol 126: 1214–1223
Sorrell DA, Marchbank A, McMahon K, Dickinson JR, Rogers HJ, Francis D (2002) A WEE1 homologue from *Arabidopsis thaliana*. Planta 215: 518–522
Sorrell DA, Chrimes D, Dickinson, JR, Rogers HJ, Francis D (2005) The *Arabidopsis* CDC25 induces a short cell length when over expressed in fission yeast: evidence for cell cycle function. New Phytol 165: 425–428

Springer PS, McCombie WR, Sundaresan V, Martienssen RA (1995) Gene trap tagging of PROLIFERA, an essential MCM2–3–5-like gene in *Arabidopsis*. Science 268: 877–880

Stevens R, Grelon M, Vezon D, Oh J, Meyer P, Perennes C, Domenichini S, Bergounioux C (2004) A CDC45 homolog in *Arabidopsis* is essential for meiosis, as shown by RNA interference-induced gene silencing. Plant Cell 16: 99–113

Sun Y, Dilkes BP, Zhang C, Dante RA, Carneiro NP, Lowe KS, Jug R, Gordon-Kamm WJ, Larkins BA (1999) Characterization of maize (Zea mays L.) Wee1 and its activity in developing endosperm. Proc Natl Acad Sci USA 96: 4180–4185

Tapia-López, García-Ponce B, Dubrovsky JG, Garay-Arroya A, Pérez-Ruíz V, Kim S-H, Acevedo F, Pelaz S, Alvarez-Buylla ER (2008) An Agamous-related MADS-box gene, XAL1 (AGL12) regulates root meristem cell proliferation and flowering transition in *Arabidopsis*. Plant Physiol 146: 1182–1192

Van't Hof J (1966) Experimental control of DNA synthesising and dividing cells in excised root tips of *Pisum sativum*. Am J Bot 53: 970–976

Van't Hof J (1985) Control points within the cell cycle. In: Bryant JA, Francis D (eds). The cell division cycle in plants. Cambridge University Press, Cambridge, UK, pp. 1–14

Völker A, Stierhof YD, Jurgens G (2001) Cell cycle-independent expression of the *Arabidopsis* cytokinesis-specific syntaxin KNOLLE results in mistargeting to the plasma membrane and is not sufficient for cytokinesis. J Cell Sci 114: 3001–3012

Wang H, Qi Q, Schorr P, Cutler AJ, Crosby WL, Fowke LC (1998) ICK1, a cyclin-dependent protein kinase inhibitor from *Arabidopsis thaliana* interacts with both Cdc2a and CYCD3 and its expression is induced by abscisic acid. Plant J 15: 501–510

Wang H, Zhou Y, Antonio J, Acosta T, Fowke LC (2007) CDK inhibitors. In: Inze D (ed) Cell cycle control and plant development. Blackwell, Oxford, UK, pp. 62–86

Wang H, Zhou Y, Fowke LC (1997) A plant cyclin-dependent kinase inhibitor geen. Nature 386: 451–452

Watanabe N, Arai H, Iwasaki J, Shiina M, Ogata K, HunterT, Osada H (2005) Cyclin-dependent kinase (CDK) phosphorylation destabilizes somatic Wee1 via multiple pathways. Proc Natl Acad Sci USA 102: 11663–11668

Weingartner M, Criqui MC, Meszaros T, Binarova P, Schmit AC, Helfer A, Derevier A, Erhardt M, Bogre L, Genschik P. 2004. Expression of a nondegradable cyclin B1 affects plant development and leads to endomitosis by inhibiting the formation of a phragmoplast. Plant Cell 1: 643–657

Wildwater M, Campilho A, Perez-Perez JM et al (2005) The retinoblastoma-related gene regulates stem cell maintenance in *Arabidopsis* roots. Cell 123: 1337–1349

Zhang K, Diederich L, John PC (2005) The cytokinin requirement for cell division in cultured Nicotiana plumbaginifolia cells can be satisfied by yeast Cdc25 protein tyrosine phosphatase: implications for mechanisms of cytokinin response and plant development. Plant Physiol 137: 308–316

Zhang K, Letham DS, John PC (1996) Cytokinin controls the cell cycle at mitosis by stimulating the tyrosine dephosphorylation and activation of p34cdc2-like H1 histone kinase. Planta 200: 2–12

Physiology

Solute Uptake in Plants: A Flow/Force Interpretation

M. Thellier(✉), C. Ripoll, V. Norris, M. Nikolic, and V. Römheld

Contents

Abstract Rates of uptake of silicic acid as a function of its concentration by the roots of various plants have been interpreted by their authors in the framework of a top-down approach based on the theoretical model of Michaelis and Menten. Although a hyperbola was fitted to the set of experimental points, this does not prove that all the simplifying assumptions underlying the Michaelis–Menten approach to transport were warranted. In consequence, the V_j^{m}- and K_j^{m}-values, which were inferred from these experiments, may not characterize the carrier molecules involved in the uptake process in the way that K_m- and V_m-values characterize classical enzymes (V_j^{m} = uptake rate when the carriers are saturated by the transported substrate, $1/K_j^{\mathrm{m}}$ = carrier affinity for this substrate). Since the experimental measurements have dealt only with solute uptake by entire biological systems (plant roots in the present case), here we adopt a "top-top" approach. The idea is

M. Thellier
Laboratory AMMIS, Faculty of Sciences of the University of Rouen,
F-76821 Mont-Saint-Aignan Cedex, France
e-mail: michel.thellier@univ-rouen.fr

U. Lüttge et al. (eds.), *Progress in Botany 70*,
DOI:10.1007/978-3-540-68421-3,

to find logical parameters that characterize these entire systems rather than look for key molecular constituents (e.g. carriers). By using the fact that the equations relating flows to forces always have a linear approximation when the system is close to equilibrium, we introduce two parameters to describe the uptake features of plant roots in the system of coordinates $\{\ln c_j^e, J_j(c_j^e)\}$: $°c_j^e$ = the particular value of the concentration c_j^e of the substrate, S_j, in the uptake medium for which the net flow, $J_j(c_j^e)$, of S_j exchange between medium and plant samples is zero and L_j = the overall conductance of the sample for S_j-uptake (= slope of the linear part of the curve $\{\ln c_j^e, J_j(c_j^e)\}$). We find that (1) the $°c_j^e$-values are of the order of magnitude expected in the experimental conditions used and (2) the greater the ability of a plant to accumulate silicon, the greater the L_j-value for the radial absorption of silicon by this plant. The flow/force approach to uptake is applicable to the absorption of any type of substrate by any type of plant. In the flow/force approach, $°c_j^e$ must equal the concentration of S_j in the growth solution of the plants prior to the experiments, which can be easily checked. It is possible to complement the interpretation of the uptake kinetics by using (1) a "symmetry criterion" to test whether uptake is active or passive and (2) an Arrhenius plot to check whether a modification of L_j corresponds to a modification of the uptake mechanism that is quantitative (e.g. number of carriers) or qualitative (e.g. post-translational modification of carriers).

1 Introduction

The uptake of various kinds of inorganic or organic solutes by plant systems (whole plants, plant tissues or cell cultures) has been the subject of many publications. In most cases, the results appear in the form of kinetic curves describing the rate of uptake of a substrate by the system under study as a function of the concentration of this substrate in the external medium under quasi-stationary conditions. By following "bottom-up" or "top-down" approaches, the authors have interpreted such measurements (carried out on the global system) in terms of parameters characteristic of underlying molecular and structural features. Here, we re-examine the bases of these classical approaches and complement them with a "top-top" approach.

To illustrate the top-top approach, we revisit existing experimental data recently published on silicon (Si) uptake by various systems of plant roots (Tamai and Ma 2003; Mitani and Ma 2005; Rains et al. 2006; Wu et al. 2006; Nikolic et al. 2007). The reasons for choosing silicon are that (1) Si is an element useful to plants, particularly under stress conditions (Epstein 1999; Ma et al. 2004, 2006; Liang et al. 2005; Ma and Yamaji 2006; Rains et al. 2006; Wu et al. 2006; Dragišić Maksimović et al. 2007; Ma et al. 2007), (2) this element is absorbed in the form of the uncharged species $Si(OH)_4$) (Raven 2003; Tamai and Ma 2003; Mitani and Ma 2005; Rains et al. 2006; Nikolic et al. 2007), which avoids problems due to an accompanying ion of opposite sign (Ripoll et al. 1984), (3) Si accumulation under natural conditions differs greatly between plant species (Epstein 1999; Tamai and Ma 2003; Mitani and Ma 2005; Ma and Yamaji 2006) and (4) precise data are now available on the nature and location within cells and tissues of the carrier molecules involved in Si uptake (Ma et al. 2004, 2006, 2007).

2 Classical Kinetic Approaches of Uptake Processes

2.1 General Experimental Conditions

Consider a substrate, S_j, which a plant system (plant roots in the present case) exchanges with its bathing medium and let J_j^{ei} and J_j^{ie} be the unidirectional inflow and outflow of S_j, with e = exterior and i = interior. The net flow of S_j, J_j, is expressed as

$$J_j = J_j^{ei} - J_j^{ie}. \tag{1}$$

Briefly, from the experimental point of view, the plants are initially grown in an appropriate growth solution. At the beginning of the kinetic experiments, they are transferred to experimental solutions containing increasing concentrations of S_j (sometimes only the separated roots are transferred to the experimental media). In some cases, the inflow is determined by measuring the tissue accumulation of S_j or the decrease of the S_j-concentration in the external medium. In other cases, especially when a radioactive tracer, S_j^*, is available, inflows can be determined by submerging the roots of non-labeled plants in media containing the S_j^*-labeled substrate and following the progressive increase of radioactivity in the plant samples. In this case, it is also possible to evaluate outflows by using plant samples previously loaded with S_j^* and following the outflow of radioactivity into unlabeled experimental media. Finally, the results are expressed in the form of curves relating the flows to the S_j-concentrations, for instance $\{c_j^e, J_j^{ei}\}$ or $\{c_j^e, J_j\}$ or $\{c_j^i, J_j^{ie}\}$, where c_j^e and c_j^i are the external and internal concentrations of S_j.

2.2 The Principles for "Bottom-Up" and "Top-Down" Approaches

A complete bottom-up interpretation of experimental data such as those obtained from a whole plant tissue system would require knowledge of the kinetic characteristics of the molecule(s) involved in the studied transport process, the topology of the arrangement (e.g. in parallel or in series) of these molecules and possible diffusion constraints. This would make it possible to build a realistic model of the system under consideration, to carry out a numerical simulation of the overall behaviour of this system and to determine the values of the main (molecular level) parameters. Investigations of this type include, for instance, the "reaction-kinetic" model (Sanders 1990), the "energy-barrier" models (White and Ridout 1995; White et al. 1999) and the "dynamic pore" model (Gradmann 1996).

However, most authors, following an original suggestion by Emmanuel Epstein and co-workers (Epstein and Hagen 1952; Epstein 1953, 1966), have reasoned differently using a top-down approach. Observing that the experimental $\{c_j^{e}, J_j^{ei}\}$ curves often increased monotonically with a negative curvature, they interpreted the cell uptake of S_j by a system of membrane carriers as formally equivalent to the catalysis of a reaction by an enzyme system of the Michaelis–Menten type. When it was possible to fit a single hyperbola to the uptake data (monophasic process), this was taken as meaning that a single carrier of S_j, C_j, was involved in the uptake process and that it was possible to describe the kinetics of functioning of C_j, under quasi-stationary conditions, by an equation of the Michaelis–Menten type

$$J_j^{ei} = (V_j^{m} c_j^{e}) / (K_j^{m} + c_j^{e}) \tag{2}$$

in which V_j^{m} and K_j^{m} were playing a role similar to that played by the parameters V_m and K_m in classical enzymology; V_j^{m} was thus considered to correspond to the maximum rate of uptake (C_j saturated by S_j) and $1/K_j^{m}$ to characterize the affinity of C_j for S_j. In more complicated cases, i.e. when two or more hyperbolas had to be used to fit a particular set of uptake results (dualphasic or multiphasic processes, respectively), this was taken as indicating that two or several different carriers (C_j^1, C_j^2, etc.) were contributing to overall uptake and that the main kinetic parameters of these carrier activities (V_j^{m1} and K_j^{m1}, V_j^{m2} and K_j^{m2}, etc.) could be obtained by using equations similar to (2) for characterizing each particular hyperbola. In cases where the experimental $\{c_j^{e}, J_j^{ei}\}$ curves exhibited a sigmoid shape, this was taken to mean that an allosteric carrier was involved (Glass 1976). Detailed theoretical analyses have shown that it is indeed possible to describe the kinetics of a carrier-mediated uptake process by use of a Michaelis–Menten model providing that a set of simplifying conditions are satisfied (see e.g. Neame and Richards 1972; Schachter 1972); the most important of these conditions are that a single type of carrier is responsible for uptake (hypothesis h1), the molecular characteristics of this carrier are independent of the location of the carrier (hypothesis h2), the binding of the carrier to its substrate is so rapid that this reaction may be considered to be at equilibrium (hypothesis h3), the stochastic changes in the conformation of the carrier are also very rapid (hypothesis h4), the carriers are arranged in parallel (hypothesis h5), the uptake process takes place under quasi-stationary conditions (hypothesis h6), and the diffusion constraints are negligible (hypothesis h7). When all these conditions are satisfied, then the resulting inflow for the whole system may be described as if it were controlled by an "equivalent carrier" the V_j^{m}-value of which is expressed by the sum

$$V_j^{m} = \Sigma^{real} V_j^{m} \tag{3}$$

(where ${}^{real}V_j^{m}$ is the saturation rate of the real, individual carrier molecules and Σ represents a summation over all these real carriers) and the K_j^{m}-value is equal to that, ${}^{real}K_j^{m}$, of the real, individual carriers, i.e.

$$K_j^{m} = {}^{real}K_j^{m} \tag{4}$$

(monophasic process described by (2)). Moreover, it is also possible to adopt this approach when two or several different types of carriers are involved, on condition that their K_j^m-values are sufficiently different from one another (dual-phasic or multiphasic processes described by two or more equations with a form similar to that of (2)). In all these cases, the measurement of the K_j^m-value of each apparent carrier of the overall system represents the ${}^{real}K_j^m$ value of the real corresponding carriers; moreover, dividing the V_j^m-value of the apparent carrier by the total number of carrier molecules involved gives an evaluation of the ${}^{real}V_j^m$-value of these carrier molecules. The Michaelis–Menten interpretation of uptake, the aim of which is to obtain the values of molecular parameters from the measurement of parameters characteristic of the overall system, is thus typical of an approach of the top-down type.

2.3 Application of the Michaelis–Menten, Top-Down Approach to Si Uptake by Plant Roots

The uptake of Si is known to involve at least two processes, the radial transport of Si from the external solution to the root cells and the release of Si from the root symplast into the xylem vessels (Mitani and Ma 2005). The radial transport of Si has been studied by exposing various plants (differing in the silicon content of their tissues under natural conditions) to external solutions containing increasing concentrations of $Si(OH)_4$. Thus the symbol j now represents Si [or $Si(OH)_4$]. The Si-inflows in the plant roots were determined by measuring the Si-accumulation in the root cells (Tamai and Ma 2003; Mitani and Ma 2005; Wu et al. 2006), or the decrease of the Si concentration in the experimental solutions (Rains et al. 2006), or the increase of radioactivity in the roots when Si was labeled with the radioactive

Table 1 Michaelis–Menten, top-down interpretation of published data of silicon uptake by the roots of various plants

References	Plant	Correlation factor	K_j^m (mM)	V_j^m
Tamai and Ma (2003)	Rice	0.981	0.58	8.33
Mitani and Ma (2005)	Rice	0.996	0.21	42.1
	Cucumber	0.990	0.23	33.5
	Tomato	0.996	0.26	16.7
Rains et al. 2006	Wheat	0.968	0.12	4.57
Wu et al. 2006	Rice (DV85)	0.98	0.59	9.9
	Rice (Kinmaze)	0.97	0.57	12.2
Nikolic et al. 2007	Rice	0.98	0.51	0.384
	Barley	0.98	0.37	0.114
	Cucumber	0.94	1.52	0.084
	Tomato	0.79	1.88	0.031

V_j^m is expressed in mg Si per g of root (dry weight) per h (Tamai and Ma 2003; Wu et al. 2006), ng Si per whole root of each plant per 8 h of uptake (Mitani and Ma 2005), µmol Si per g of roots (fresh weight) per h (Rains et al. 2006), mmol Si per g of roots (dry weight) per h (Nikolic et al. 2007)

isotope ^{68}Ge in the external medium (Nikolic et al. 2007). The authors followed what we have termed above the top-down approach. They estimated the values of the parameters V_j^m and K_j^m either by direct graphical determination on the $\{c_j^e, J_j^{ei}\}$ curves or by using linear graphs, e.g. in the $\{1/c_j^e, 1/J_j^{ei}\}$ system of coordinates. For the sake of homogeneity, we have recalculated the parameter values from the authors' data by use of linear graphs in the $\{c_j^e, c_j^e/J_j^{ei}\}$ system of coordinates (which is usually considered to be the most reliable method of linearisation). The results, which are not very different from those originally announced by their authors, are given in Table 1.

In all cases, the fitting of a straight line to the uptake data in the system of coordinates $\{c_j^e, c_j^e/J_j^{ei}\}$ appeared to be satisfactory (correlation factors usually close to 0.98), which means that a unique Michaelis–Menten model (2)) can explain the uptake data for the radial transport of silicon over the range of $Si(OH)_4$-concentrations used in the experiments. Moreover, when considering plants ranging from those that accumulate much Si to those that do not, the V_j^m values were observed to decrease systematically (Mitani and MA 2005; Nikolic et al. 2007). This is interpreted by the authors as suggesting that the density of Si-carriers on the root cell membranes differs in the plant species in the order rice > barley > cucumber > tomato. In contrast, the different authors disagree over the K_j^m-values and therefore disagree over the carrier affinity for $Si(OH)_4$. According to Mitani and Ma (2005), the K_j^m-value is practically invariable and in the range of 0.2–0.3 mM whatever the plant species, while it is either significantly lower (Rains et al. 2006) or significantly higher (Tamai and Ma 2003; Wu et al. 2006; Nikolic et al. 2007) especially in plants that do not accumulate much Si under natural conditions (Nikolic et al. 2007).

2.4 Re-Examination of the Applicability of the Michaelis–Menten Top-Down Approach

If it could be proven that a Michaelis–Menten equation is actually the correct theoretical model for the above measurements of Si uptake, then this top-down approach would be extremely powerful. It would indeed permit the values of important molecular parameters (number, affinity and saturation rate of the carrier molecules involved) to be derived from measurements performed with the overall root system.

Recent investigations (Ma et al. 2006; Ma and Yamaji 2006) have shown that a gene (*Lsi1*) is responsible for Si uptake in rice roots and that the encoded protein is localised on the plasma membrane of the distal side of both the exodermis and endodermis cells (where solute-impermeable Casparian strips exist). Such data about the existence and location of a Si carrier are consistent with some of the simplifying conditions required for the top-down approach (especially hypotheses h1, h2 and h5 in Sect 2.2) and are not in contradiction with the Michaelis–Menten top-down approach and the molecular interpretation of the parameters K_j^m and V_j^m.

However, despite these remarkable data, it is not possible to be sure that all the simplifying conditions listed in Sect 2.2 are actually satisfied. Moreover, the authors have only shown from the kinetic measurements (see Sect 2.3) that a Michaelis–Menten hyperbola can be fitted to the experimental data: they have not shown that the Michaelis–Menten model is the only – let alone the best – theoretical model. Importantly, the K_j^{m} and V_j^{m} parameters are in fact complicated functions of the internal concentrations and of the rate constants of the molecular events involved in the transport process. For instance, it is only in the rare (and not easily demonstrated) case when the transmembrane carriers are totally symmetrical (i.e. when inverting the concentrations on the two faces of the membrane reverses the direction of the substrate flow without changing its value) that $1/K_j^{\mathrm{m}}$ represents the affinity of the carrier for its substrate. Finally, it has also been shown that more or less complex natural or artificial transport systems exhibit monophasic or dualphasic kinetic behaviour without meeting the Michaelis–Menten requirements (Vincent and Thellier 1983; Vincent et al. 1988a, b; Thellier et al. 2006); in these cases, the apparent values of K_j^{m} and V_j^{m} that can be obtained by fitting hyperbolas to the uptake data no longer have a simple biological significance.

In brief, a full exploitation of the Michaelis–Menten approach would require being able to numerically simulate the functioning of the carrier network in all its complexity. In the absence of a complete knowledge of the network topology, it is likely that the estimated K_j^{m} and V_j^{m} do not have the molecular meaning often attributed to them and that, even when they have, it is usually not possible to be sure of this.

3 Flow/Force, A "Top-Top" Approach

In general, the parameters K_j^{m} and V_j^{m} not only lack a clear molecular significance but are also hard to relate to the properties of the overall system. This is because the Michaelis–Menten top-down approach was formulated for molecular analyses rather than for characterising overall properties. As an alternative, we propose here to follow an approach we term "top-top" that is less ambitious in that it ignores the microscopic complexity (possible existence of several carriers and/or channels, interference by the uptake of other substrates, role of the cell wall, etc.), but that may be more realistic in that it is based on the flow/force relationships of non-equilibrium thermodynamics (for an introduction to non-equilibrium thermodynamics, see e.g. Katchalsky and Curran 1965). Our aim is to treat the problems of uptake kinetics entirely (i.e. from the point of view of both measurements and interpretation) at the level of the overall system (plant roots in the present case). Non-equilibrium thermodynamics provides unambiguous overall parameters for the transport processes which make it possible to compare plant systems under variable physiological conditions. The top-top approach is an extension of flow/force modelling and experiments that have been described in previous publications (e.g. Thellier 1971; Ripoll et al. 1984; Thellier et al. 1993) and that may be summarised as follows.

3.1 The Formalism

Consider a set of plants with roots bathed in a nutritive solution containing a constant concentration, $°c_j^e$, of a substrate S_j (e.g. silicic acid). After a time sufficient for equilibration of the internal with the external medium, the balance of the exchanges of S_j is obviously zero (insofar as additional absorption of S_j corresponding to growth can be taken as negligible for short periods of measurement). This is written

$$J_j(°c_j^e) = 0 \tag{5}$$

in which $J_j(°c_j^e)$ is the net rate of exchange of S_j when the external concentration of S_j is $°c_j^e$. Moreover, the chemical potential of S_j in the external solution is then equal to $\mu_j(°c_j^e)$ with

$$\mu_j(°c_j^e) = RT\ln(°\gamma_j^e\, °c_j^e) \tag{6}$$

in which R is the gas constant, T is the absolute temperature and $°\gamma_j^e$ is the activity coefficient of S_j in the nutritive solution.

At the beginning of each uptake experiment, the external concentration of S_j is changed to a value c_j^e either larger or smaller than $°c_j^e$ (uptake solution), all other features of the external solution remaining unchanged. The chemical potential of S_j in the external solution thus becomes equal to $\mu_j(c_j^e)$ with

$$\mu_j(c_j^e) = RT\ln(\gamma_j^e c_j^e) \tag{7}$$

in which γ_j^e is the activity coefficient of S_j in the uptake solution. This change in the chemical potential of S_j in the external medium causes the initial net rate of exchange of S_j between plants and medium, $J_j(c_j^e)$, to become non-zero. This may be written formally

$$J_j(c_j^e) = \lambda_j(\mu_j(c_j^e) - \mu_j(°c_j^e)) = RT\lambda_j(\ln((\gamma_j^e c_j^e)/(°\gamma_j^e\, °c_j^e))) \tag{8}$$

in which λ_j relates the "flow" of S_j, $J_j(c_j^e)$, to the "force" acting on the exchanges of S_j as a consequence of the difference of chemical potential, $RT(\ln(\gamma_j^e c_j^e) - \ln(°\gamma_j^e\, °c_j^e))$. Since the composition of the uptake solution is not very different from that of the nutritive solution, one may assume that the activity coefficient of S_j has very similar values in the two solutions

$$\gamma_j^e \# °\gamma_j^e \tag{9}$$

and therefore

$$J_j(c_j^e) = L_j \ln((c_j^e)/(°c_j^e)) \tag{10}$$

in which L_j groups several parameters

$$L_j = RT\lambda_j. \tag{11}$$

Moreover, the quantity $RT \ln(1/(^{\circ}c_j^{e}))$ may be considered to correspond to the resulting effect of all terms other than the external concentration of S_j (e.g. internal concentration of S_j, c_j^{i}, internal activity coefficients, possible couplings with metabolic reactions and/or with transport of other substances, etc.) that contribute to the force driving the uptake of S_j. In practice, the measurement of $J_j(c_j^{e})$ has to be carried out in a time short enough for these contributions to the force (especially c_j^{i}) not to be changed significantly during the measurements. Moreover, it is clear in (10) that $J_j(c_j^{e})$ will be positive (net entry of S_j into the plant roots) when $c_j^{e} > {}^{\circ}c_j^{e}$), while it will be negative (net exit of S_j from the plant roots) when $c_j^{e} < {}^{\circ}c_j^{e}$.

According to non-equilibrium thermodynamics, the parameter λ_j, and therefore also L_j, are equivalent to conductance terms that may be assumed to keep a constant value as long as the system remains close enough to equilibrium (i.e. as long as $J_j(c_j^{e})$ is not too large). In this sense, (10) may be considered to be the equivalent for an uptake process of Ohm's law for an electric process.

By plotting the data of previous authors (e.g. Andrews 1966; Lüttge and Laties 1966; Torii and Laties 1966; Rains and Epstein 1967) in the system of coordinates {ln (c_j^{e}), $J_j(c_j^{e})$} it was indeed observed (Thellier 1970) that a straight line could often be fitted to the experimental data over up to two orders of magnitude (or even more) of the values of c_j^{e}, when the c_j^{e} and $J_j(c_j^{e})$ values were not too high. For higher c_j^{e} values (i.e. beyond the domain of linearity of the plot), we tried to model the non-linear behaviour by considering it as formally equivalent to that of an electrical varistor (Thellier 1969), but this had no theoretical justification thus making biological interpretation difficult. When there is no clear linear region, even for very small values of $J_j(c_j^{e})$, this may correspond either to the linear region being very short or to rapid adaptation of the system at each imposed value of c_j^{e} (Falkner et al. 1989) and a consequent modification of parameters L_j and $\ln(1/{}^{\circ}c_j^{e})$ during the experiment.

3.2 *Optimal Design of Experiments for the Flow/Force Approach*

For an optimal application of the formalism developed above, the experiments should be conducted as follows. To allow equilibration, the plants are grown for a long period of time (e.g. for several weeks following germination) in a nutritive solution (the growth medium) containing the substrate, S_j, at a given concentration, ${}^{\circ}c_j^{e}$. The composition of this nutritive solution is maintained constant by continuous (or at least frequent) renewal. The plants are old enough when the uptake experiments are performed for their growth to be negligible during measurements of S_j uptake. The uptake experiments entail maintaining the plants in their growth medium but changing concentration ${}^{\circ}c_j^{e}$ to c_j^{e} in this external medium; this may be achieved by adding to the external medium an appropriate amount of S_j ($c_j^{e} > {}^{\circ}c_j^{e}$) or by diluting the external medium with an appropriate volume of a solution lacking S_j but otherwise identical in composition to the growth medium of the plants ($c_j^e < {}^\circ c_j^e$). The initial net rate of exchange of S_j between plants and medium, $J_j(c_j^e)$, which corresponds to the difference between the inflow, $J_j^{ei}(c_j^e)$, and the outflow, $J_j^{ie}(c_j^e)$, of S_j

$$J_j(c_j^e) = J_j^{ei}(c_j^e) - J_j^{ie}(c_j^e) \tag{12}$$

is measured in a time short enough for the terms in (12) to correspond to actual initial rates of transport. Equation (12) is a slightly more explicit form of (1). There are different possibilities for carrying out the measurement of $J_j(c_j^e)$. When appropriate radioactive labels exist, the best approach consists in (1) measuring $J_j^{ei}(c_j^e)$ by determining the accumulation into unlabeled plants of radioactively labeled S_j from the experimental solution, (2) measuring $J_j^{ie}(c_j^e)$ by determining the outflow of S_j from plants previously enriched with radioactively labeled S_j into an unlabeled experimental solution, and (3) evaluating $J_j(c_j^e)$ using (12).

3.3 Principle of the Interpretation of Experimental Data

The predictions of the flow/force approach for experiments as described in Sect 3.2 are that the curve $\{\ln(c_j^e), J_j(c_j^e)\}$ ((10)) intersects the abscissa at the point $\{\ln({}^\circ c_j^e), J_j(c_j^e) = 0\}$ and that this curve is linear on either side of intersection providing $J_j(c_j^e)$ (which can be positive or negative) is close to 0 (Fig. 1).The value of parameter L_j

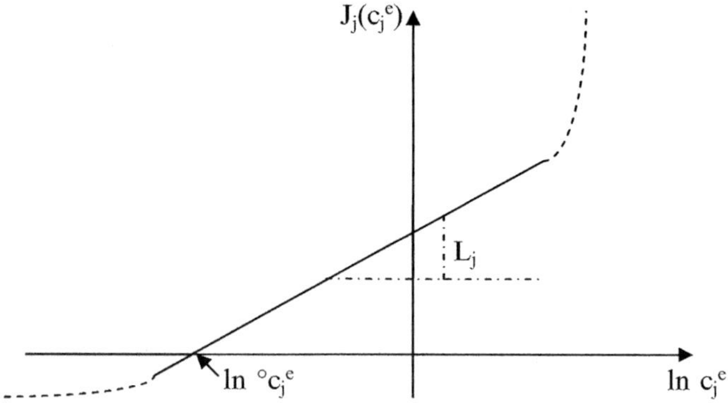

Fig. 1 Theoretical dependence of $J_j(c_j^e)$ (the rate of uptake of a solute S_j) upon ln c_j^e (where c_j^e is the concentration of S_j in the external medium of the plants during the uptake measurements). The theoretical model is (10). The curve {ln c_j^e, $J_j(c_j^e)$} admits a linear approximation close to the equilibrium situation (i.e. when $J_j(c_j^e)$ is not too different from zero), then it becomes non-linear (*dotted lines*). The slope of the linear part of the curve is a measurement of the overall conductance, L_j, of the plant system under study for S_j. The rate of uptake, $J_j(c_j^e)$, becomes equal to zero when $c_j^e = {}^\circ c_j^e$ (where ${}^\circ c_j^e$ is the concentration of S_j in the growth solution of the plants prior to the uptake experiments)

is then equal to the slope of the linear part of the curve $\{\ln(c_j^e), J_j(c_j^e)\}$ in the vicinity of the point $\{\ln(°c_j^e), J_j(c_j^e) = 0\}$.

3.4 *Application of the Flow/Force Approach to Si Uptake by Plant Roots*

Let us assume that the kinetic experiments of the previous authors may be reinterpreted in the terms of our flow/force approach, although these experiments were not carried out under the exact experimental conditions as described in Sect 3.2. The index j now holds for silicic acid in the symbols $°c_j^e$, c_j^e, $J_j(c_j^e)$, L_j, etc., occurring in (5)–(12).

As an example, in Fig. 2 we have plotted the original data of Tamai and Ma (2003) [uptake of silicic acid by rice (*Oryza sativa* L. cv. Oochicara) roots] in the system of coordinates $\{\ln(c_j^e), J_j(c_j^e)\}$ and have fitted these values to a straight line (correlation factor = 0.98). This fit is just as good as that obtained in the system of coordinates $\{(c_j^e), J_j(c_j^e)\}$ to a Michaelis–Menten hyperbola. The computed values of the main parameters here are $°c_j^e = 0.056$ mM and $L_j = 1.83$ mg Si per g of root (dry weight) per h.

When the original data obtained by various groups of authors were plotted in the system of coordinates $\{\ln(c_j^e), J_j(c_j^e)\}$ (not shown), the computed values of the main parameters for fitting a straight line to the data are as indicated in Table 2. The correlation factors for the fitting of a linear model are again close to 0.98. The $°c_j^e$-values

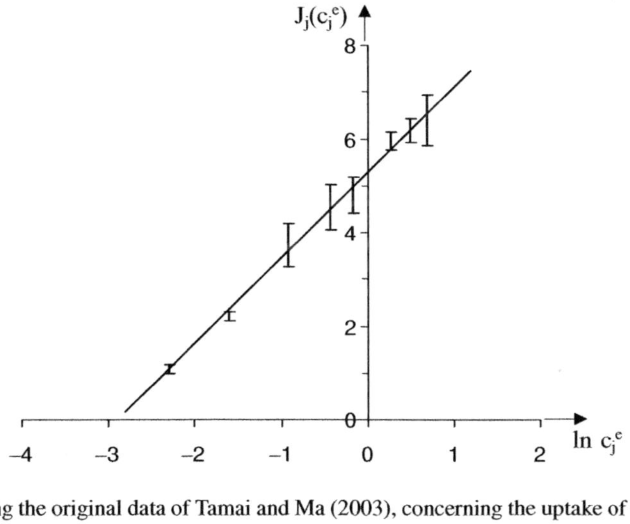

Fig. 2 Plotting the original data of Tamai and Ma (2003), concerning the uptake of silicon by rice roots, in the system of coordinates $\{\ln c_j^e, J_j(c_j^e)\}$. A straight line can be adjusted to the experimental data (correlation factor = 0.98). The equation of this straight line is $J_j(c_j^e) = 1.83 \ln c_j^e + 5.26$, when c_j^e is measured in mM and $J_j(c_j^e)$ in mg Si per g of root (dry weight) per h. The estimated values of $°c_j^e$ and L_j are 0.056 mM and 1.83 mg Si per g of root (dry weight) per h

are always low (of the order of a few tens of µM in the experiments reported in Tamai and Ma 2003, Mitani and Ma 2005, Wu et al. 2006 and Nikolic et al. 2007); this is consistent with the fact that, in all these experiments, the growth medium of the plants was without any addition of silicic acid and thus contained only contaminant silicon. Moreover, the $^{\circ}c_j^{e}$-value was especially low in the experiments by Rains et al. (2006), in which measures were taken to eliminate silicon from the growth solution; significantly, this $^{\circ}c_j^{e}$-value (8 µM) then approaches the Si-concentration in the growth solution (≤4 µM). Since different authors have expressed the rates of Si uptake using different systems of units, the L_j-values are also expressed with different units and thus cannot readily be compared. However, it is clear from Table 2 that (1) the L_j-values are fairly close to each other (differing by only 10%) for two different varieties of rice plants (data by Wu et al. 2006) while (2) they are very different from one another for different plant species and they decrease in parallel with the decreasing ability of these plants to accumulate silicon under natural conditions (Mitani and Ma 2005 and Nikolic et al. 2007).

Despite the likely complexity of the molecular events involved, the overall conductance for the radial transport of Si by a plant root system thus plays a key role in the control of Si accumulation by this plant.

3.5 Suggestions for Further Experiments

For future kinetic studies of uptake processes, it would be interesting to carry out these investigations exactly as indicated in Sect 3.2 in order to facilitate the flow/force interpretation of the measurements.

To examine the possible involvement of an active process in the uptake of a substrate under study (coupling with a metabolic reaction or with the transport of

Table 2 Proposed flow/force, top-top interpretation of the same published data of silicon uptake by the roots of various plants as in Table 1

References	Plant	Correlation factor	$^{\circ}c_j^{e}$ (mM)	L_j
Tamai and Ma (2003)	Rice	0.98	0.056	1.83
Mitani and Ma (2005)	Rice	0.98	0.020	8.78
	Cucumber	0.97	0.018	6.69
	Tomato	0.97	0.020	3.34
Rains et al. 2006	Wheat	0.97	0.008	0.94
Wu et al. 2006	Rice (DV85)	0.98	0.061	2.23
	Rice (Kinmaze)	0.96	0.059	2.77
Nikolic et al. 2007	Rice	0.97	0.040	0.080
	Barley	0.98	0.029	0.023
	Cucumber	0.96	0.061	0.013
	Tomato	0.88	0.064	0.004

L_j is expressed in mg Si per g of root (dry weight) per h (Tamai and Ma 2003; Wu et al. 2006), ng Si per whole root of each plant per 8 h of uptake (Mitani and Ma, 2005), µmol Si per g of roots (fresh weight) per h (Rains et al. 2006), mmol Si per g of roots (dry weight) per h (Nikolic et al., 2007)

another substrate), it is not enough to study the effect of metabolic inhibitors. The reason is that an observed effect might be a side-effect of the action of the inhibitor. The classical way for checking unambiguously the possible involvement of an active transport consists in using Ussing's "flux-ratio" equation (Ussing 1949, 1971). However, with biological specimens other than epithelia, the measurements for determining the variables of this equation require rather intricate morphometric and compartmental analysis studies (Gaudinet et al. 1984). Our group has proposed a "symmetry criterion" (Lassalles and Thellier 1977) that can be used instead of the flux-ratio equation when studying the uptake of substrates by cell or tissue samples. Since it is likely that the same transport process is involved in the uptake of both Si and Ge (Rains et al. 2006; Nikolic et al. 2007), the "symmetry criterion" is particularly appropriate for determining the active or passive character of both Si and Ge transport in plant roots. In this case, the experimental investigation would simply consist of measuring the net flow of Si at a constant concentration of Si and a variable concentration of Ge in the uptake medium and, conversely, in measuring the net flow of Ge at a constant concentration of Ge and a variable concentration of Si in the uptake medium (Lassalles and Thellier 1977).

We have seen above that the difference between plants that have a high natural content of Si in their tissues (e.g. rice) and those known as Si-excluding species (e.g. tomato) is that they exhibit high and low L_j-values, respectively. A priori, this difference in radial conductance could be either quantitative (e.g. concentration of Si-carriers) or qualitative (e.g. catalytic efficacy of these carriers). In the case of other substrates taken up by other plant species, studying the dependence of log L_j on $1/T$ (Arrhenius plot) helps distinguish between qualitative and quantitative differences (Thellier et al. 1971): parallel straight lines correspond to a purely quantitative difference whilst non-parallel straight lines correspond to a qualitative difference. It would be interesting to conduct new experiments to determine the Arrhenius dependence of the differences in the L_j-values that have been measured in the uptake of silicon by the roots of different plants. Moreover, since L_j is in fact the product of several parameters (see (11)), one of which is the absolute temperature T, it might be preferable to study the dependence of $\log(L_j/T)$ on $1/T$ rather than that of log L_j on $1/T$.

4 Conclusion

Although it is possible to fit a hyperbola to the experimental points representing the rate of uptake of silicic acid by plant roots as a function of the concentration of this substrate in the external medium, it is virtually impossible to decide whether all the simplifying assumptions required to use a Michaelis–Menten model to interpret uptake kinetics are really founded. Under such circumstances, (1) it is not certain that the K_j^{m}- and V_j^{m}-values that are obtained from the measurements have a molecular significance (i.e. a significance equivalent to that of parameters K_m and V_m in classical enzymology) and, even if they have, it is usually impossible to know that this is actually the case, and (2) K_j^{m} and V_j^{m} should rather be considered as phenomenological parameters of the overall system. However, kinetic studies of uptake processes can be interpreted providing that the measurements, which are carried out on complex plant systems, are also interpreted relative to the whole systems (top-top approach). By taking advantage of the linear dependence close to equilibrium of flows on forces, the values of two main parameters, $°c_j^e$ and L_j, may be determined. The first of these parameters makes it possible to check the validity of the approach since the measured $°c_j^e$-value must be equal to the concentration of the substrate, S_j, in the growth medium before the start of the experiment. The second parameter, L_j, represents the combined effect of the molecular processes on the overall conductance of the tissue system for the transport of the substrate. For different plant species, the conductance of the roots for the radial transport of Si decreases in parallel with the tissue accumulation of Si by these plants under natural conditions. Although no experimental data are available at present, (1) the use of Arrhenius plots might make it possible to check whether the origin of species differences in L_j-values is quantitative (e.g. differences in the numbers of carriers in the root cell membranes) or qualitative (e.g. differences in the catalytic efficiencies of carriers), and (2) the use of the symmetry-criterion would permit one to decide unambiguously whether uptake is an active process.

References

Andrews CS (1966) A kinetic study of phosphate absorption by excised roots of *Stylosanthes humilis*, *Phaseolus lathyroides*, *Desmodium uncinatum*, *Medicago sativa* and *Hordeum vulgare*. Aust J Agric Res 17:611–624

Dragišić Maksimović J, Bogdanović J, Maksimović V, Nikolic M (2007) Silicon modulates the metabolism and utilization of phenolic compounds in cucumber (*Cucumis sativus* L.) grown at excess manganese. J Plant Nutr Soil Sci 170:739–744

Epstein E (1953) Mechanism of ion absorption by roots. Nature 171:83–84

Epstein E (1966) Dual pattern of ion absorption by plant cells and by plants. Nature 212:1324–1327

Epstein E (1999) Silicon. Annu Rev Plant Physiol Plant Mol Biol 50:641–664

Epstein E, Hagen CE (1952) A kinetic study of the absorption of alkali cations by barley roots. Plant Physiol 27:457–474

Falkner G, Falkner R, Schwab AJ (1989) Bioenergetic characterization of transient state phosphate uptake by the cyanobacterium *Anacystis nidulans*: theoretical and experimental basis for a sensory mechanism adapting to varying environmental phosphate levels. Arch Microbiol 152:353–361

Gaudinet A, Ripoll C, Thellier M, Kramer D (1984) Morphometric study of *Lemna gibba* in relation to the use of compartmental analysis and the flux-ratio equation in higher plant cells. Physiol Plant 60:493–501

Glass ADM (1976) Regulation of potassium absorption in barley roots: an allosteric model. Plant Physiol 58:33–37

Gradmann D (1996) Selectivity of ion channels: competitive catalysis versus independent electrodiffusion. J Exp Bot 47:1733–1736

Katchalsky A, Curran PF (1965) Nonequilibrium thermodynamics in biophysics. Harvard University Press, Cambridge, MA

Lassalles JP, Thellier M (1977) Discussion on the symmetry of macroscopic coefficients in the formulation of the cellular fluxes: possible application to a "symmetry criterion" for the study of active and passive transports. J Theor Biol 68:53–63

Liang YC, Sun WC, Si J, Römheld V (2005) Effects of foliar- and root-applied silicon on the enhancement of induced resistance to powdery mildew in *Cucumis sativus*. Plant Pathol 54:678–685
Lüttge U, Laties GG (1966) Dual mechanisms of ion absorption in relation to long distance transport in plants. Plant Physiol 41:1531–1539
Ma JF, Yamaji N (2006) Silicon uptake and accumulation in higher plants. Trends Plant Sci 11:392–397
Ma JF, Mitani N, Nagao S, Konishi S, Tamai K, Iwashita T, Yano M (2004) Characterization of the silicon uptake system and molecular mapping of the silicon transporter gene in rice. Plant Physiol 136:3284–3289
Ma JF, Tamai K, Yamaji N, Mitani N, Konishi S, Katsuhara M, Ishiguro M, Murata Y, Yano M (2006) A silicon transporter in rice. Nature 440:688–691
Ma JF, Yamaji N, Mitani N, Tamai K, Konishi S, Fujiwara T, Katsuhara M, Yano M (2007) An efflux transporter of silicon in rice. Nature 448:209–212
Mitani M, Ma JF (2005) Uptake system of silicon in different plant species. J Exp Bot 56:1265–1261
Neame KD, Richards TG (1972) Elementary kinetics of membrane carrier transport. Blackwell, Oxford
Nikolic M, Nikolic N, Liang Y, Kirkby EA, Römheld V (2007) Germanium-68 as an adequate tracer for silicon transport in plants. Characterization of silicon uptake in different crop species. Plant Physiol 143:495–503
Rains DW, Epstein E (1967) Sodium absorption by barley roots: its mediation by mechanism 2 of alkali cation transport. Plant Physiol 42:319–323
Rains DW, Epstein E, Zasoski RJ, Aslam M (2006) Active silicon uptake by wheat. Plant and Soil 280:223–228
Raven JA (2003) Cycling silicon – the role of accumulation in plants. New Phytol 158: 419–421
Ripoll C, Demarty M, Thellier M (1984) Formulation of transmembrane ionic fluxes. In: Cram J, Janacek K, Rybova R, Sigler K (eds) Membrane transport in plants. Publishing House of the Czechoslovak Academy of Sciences, Praha, pp. 21–26
Sanders D (1990) Kinetic modeling of plant and fungal membrane transport systems. Annu Rev Plant Physiol Plant Mol Biol 41:77–107
Schachter H (1972) The use of the steady-state assumption to derive kinetic formulation for the transport of a solute across a membrane. In: Hokin LE (ed) Metabolic transport, vol VI. Academic Press, New York London, pp. 1–15.
Tamai K, Ma JF (2003) Characterization of silicon uptake by rice roots. New Phytol 158:431–436
Thellier M (1969) Interprétation électrocinétique de l'absorption minérale chez les végétaux. Bull Soc Franç Physiol Veg 15:127–139
Thellier M (1970) An "electrokinetic" interpretation of the functioning of biological systems and its application to the study of mineral salt absorption. Ann Bot 34:983–1009
Thellier M (1971) Non-equilibrium thermodynamics and electrokinetic interpretation of biological systems. J Theor Biol 31:389–393
Thellier M, Thoiron B, Thoiron A (1971) Electrokinetic formulation of overall kinetics of *in vivo* processes. Physiol Vég 9:65–82
Thellier M, Ripoll C, Vincent JC, Mikulecky D (1993) Interpretation of the fluxes of substances exchanged by cellular systems with their external medium. In: Greppin H, Bonzon M, Degli Agosti R (eds) Some physico-chemical and mathematical tools for understanding of living systems. The University, Geneva, pp. 243–264
Thellier M, Legent G, Amar P, Norris V, Ripoll C (2006) Steady-state kinetic behaviour of functioning-dependent structures. FEBS J 273:4287–4299
Torii K, Laties GG (1966) Dual mechanism of ion uptake in relation to vacuolation in corn roots. Plant Physiol 41:863–870
Ussing HH (1949) The distinction by mean of tracers between active transport and diffusion. Acta Physiol Scand 19:43–56

Ussing HH (1971) The interpretation of tracer fluxes in terms of active and passive transports. Physiol Vég 9:1–19

Vincent JC, Thellier M (1983) Theoretical analysis of the significance of whether or not enzyme or transport systems in structured media follow Michaelis–Menten kinetics. Biophys J 41:23–28

Vincent JC, Alexandre S, Thellier M (1988a) How a soluble enzyme can be forced to work as a transport system: description of an experimental design. Arch Biochem Biophys 261:405–408

Vincent JC, Alexandre S, Thellier M (1988b) How a soluble enzyme can be forced to work as a transport system: theoretical interpretation. Bioelectrochem Bioenerg 20:215–222

White PJ, Ridout M (1995) The K^+ channel in the plasma membrane of rye roots has a multiple residency pore. J Membrane Biol 143:37–49

White PJ, Biskup B, Elzenga JTM, Homann U, Thiel G, Wissing F, Maathuis JM (1999) Advanced patch-clamp techniques and single-channel analysis. J Exp Bot 50:1037–1054

Wu QS, Wan XY, Su N, Cheng ZJ, Wang JK, Lei CL, Zhang X, Jiang L, Ma JF, Wan JM (2006) Genetic dissection of silicon uptake ability in rice (*Oryza sativa* L.). Plant Science 171:441–448

Oscillations, Synchrony and Deterministic Chaos

D. Lloyd

Contents

Abstract The coherence and robustness of biological systems is an astonishing phenomenon that depends on oscillations, synchronous behaviour and, in some instances, deterministic chaos. Understanding of dynamic interactions on an extended range of timescales involves homeodynamic rather than homeostatic concepts. Thereby, oscillations produce highly complex processes of intracellular as well as intercellular synchrony and have led to the evolutionary emergence of responsiveness, motility, and developmental change. "Creative destruction" is as important as biosynthesis in the elaboration and maintenance of cellular structure. Synchronisation of oscillators even of the simplest physical kind is still incompletely understood. In a huge population of oscillators it involves the idea of coupling strength and sudden cohesion of a small cluster of oscillators as the spread of natural frequencies is decreased below a threshold. This represents an example of Prigogine's theory of time order in which spontaneous transitions give spatio-temporal patterns in non-equilibrium open systems. Clearly, the ordered complexity of biological systems out-performs the simplicity of physical or chemical systems and its basic understanding remains a challenge despite recent successes in imaging and the increasing power of analytical chemistry. Network dynamics provides a promising tool for handling large data sets in such a way as to provide interpretation of behaviour of the whole system.

D. Lloyd
Microbiology (BIOSI 1), Main Building, Cardiff University, P.O. Box 915, Cardiff CF10 3TL, Wales, UK
e-mail: lloydd@cardiff.ac.uk

U. Lüttge et al. (eds.), *Progress in Botany 70*,
DOI:10.1007/978-3-540-68421-3,

1 Introduction

Time and change characterise the natural world, but in the biological sciences, by comparison with spatial measurements, time is a somewhat neglected parameter. Structural analyses of great depth and elegance have taken our spatial understanding to atomic dimensions, where distances are measured in Å. To obtain temporal measurements appropriate to this spatial scale, dynamics on an attosecond time-scale (10^{-18}s) are required in order to visualise physico-chemical mechanisms (Baum and Zewail 2006). For certain specific reactions of molecular components obtained from biological sources (e.g. the formation of carboxyhaemoglobin by the oxygenation of haemoglobin), probing of picosecond reactions are important (Brunori et al. 1999). In plants, femtosecond lifetimes of excited states of chlorophyll are key to the photosynthetic light reaction. These considerations underline the extreme range of dynamic interactions that are necessitated for an understanding of the living organism, for if we include the long history of evolutionary change (Fenchel 2002), an upper limit to our studies would extend over about 3.8×10^9 years (Fig. 1).

When the dynamic range of biological processes is to be considered, we must be aware that the system as it performs *in vivo* is a heterarchy with interactions of great complexity that occur, not merely within a level but between levels, and often across widely-separated time domains. The living state is better considered to be homeodynamic rather than homeostatic (Yates 1992; Lloyd et al. 2001). The importance of making measurements on appropriate time scales has been emphasised: "*This situation is analogous with that involved in deciding whether a mountain is a rigid or a flexible structure. Evidently over a period of a human lifetime, of the order of tens of years, the mountain would not change appreciably its shape, but this conclusion could not be extended as valid over periods of geological time, of the order of 10^8 years. The latter time interval would correspond to 1 s on the molecular time scale if the year is 1ns. We must therefore guard against deriving rigorous conclusions as the "molecular rigidity" from observations on the nanosecond time scale*" (Weber 1976).

That the living state is composed of a highly complex system with multiple interactions lies at the basis of its non-linear behaviour (Kaiser 2000). As a thermodynamic system open for fluxes of matter and energy and operating some distance, although not far, from equilibrium (Prigogine and Nicolis 1971; Yates 1992; Schneider and Sagan 2005) it is constructed in a manner where its dynamics will be predominantly oscillatory (rather than steady state, Fig. 2). In such systems, even in the inanimate world, spatial inhomogeneities (patterns) and dynamic structures (spiral scroll waves) often emerge (Winfree 2001). In life, emergence has proceeded to the extremely rich levels of dynamic complexity that we recognise as responsiveness, motility and the processes of growth and developmental change.

In order to achieve coherence of performance within living systems, an extensive pattern of control systems operates. Nowhere has this been more vividly described than by Reich and Sel'kov (1981).

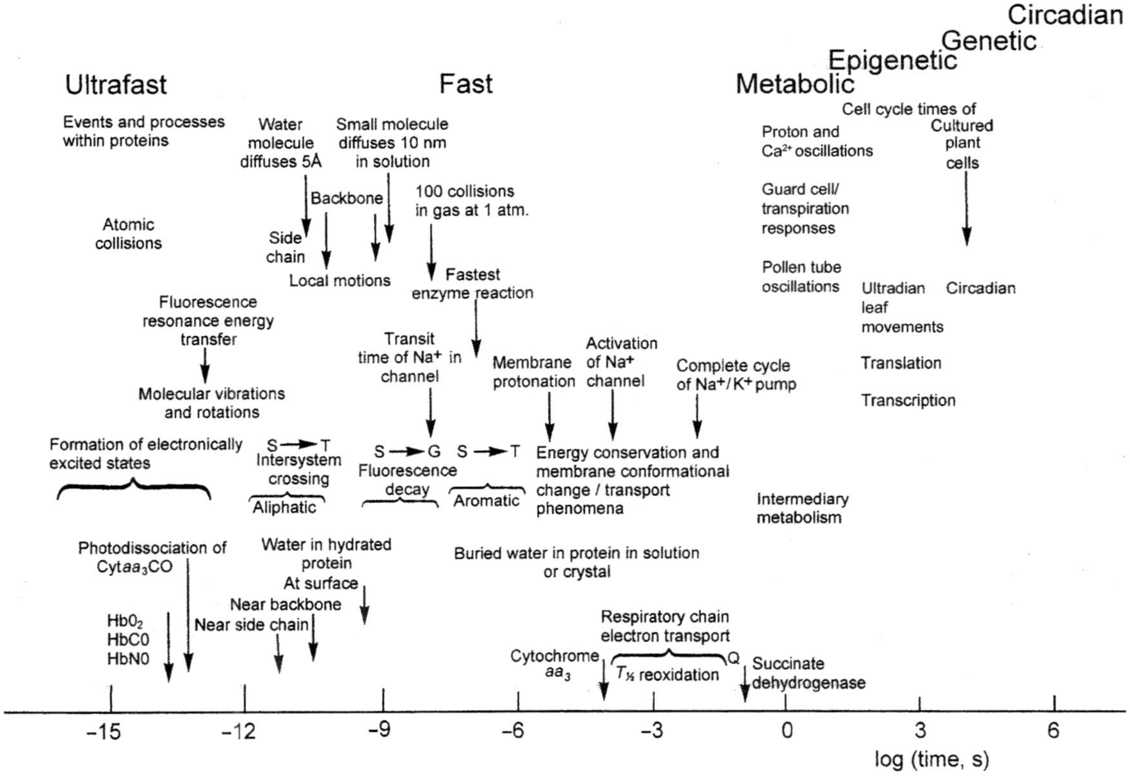

Fig. 1 The time domains of living systems. Approximate relaxation times (to a return of 1/e following perturbation) are plotted against time in seconds (log scale)

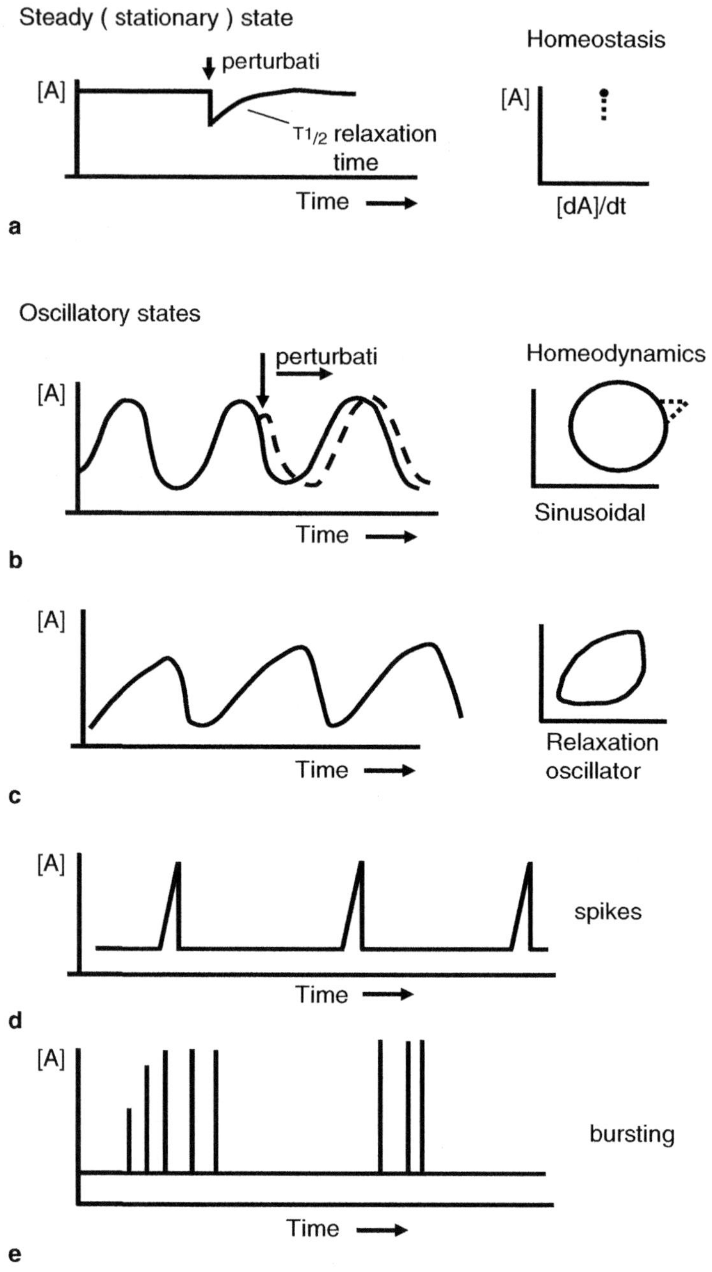

Fig. 2 (a) Steady-state operation of a system yields a time-independent output with formation and decay of the constituent precisely balanced. Perturbation away from the steady state leads to exponential restoration. The relaxation time (see Fig. 1) for this process can be evaluated. (b) The homeostatic concept indicated in (a) is most likely to be in fact homeodynamic, where a control system exerts an oscillatory output. One possible outcome is a phase-shift of the oscillator. (c) The sinusoidal output is only one of many possible waveforms. (d) In biological systems, intermittent increases (spikes) are often evident. (e) Episodic activation above a baseline is termed “bursting”

"A whole factory indeed. The number of metabolites partaking is counted by thousands. They are all present, from a few up to several million molecules in number, in that tiny volume of 10^{-12} cm^3 which we call a (bacterial) cell. They form more or less a mixture, in aqueous solution, though not a perfect one in the physiochemical sense. The organization is mysterious. ... [The metabolic map] ... looks like a cobweb. The first task, obviously, is to unravel the tangle. To make the network more transparent while retaining its principal multicomponent structure. . ...

... In the cell, nearly all the routes are in fact frequented very heavily and simultaneously. If reactions buzzed, we would hear an infernal concert. Everything crooning as in a beehive. We would observe individual molecules bustling to and fro, changing their binding state, or the compartment where they are, becoming engaged in chemical conversions, leaving or entering the cell. If one puts a handy label into it, for example a radioactive molecule, then seconds to minutes later everything carries a mote. The label spreads instantaneously into the most remote corner, and quite unexpected substances will be tagged a few moments later ... Apparently of all the billions of molecules, very few survive a dozen quiet seconds without undergoing chemical or physical conversion. Yet law and order appears to reign in that seemingly chaotic rush. The average content of metabolites remains unchanged for a long time. Input and output rates are stationary. The whole motion pattern is steady, re-established after perturbation and obstinately defended. The experimenter is often perplexed by the precise status maintenance of a system that may undergo instantaneous and dramatic irreversible change if handled unskilfully. This leads us to the second obvious task of analysis to explain how a system in that bustle can know, find and defend its 'normal' state and maintain homeostasis. To clarify how continuous motion retains coincidence and simultaneity.

On the map, we can discern very many different possible routes. Some of them short-circuit each other to apparent futility. Every function is a cross-way where several enzymes tussle for the common substrate or cofactor. Yet if the motion could be made visible, it would look like a fish shoal: masterly coordinated and capable of instantaneous reorientation being responsive to the slightest specific signal.

Apparently there is co-ordination without confusion and appropriateness of response without error. In some cases the result is close to perfection – amazing, since everything goes on more or less in a broth of thousands of interacting molecules and enzymes. Is there a highly efficient set of commands, collection of interactive regulatory feedback signals which can be demonstrated – literally a second cobweb stretched over the first one of chemical pathways, or just a spontaneous co-ordination and expediency of motion like that of droplets in a stream? ... Thus a third objective is to describe the kinetic functioning of well coordinated and stable flow in a system of potentially chaotic diffuseness".

The complexity of time structure is to some extent made more understandable once we have better appreciated the inherent importance of cycles (Krebs 1981), hypercycles (Eigen et al. 1991), oscillations (Hess and Boiteux 1971; Chance et al. 1973) and also more complex temporal behaviour, e.g. deterministic chaos (Lorenz 1963; Lloyd and Lloyd 1995; Aon et al. 2000; Roussel and Lloyd 2007) to the functioning of the living organism.

2 Dynamic States in Living Organisms

Decay to an equilibrium state represents the elementary thermodynamic process still predominately studied and taught in chemical reaction kinetics. Thus, equilibria are of limited interest to biologists other than to those that study single reactions (e.g. of purified enzymes). Extension of these principles to functional biological systems led historically to the principle of "dynamic equilibrium in a polyphasic system" as described by Hopkins (1913). Claude Bernard (1927) based his ideas of homeostasis, the constancy of the internal environment, on the concept of an organism having control systems that perform to ensure an enduring stability that lasts for its whole lifetime. Taken further by Cannon (1932) and Wiener (1961), processes of feedback became central to an understanding of control and also to the "cybernetic" principles of signal function.

Biological organisation viewed as an open thermodynamic system in which a continual flux of energy and matter maintain the illusion of a permanent living state was first enunciated by von Bertalanffy (1952), who developed earlier comparisons with Heraclitus' flowing stream and Roux's combustion of matter in the candle flame. Further elaboration of these ideas led to a fundamental concept:

> *"What are called structures are slow processes of long duration, functions are quick processes of short duration."*

Measurement of protein turnover rates required new techniques, and the post-World War II availability of radioisotopes facilitated experiments that had earlier been pioneered using stable heavy isotopes of carbon and nitrogen by Schoenheimer (1942). Rittenberg (1948) was able to demonstrate that the apparent stability of components in living cells simply represents a balance between synthesis and degradation. It became evident that "the approach to equilibrium is a sign of death."

Now we realise that until the late 1970s the emphasis on biosynthetic processes of growth led to a neglect of macromolecular degradation as an equally significant component in the maintenance of cellular structure (Edwards and Lloyd 1980; Luzikov 1981, 2004). Creative destruction is a key concept in biology and the underestimation of protein turnover rates continues despite earlier contrary entreaties (Wheatley 1989; Beynon 2005).

3 Oscillations in Biological Systems

Conditions that favour oscillations in physical, chemical, or biological systems include non-linear behaviour, non-equilibrium states and flux of energy through an open system. Clearly, these are all fundamental characteristics of living systems, and on every time scale biological oscillations occur and provide the dynamic signatures of life (Lloyd and Rossi 1992; Yates and Yates 2008). The recent developments in our understanding of circadian and ultradian rhythms in plants have been reviewed (Lüttge 2003; Lüttge and Hütt 2004) in this series and in a collection of reviews (Mancuso and Shabala 2007). In this article, I will be concentrating on general principles that emerge mainly from my own experience of working with yeast.

Table 1 Functions of oscillations, rhythms and clocks

A. Temporal organisation		
	1. Separation of incompatible processes	
	2. Co-ordination of events, processes and states (entrainment, synchronisation)	
		a. Intracellular processes
		b. Co-ordination of tissues or organs
		c. Environmental matching (tidal, circadian, lunar, annual)
		d. Predictive functions
	3. Faster responsiveness	
B. Energetic advantage		
C. Spatio-temporal organisation		
	1. Signalling	
	2. Developmental processes	
		a. Embryonic
		b. Growth

Figure 2 indicates some examples of the time-dependent non-linear behaviour of biological constituents; alongside the amplitude (e.g. concentration) versus time plots are the corresponding phase plane diagrams in which the amplitude is plotted versus the first derivative of the amplitude. For steady (stationary) state behaviour (Fig. 2a), the amplitude is unchanging, the phase portrait is a point, and a small perturbation results in a subsequent relaxation (i.e. a return) to the steady state. The characteristic relaxation time of the system is expressed as the interval required for it to return to some fraction of its initial deviation (e.g. $t_{½}$, $t_{90\%}$, etc.). This behaviour represents the simplest dynamic state (the homeostatic paradigm).

Oscillatory states of the simplest kind are represented (Fig. 2b) by a simple harmonic oscillator which shows sinusoidal waveform and where no damping occurs (i.e. the amplitude is constant): the phase portrait is a circle. Perturbation may phase-shift the oscillator (as shown) or it may alter the amplitude: further examples (Chance et al. 1967) of modulated waveforms generated in spectrophotometric recordings of NADH absorption of cell-free extracts of yeast under anaerobic conditions. Departures from a simple sinusoidal motion in which a fast rise-time precedes a slow return, or vice versa (Fig. 2c), are termed relaxation oscillators. Other waveforms commonly encountered in biology include spikes (Fig. 2d) and the periodic phenomenon of “bursting” (Fig. 2e).

Self-sustained oscillators (that are driven by an energy source and are not damped) may be extremely robust; these rhythms sometimes continue for the lifetime of the organism. Biological clocks are temperature-compensated rhythms (Hastings and Sweeney 1957; Lloyd et al. 1982). Circadian clocks (τ~24 h, match all the physiological characteristics of the organism to the daily rotation of the earth and the consequent cycles of day and night. Current attempts to explain this biological clock on the basis of a few “canonical” circadian genes are simplistic (Lüttge 2003) as the entire cellular network has to be involved. Ultradian clocks (that cycle many times in a day) underpin the circadian control circuitry and preceded it (Lloyd 1997). Circadian rhythms were an “evolutionary afterthought” (Yates and Yates 2008). Functions of oscillations, rhythms and clocks are shown in Table 1.

4 Synchronisation of Oscillators

Many of the most challenging aspects of dynamic behaviour involve the collective outputs from large ensembles of oscillators interconnected in complex networks (Hütt and Lüttge 2005). Understanding of cellular biochemistry and the molecular biology of transcriptional controls, as well as the superstructural kinetics of the processes of assembly of membrane and organelles, and of the chromosomal dynamics, as well as the cell division cycle, depends on the non-linear behaviour of an exceedingly complex interacting web of components.

Even for the simplest physical model, a network of identical oscillators, the set of rules that govern the tendency of the system to fall into synchrony has long been a matter of some conjecture, as in the well-known original observations of Huygens in 1665.... "*that two clocks hanging next to one another separated by one or two feet keep an agreement so exact that the pendulums always oscillate together without variation. . . .. mixing up the swings of the pendulums, I have found that within a half hour they always return to consonance and remain so constantly for as long as I let them go*".

From the very small, the concerted electron movement in superconductors, to the very large, the motion of planets in orbital synchrony, the universal tendency of physical systems to become coordinated is still incompletely understood. Starting from a pair of coupled oscillators, Wiener (1961) progressed to the problems of coherent behaviour of large-scale populations, and to speculating on the synchronous behaviour of power generators of the US powergrid. He also looked further, into the biology of acoustic signalling of frogs and crickets, the synchronised flashing of fireflies and the large-scale coordinated behaviour of the brain. Within the brain, it was suggested there must be a master clock beating out a time signal that keeps all other signals entrained by a process of "frequency pulling". Winfree (1967) considered the individual oscillators as somehow in communication with one another by signals that change with their relative positions as they progress through their cycles. Each can advance or delay; depending on the signal they receive from one another. For a multi-oscillator system the location of one individual is determined by its speed and therefore its position in the cycle (its phase), its sensitivity to incoming signals, and the global influence of all the other oscillators. The set of non-linear differential equations required to solve the problem becomes impossibly difficult for any number of oscillators greater that two, and so computer simulation becomes necessary. When the oscillators act independently the outcome is incoherence (i.e. all the individuals remain out of phase). When just a minority "happen" to come into coherence, this minority population grows bigger to generate self-synchrony as an emergent property. Winfree (1967) next considered the results of different breadths of distributions of phase behaviour within the initially minority population. He discovered that a critical threshold of homogeneity was necessary to generate global coherence. Analogies between this system and that of a phase transition point led to a new way of thinking and unifying two seemingly disparate phenomena. Non-linear dynamics, working well for systems with a small number of individual entities and a few variables could be allied with the methods of statistical mechanics.

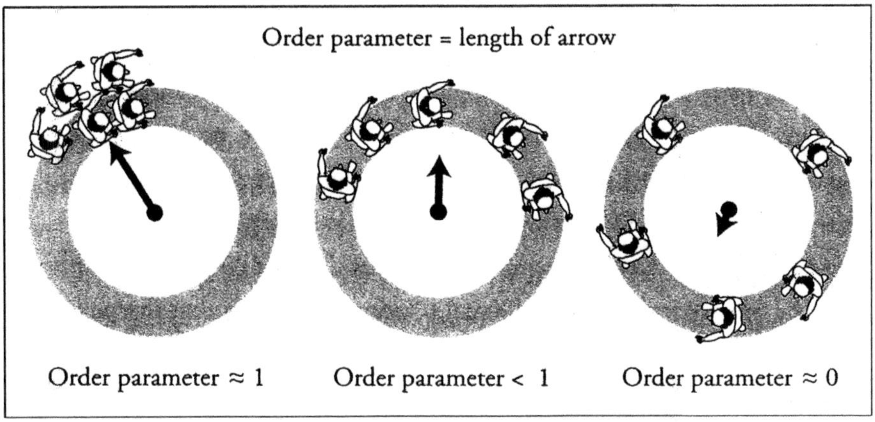

Fig. 3 Coupling and synchronisation of oscillators (Strogatz 2003)

Kuramoto (1975, 1984) like Wiener, considered a huge population of oscillators with a natural distribution of frequencies, but developed the idea of coupling strength (the sine function of the phase angle between a pair of oscillators) (Fig. 3).

Kuramoto and Nishikawa (1987) seized on the concept of Winfree (1967) that the sudden cohesion of a small cluster of oscillators occurs as the spread of natural frequencies is decreased below a threshold. They realised that this represents an example *par excellence* of Prigogine's theory of time order (Glandsdorff and Prigogine 1971) in which spontaneous transitions occur to give spatio-temporal patterns in non-equilibrium open systems. Further refinements of the Kuramoto phase oscillator model have resulted in useful predictions for a wide class of oscillator models (Kori and Kuramoto 2001) in which the introduction of a time delay and symmetry-breaking perturbations can lead to slow switching. These principles are particularly applicable to an assembly of relaxation oscillators. One important feature of this slow switching is a system involved in a much longer time scale in which collective order varies periodically by comparison with the intrinsic period of the oscillation. Relevance to rhythm splitting (e.g. 12-h rhythms in circadian systems, Iglesia et al. 2000), and the generation of long-period biological rhythms is evident (Lara-Aparicio et al. 2006). Technical aspects of the Kuramoto model have been further resolved (Crawford and Davies 1999) and have led to new lines of inquiry in kinetic theory, fluid mechanics and plasma physics (Strogatz 2000, 2003). Analogies with the behaviour of positive ions and electrons in rarefied hot plasmas has led to an understanding of the stability characteristics of coupled oscillators (Strogatz and Mirello 1991; Strogatz et al. 1992) The relaxation decay of electrostatic waves in ionized gases ("Landau damping", Landau 1946; Sagan 1994) provides insights into the decay to incoherence in the Kuramoto model.

It is still not certain whether the synchronisation of biological oscillators is entirely dependent on the rules (Fig. 3) that have been established for physical systems. However, chemical oscillators do fall into coherence once coupling exceeds a threshold (Kiss et al. 2002). The degree of synchrony (i.e. the order parameter) increased in this system as the coupling strength was augmented in a precise relationship as predicted by Kuramoto (1975, 1984).

5 Synchrony in Biological Systems

Rhythmicity is a property so central in biology that it should be listed as one of life's essential characteristics. The heartbeat depends upon networks of pacemaker cardiomyocytes. When these cells are cultured, they too beat. Networks of interconnected neurons produce the 100 ms α – wave electrical output of the human brain. The 90-min basic rest-activity cycle (BRAC) most easily monitored as REM–non-REM episodes in sleeping subjects (Aserinsky and Kleitman 1955) is an example of an ultradian rhythm produced as the collective summation of activities of the whole organism; likewise, the numerous circadian physiological functions that have become entrained to the 24 h night-day cycle on our rotating planet Earth.

Even single-celled organisms studied in isolation exhibit ultradian (Aon et al. 2007a, b) and circadian (Sweeney et al. 1967) phenomena as a consequence of intracellular coherence and self-optimisation of function. In each cell, a massively interconnected web of metabolic, transcriptional and informational processes and events interweave in a heterarchical whole. Parallel processing, divergence and convergence are synchronised on a time-line and to a strict temporal program (Klevecz 1992; Klevecz et al. 2004; Lloyd and Murray 2005, 2006, 2007) provided by the recurrent insistence of a series of clocks performing interactively on every time-scale from nanoseconds to days, or even months or years.

Clearly, the ordered complexity of biology out-performs the simplicity of physical or chemical oscillators. Its coherent behaviour remains a huge challenge to our understanding, despite enormous recent technical successes brought about by fast optical imaging methods and the high-throughput facilities provided by analytical chemistry and molecular biology.

For the synchronous behaviour of populations of organisms, an enormous literature provides a plethora of mechanisms for interactions by signal transmission and reception. Even in the best-studied cases (e.g. acoustic interaction of crickets, Nityananda and Balakrishnan 2007) or the synchronous flashing of fireflies (Smith 1935; Buck and Buck 1976; Buck 1988; Trimmer et al. 2001; Kim 2004), complete explanations are still awaited (Strogatz 2003).

6 Synchronous Single Cell and Unicellular Behaviour and the Evolution of Multicellularity

It has been known for several years that the ultradian rhythm of protein synthesis in isolated hepatocytes first discovered by Brodsky (1975) is involved in cell–cell communication (Brodsky 2005). Gangliosides and/or catecholamines and Ca^{2+} dependent protein kinase-mediated protein phosphorylation provides the phase-modulation of individual cells for harmonious cell–cell interaction during the establishment of synchrony. It is evident that these co-operative phenomena occur not only in isolated cells, but are a central feature of tissue and organ function.

The heart can be viewed as a collection of electrically coupled relaxation oscillators (Van der Pol and Van der Mark 1928). Spontaneous metabolic oscillations in sarcolemmal K^+ currents, action potentials and Ca^{2+} transients (O'Rourke 2000; O'Rourke et al. 1994) create order. Subcellular cytological investigations revealed spatiotemporal patterns that underpin these phenomena; especially important are propagated mitochondrial redox waves (Romashko et al. 1998). Oscillation of the mitochondrial energy state results from the interplay between mitochondrial reactive O_2 species (ROS) production and the ROS scavenging systems of the cell (Aon et al. 2003). The network of cardiac mitochondria acts as a system of coupled oscillators (Aon et al. 2006a, b). More recently, analogous processes have been described in yeast, whereby spontaneous self-synchrony occurs both in continuously stirred aerobic suspensions of growing organisms (Satroutdinov et al. 1992) at high cell densities (5×10^8 organisms/ml) or in glucose-perfused yeasts tethered on a polylysine coated slide under aerobic conditions (Aon et al. 2007).

In the first case, messenger substances responsible for internal coherence and cell–cell interactions include H_2S and acetaldehyde (Murray et al. 2003) as shown by phase response curves generated on addition of these compounds at set times during their 40 min ultradian cycles.

Cell division synchrony is also spontaneously produced in these cultures, as about 8% of the total population enter S phase in each ultradian cycle under the stringent timekeeping regime of the ultradian clock (Klevecz et al. 2004).

The synchronous (ca. 2 min period) behaviour observed in the attached monolayer of yeasts is revealed by laser scanning 2-photon excitation microscopy (Aon et al. 2007a). Using this technique, several mitochondrial properties could be simultaneously monitored in real-time (i.e. NAD(P)H redox state, inner mitochondrial membrane electrochemical-potential and levels of reactive O_2 species). All these variables showed oscillations with similar periods. Observation of single yeast cells and discrimination between changes in mitochondrial and cytosolic nicotinamide nucleotide pools was possible. The intracellular rhythm appears to be initiated in a single mitochondrion before spreading throughout these respiratory organelles of a single organism. The rhythmic activity thence propagates across the entire layer of yeasts. This is a most striking pheonomenon, whereby the redox balance of the entire ensemble alternates through a cycle of more oxidised to more reduced intracellular states. Thus, even in an organism that we usually regard as living an independent existence, loss of autonomy amongst a crowd appears inevitable (Fig. 4).

Synchronised cycling gene expression is necessary for the generation of heteromeric symmetry, and one of the prime cases in the study of molecular regulation of the development of segmentation and the production of biopolarity is *Drosophila*. Another example is provided by the processes involved in segmentation of the vertebrate body axis early in embryonic development (Maroto and Pourquié 2001; Pourquié 2003). Mechanisms that converge to orchestrate this spatial architecture are only beginning to be understood. A key role for a community effect involves cell–cell communication by an autonomous system as oscillatory expression continues on schedule in cells isolated from the mesenchymal presomatic mesoderm, which lies either side and juxtaposed to the neural tube (Maroto et al. 2005). Both

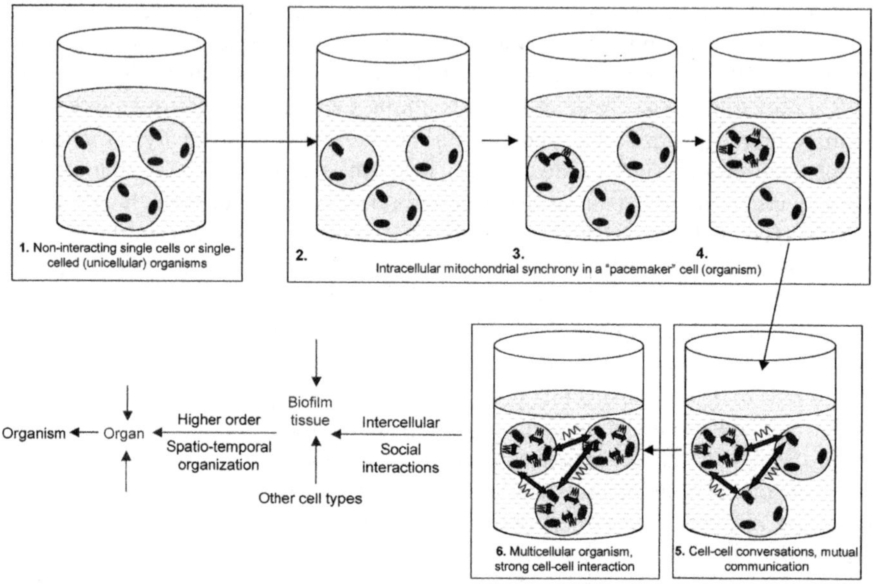

Fig. 4 The evolution of single-cell eukaryotes and of multicellular higher organisms has utilised ultradian rhythms for intracellular and intercellular synchronisation of redox state and hence coherent performance (courtesy of Victoria L. Gray)

experiments and simulations reveal how these cellular oscillators, driven by the negative feedback loop of Hairy transcription factor, are linked through Notch-dependent intercellular coupling to maintain synchrony. This coupled oscillator system forms a robust mechanism resistant to the effects of developmental noise (stochastic gene expression, active cell proliferation; Ishimatsu et al. 2007)

We can conclude from these data from four very disparate biological situations that there is an underlying uniformity (Brodsky and Lloyd 2008). Thus, ultradian clock-generated rhythms are a basic signature of life (Yates and Yates 2008), and the evolutionary development of cellular co-operation leading to the multicellular organisation of tissues and the harmonious functioning of the whole organism is based on ancient mechanisms. Hepatocytes, cardiomyocytes, yeast and the vertebrate segmentation clock all provide highly convenient systems for studies of both intra- and inter-cellular processes of co-ordination and coherence. Fundamental to these concerted higher-order functions are the basic principles of biological time-keeping and consequent spatio-temporal organisation.

7 Deterministic Chaos

The co-operative behaviour so evident in biology results in the whole being more than the sum of its parts; simple addition of effects rarely occurs and outputs are not proportional to inputs in any elementary relationship. However, it is now widely accepted that complex dynamic behaviour can be produced by simple deterministic systems that obey simple rate-determining equations and can therefore be modelled easily. That this is so provides encouragement for the investigation of biological complexity in unicellular organisms (Lloyd 2007a, b) and in plants (Prytz 2001; Hütt and Lüttge 2005, 2007).

Chaotic systems are those that show sensitive dependence on initial conditions (Schuster 1988). If we follow the evolution of two identical systems over time, any small difference initially becomes amplified as time goes by. The growing divergence limits the long-term predictability of the system. The average rate of trajectory divergence is known as the largest Lyapunov exponent.

If our system can be completely specified by *N* variables, then the evolution of the system can be represented as a trajectory in an *N*-dimensional space (the phase space). After any transient behaviour has occurred the motion takes place on a set of lower dimensions than the phase-space, and this set is known as the attractor of the system.

The logistic map (May 1976), $x_{n+1} = \mu x_n (1-x_n)$ gives a sequence of x_n values by successively applying the map to obtain the iterates of the map (Fig. 5a). As the strength of non-linearity (parameter μ) is increased, different long-term dynamic behaviours occur. For $\mu < 3$ stable equilibrium is observed. This is then replaced by stable period 2 orbit, in which x_n takes one of two different values, depending on whether *n* is odd or even. This new orbit is created in a period-doubling bifurcation. Further increases in μ leading to successive period bifurcations occur with smaller

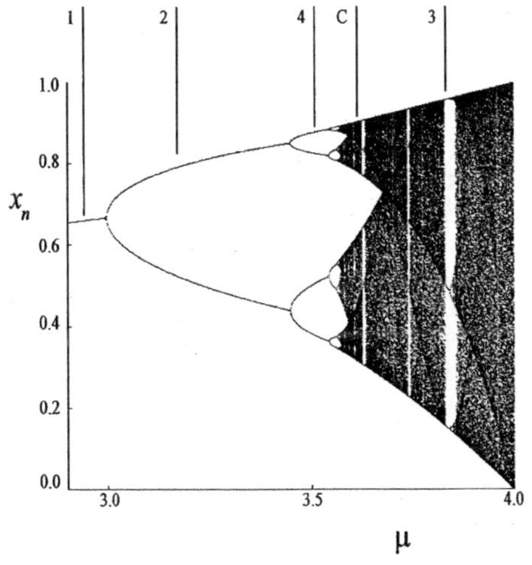

Fig. 5 **a** Graph of the relation between x_{n+1} and x_n. Notice how the points on the x_n axis are stretched apart by the map. Almost every point on the x_{n+1} axis has two pre-images on the x_n axis; the interval is folded over by the map. **b** Bifurcation diagram for the logistic map. For each μ value, the map was iterated 1,000 times, and the final 400 iterates are plotted vertically (Lloyd and Lloyd 1995). Parameter regions that exhibit different dynamic behaviours are as follows: stable equilibrium (*1*), stable period two orbit (2), period four (*4*), chaos (**c**), and a periodic window within the chaotic region where period three orbit can be seen (*3*)

increases in μ and, at a critical parameter near 3.57, the bifurcations accumulate (Feigenbaum 1983). Beyond this point, chaotic behaviour is observed, although there are also infinitely many parameter "windows" in which stable periodic behaviour occurs (Fig. 5b).

An example of chaotic behaviour in biology is provided by a continuous culture of yeast (Roussel and Lloyd 2007). Membrane inlet mass spectrometry allows very rapid simultaneous measurement (one point every 12 s) of several dissolved gases (O_2, CO_2 and H_2S). We observed a slow, cell-cycle dependent mode (τ~12 h), the ultradian clock output (τ~40 min) and a faster metabolic oscillation (τ~4 min) (Fig. 6). The chaotic metabolic attractor (Fig. 7) for this system had a positive leading Lyapunov exponent with a value of $0.752 \pm 0.004\,h^{-1}$ (95% confidence), indicating its sensitive dependence on initial conditions. If by contrast this value had been

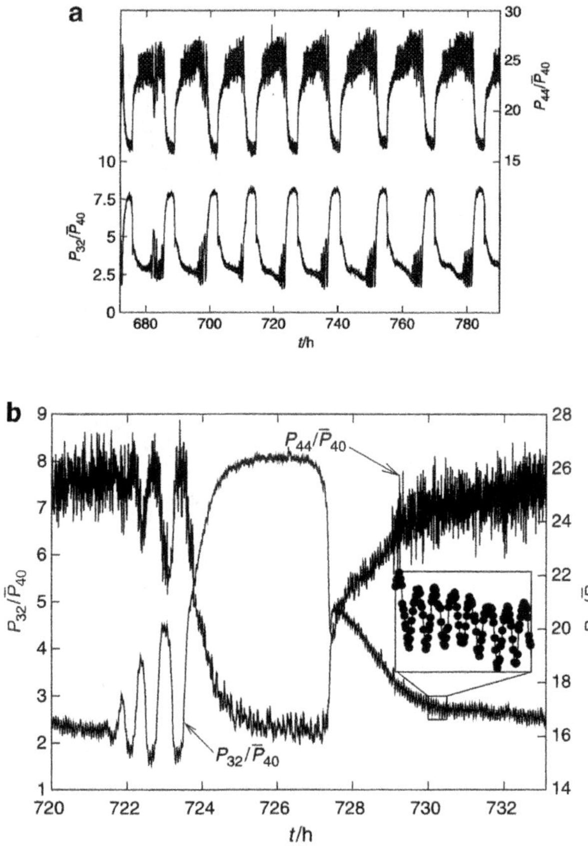

Fig. 6 Dissolved O_2 and CO_2 (mass spectrometric signals at *m/z* 32 and 44 normalised with respect to argon (*m/z* 40) vs time in a self-synchronized continuous culture of yeast; Roussel and Lloyd 2007). **a** The gas concentrations monitored during an interval that begin 672 h after the continuous culture was initiated. **b** One period of the oscillatory outputs shown in (**a**). The *inset* shows the individual data points for *m/z* 32(O_2) for a 30-min span starting at 730 h

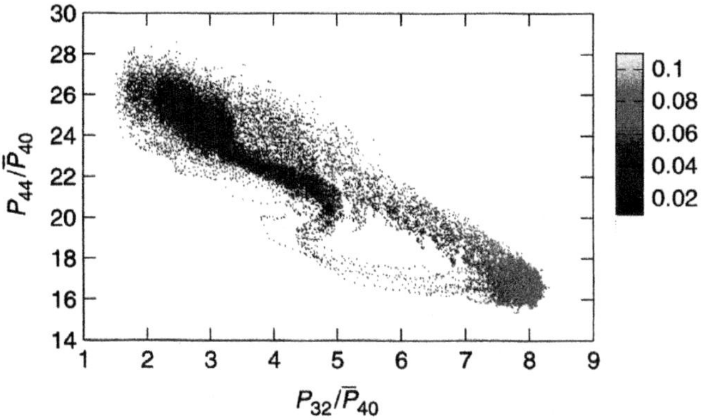

Fig. 7 A chaotic attractor, i.e. the set of metabolic states obtained from dissolved O_2, CO_2 and H_2S (indicated on the *grey scales*) measured every 15 s in the self-synchronised continuous yeast culture. The experiment lasted 3 months and 37,374 data points were collected (Roussel and Lloyd 2007)

zero, the dynamics would have been classified as being quasiperiodic. These results may mean that one of the variables was chaotic, or more likely that these oscillators (and possibly others) interact. Multiple interacting oscillators are thus involved in, and perhaps critical to, cell function (Gilbert and Lloyd 2000; Lloyd and Murray 2005; Murray and Lloyd 2007). Increasingly stressful conditions (e.g. progressively lowering the pH of a continuous culture of yeast) also drive its dynamics into a chaotic regime (Murray and Lloyd 2007). Observations of biological chaos encompass a diversity of situations: a considerable literature on the dynamical outputs of the brain and heart as well as in experiments with cultured mammalian cells and yeasts suggest its widespread occurrence (Visser et al. 1990; Davey et al. 1996; Lloyd et al. 2001; Gilbert et al. 2000). Models of interacting oscillators also predict chaotic solutions for some values of variables. Thus, the interplay between the ultradian clock (Lüttge and Hütt 2004) and the cell division cycle oscillator (Lloyd and Volkov 1990, 1991) can produce quantised increments of cell cycle duration or can give chaotic outputs (Lloyd et al. 1992). Thereby, it is conceivable that the inherent variability of cell division times arises from deterministic principles rather than from stochastic environmental influences. Because the latter are, however, omnipresent, the spread of biological properties must arise from a combination of both noise and chaos.

Over the past decade, methods for the control of chaotic performance have been extensively researched and have proved to be a rich source of new and useful electronic and mechanical devices. Of especial interest in biology is the discovery that a controlled chaotic attractor can be harnessed as a tuneable multi-oscillator (Ott et al. 1990). This is because a chaotic attractor is composed of many periodic orbits, and as the system is ergodic, all possible orbits are visited. We have suggested that

Table 2 Possible functions of chaos in biology: the generation of diversity

A. Innovation
B. Behavioural unpredictability
C. Prevention of entrainment
D. Dissipation of disturbance
E. Generation of multi-oscillatory functioning (by controlled chaos)

if such a situation exists biologically (Lloyd and Lloyd 1993) it would provide a means of providing a variable range of outputs (i.e. simultaneous performance on multiple time-scales, e.g. the generation of ultradian and circadian rhythms; Lloyd and Lloyd 1995). The control principles required involve feedback with delay circuitry (Pyragas 1992), already well-known in metabolic and transcriptional control circuits. Possible functions of chaos are summarised in Table 2.

The relationship between chaotic performance and fractal architecture has been extremely usefully documented (Yates 1992). In the multi-oscillatory state of the yeast continuous culture described in (Fig. 6) we have recently discovered a self-similar temporal relationship between the 12 h, 40-min and 4 min oscillators. Thus, fractal correlation of dynamics on differing time scales is evident and justifies use of the term "heterarchical organisation" (Yates 1992; Lloyd et al. 2001).

8 Conclusions: Coherence of the Cell and of the Organism

The living organism as a multi-oscillatory system (Lloyd and Gilbert 1997; Gilbert and Lloyd 2000) depends for its organisation (order with a purpose; Pittendrigh Pikovsky 1993) and coherence on rhythms and biological clocks. This temporal order encompasses domains from the very fast to the very slow in a network of interactions that includes electron and proton transport and ligand binding of small molecules, enzyme-catalysed reactions of metabolism, transcriptional and translation of macromolecules, replication of DNA, chromosome dynamics, the events of the cell division cycle, cell–cell interactions and the macroscopic events and results of development and cytodifferentiation to give tissues, organs and gross anatomical features. The usual way of arranging these events and processes on a hierarchical diagram of organisation is inadequate. (Yates 1982, 1987, 1992; Lloyd et al. 2001) as networks of co-operativity are required and these sometimes span very widely separated domains of time not easily represented in a multi-layered representation.

Table 1 summarises the functions of oscillations, rhythms and clocks in the integrated performance of living organisms. Perhaps a clear necessity for temporal separation of incompatible processes came about early in evolution. The vulnerability of DNA during its replicative stage to free-radical damage by photochemically generated species in the light, and to the univalent reduction products of O_2 consumption, provided the driving force for both ultradian and circadian cycles. The O_2-sensitivity of nitrogenase represents another well-established example of temporal partitioning in the non-heterocystous cyanobacteria (e.g. *Oscillatoria* spp.) where spatial separation between vegetative cells and heterocysts (e.g. as in *Anabena cylindrica*) does not occur (Mitsui et al. 1986).

Co-ordination of events or processes that occur in separate locations in the cell or in the organism requires a time-line (the term used in operational research) or a time-base (as in a computer) upon which all the parallel processing is ordered in preparation for convergent integration or divergent signalling functions. Well-explored examples of biological signalling depending on amplitude- or frequency-modulated oscillations (e.g. of Ca^{2+} Moreno et al. 2007; Beilby 2007, or reactive oxygen species, Cortassa et al. 2004) provide not only temporal coherence but spatial information. The segmentation clock in somitogenesis is the precursor process for the formation of the vertebral column, ribs and skeletal muscles in mice, zebrafish and spiders alike (Peel and Akam 2003). The Notch and Wnt signalling molecules involved are expressed in oscillatory fashion; this results in wave-like propagation depending on a negative-feedback loop (Masamizu et al. 2006). Cell–cell communication is essential not only for synchronisation but for stabilisation of cellular oscillators in this system, as previously indicated for the hepatocyte-mediated integrated functions in liver, and the co-ordinated beating of cardiomyocytes in heart.

Spatio-temporal pattern formation as a consequence of non-linear dynamics has been explored in both experiments and models of the obligate crassulacean acid metabolism of *Kalanchoë daigremontiana* (Hütt and Lüttge 2002, 2007). In this process, a cycle involves switching from vacuolar malate loading (by CO_2 fixation via phospho*enol*pyruvate carboxylase) to vacuolar malate remobilisation, when malate is decarboxylated and the CO_2 so obtained, refixed and assimilated via ribulose-1, 5-bisphosphate carboxylase-oxygenase.

These changes free-run as an endogenous rhythm with circadian oscillations of CO_2 exchange and malate levels in continuous light and under strictly constant conditions, but direct involvement of a circadian-controlled gene is excluded and the primary determinant is based on post-translational functions in a highly non-linear dynamic system. Furthermore, consideration of the patchiness of the leaf is necessary, in which each cell may be modelled as an independent oscillator in an ensemble of weakly coupled oscillators (Lüttge and Hütt 2007). Loss of synchronisation of discrete leaf areas has been confirmed by non-invasive photographic imaging of chlorophyll fluorescence as a measure of photosynthetic activity in leaf patches (Rascher et al. 2001).

The evolution of timekeeping may have had a prebiotic origin, as chemical clocks can be as simple as a hydrated cupric ion that undergoes cyclic changes of redox state in a self-sustained manner (Jiang et al. 2006; Morré et al. 2007, 2008) or as complex as the well-studied Belousov–Zhabotinsky reaction (Toth and Taylor 2006) with its 20-odd defined chemical intermediates. Single purified enzymes that can also show clock-like dynamics include malate dehydrogenase (Queiroz-Claret et al. 1988), the plasma membrane NADH oxidases (NOX systems, Morré et al.

2002) and the "oscillophore" component of the glycolytic oscillator (Yuan et al. 1990) studied *in vitro*.

The earliest *in vivo* control systems may have originated as feedback circuitry in central metabolic networks that functioned to balance bioenergetic functions of glycolysis, gluconeogenesis and redox state by production of oxidised products of pyruvate at the hub of metabolism.

This deeply embedded redox core is still central to the maintenance of intracellular coherence in all cells (Lloyd and Murray 2007). A series of investigations on a spontaneously synchronised population of yeast in a continuous culture pinpoint a redox cycle driven by NAD(P)H and involving dithiol-disulphide interconversions that recur about every 40 min. The complex network of biochemical reactions (metabolic and transcriptional) has been partially elucidated (Murray et al. 2007): in its derived form it consists of 1,650 nodes with 3,528 edges and includes 96 transcriptional regulators and 287 metabolites. Within this cohort, the most oscillatory metabolites were contained at 27% of the nodes. More than 70% of the oscillating network remains unaccounted for. Both NAD(P)H and ATP show maximum production in the reductive phase. Network dynamics as a tool for handling the large data sets in a way that gives us an interpretation of the whole system provides great promise (Hütt and Lüttge 2005). The topology, architecture and connectivity properties of the large collections of non-linear dynamic systems that constitute the networks of biochemical reactions within the cell are becoming more and more tractable. Signal functions, the cellular network of the higher organism and the dynamics of populations of organisms in an ecosystem, can all be dealt with using network theory. Among the questions that can be asked are those of system robustness, perturbability, responsiveness to specific stimuli and sensitivity to external noise. We may also discover more about the tendencies of cells and organisms to fall into spontaneous synchrony (Wiley et al. 2006).

References

Aon MA, Cortassa S, Lloyd (2000). Chaotic dynamics and fractal space in biochemistry: simplicity underlies complexity? Cell Biol Int 24: 581–587

Aon MA, Cortassa S, Marban E, O'Rourke (2003). Synchronized whole-cell oscillations in mitochondrial metabolism triggered by a local release of reactive oxygen species in cardiac myocytes. J Biol Chem 278: 4475–44744

Aon MA, Cortassa S, O'Rourke B (2006a). The fundamental organization of cardiac mitochondria as a network of coupled oscillators. Biophys J 91: 4317–4327

Aon MA, Cortassa S, O'Rourke B (2006b). Mitochondrial oscillations in physiology and pathophysiology. Landes Bioscience, Austin

Aon MA, Cortassa S, Lemar KM, Hayes AJ, Lloyd D (2007). Single and cell population respiratory oscillations in yeast: a 2-photon scanning laser microscopy study. FEBS Lett 581: 8–14

Aserinsky E, Kleitman N (1955). Regularly occurring periods of eye motility and concomitant phenomena during sleep. Science 118: 273–274

Baum P, Zewail AH (2006). Breaking resolution limits in ultrafast electron diffraction and microscopy. Proc Nat Acad Sci USA 103: 16105–16110

Beilby MJ (2007). Modeling oscillations of membrane potential difference. In Mancuso S and Shebala S (eds.) Rhythms in Plants. Springer, Berlin
Bernard C (1927). Introduction à l'Etude de la Medicine Experimentale. Green HC (trans.) McMillan, New York
Beynon JR (2005). The proteome as a dynamic entity – is yesterday's proteome the same as today's. Biochem Soc Trans, abstract SA038
Brodsky VY (1975). Protein synthesis rhythm. J Theor Biol 55: 167–200
Brodsky VY (2005). Direct cell–cell communication: a new approach derived from recent data on the nature and self-organization of ultradian (circahoralian) intracellular rhythms. Biol Rev Cambs Philos Soc 81: 143–162
Brodsky VY, Lloyd D (2008). Self-organised intracellular ultradian rhythms provide direct cell-cell communication. In Lloyd D, Rossi ER (eds.) Ultradian Rhythms from Molecules to Mind: A New Vision of Life. Springer, Berlin
Brunori M, Cutruzzola F, Savino C, Travaglini-Allocatelli C, Voellone B, Gibson QH (1999). Does picosecond protein dynamics have survival value? Trends Biochem Sci 24: 253–255
Buck J. (1988). Synchronous rhythmic flashing of fireflies II. Q Rev Biol 63: 265–289
Buck J, Buck E (1976). Synchronous fireflies. Sci Am 234 (May): 74–85
Cannon WB (1932). The Wisdom of the Body. Knopf, Norton, New York
Chance B, Pye K, Higgins J (1967). Waveform generation by enzymatic oscillators. IEEE Spectrum 4: 79–86
Chance B, Williamson G, Lee IY, Mela L, DeVault D, Ghosh A, Pye EK (1973). Synchronization Phenomena in Oscillations of Yeast Cells and Isolated Mitochondria. Academic Press, New York
Cortassa S, Aon MA, Winslow RL, O'Rourke B (2004). A mitochondrial oscillator depends on reactive oxygen species. Biophys J 87: 2060–2073
Crawford JD, Davies KTR (1999). Synchronisation of globally coupled phase oscillations: singularities and scaling for general couplings. Physica D 125: 1–46
Davey HM, Davey CL, Woodward AM, Edmonds AN, Lee AW, Kell DB (1996). Oscillatory, stochastic and chaotic growth rate fluctuations in permittistatically controlled yeast cultures. BioSystems 39: 43–61
Edwards SE, Lloyd D (1980). Oscillations in protein and RNA during synchronous growth of *Acanthamoeba castellanii*: evidence for periodic turnover of macromolecules during the cell cycle. FEBS Lett 109: 21–26
Eigen M, Biebricher CK, Grebinoga M, Gardiner WC (1991). The hypercycle. Coupling of RNA and protein biosynthesis in the infection cycle of an RNA bacteriophage. Biochem 30: 11005–11018
Fenchel T (2002). The Origin and Early Evolution of Life. Oxford University Press, Oxford
Feigenbaum MJ (1983). Universal behaviour in non-linear systems. Physica D 7: 16–39
Gilbert DA and Lloyd D (2000). The living cell: a complex autodynamic multi-oscillator system? Cell Biol Int 24: 569–580
Gilbert DA, Visser G, Ferreira GMN, Hammond KD (2000). Transient chaos in intracellular dynamics? Cell Biol Int 24: 589–591
Glandsdorff P, Prigogine I (1971). Thermodynamic Theory of Structure, Stability and Fluctuations. Wiley, New York
Hastings JW, Sweeney B (1957). On the mechanism of temperature compensation in a biological clock. Proc Natl Acad Sci USA 43: 804–811
Hess B, Boiteux A (1971). Oscillatory phenomena in biochemistry. Ann Rev Biochem 40: 237–258
Hopkins FG (1913). The dynamic side of biochemistry. Brit Med J 2: 13–24
Hütt M-T, Lüttge U (2002). Non linear dynamics as a tool for modelling in plant physiology. Plant Biol 4: 281–297
Hütt M-T, Lüttge U (2005). Network dynamics in plant biology: current progress and perspectives. Prog Bot 66: 277–309

Hütt M-T, Lüttge U (2007). Noise-induced phenomena and complex rhythms: theoretical considerations, modelling and experimental evidence. In Mancuso S and Shabala S (eds.) Rhythms in Plants. Springer-Verlag, Berlin, Heidelberg, New York
Iglesia H, Meyer J, Carpino A, Schwartz W (2000). Antiphase oscillation of the left and right suprachiasmatic nucleus. Science 290: 799–801
Ishimatsu K, Horikawa K, Takeda H (2007). Coupling cellular oscillations: a mechanism that maintains synchrony against developmental noise in the segmentation clock. Dev Dyn 236: 1416–1421
Jiang Z, Morré DM, Morré DJ (2006). A rate for copper in biological timekeeping. J Inorg Biochem 100: 2140–2149
Kaiser F (2000). External signals and internal oscillation dynamics: principal aspects and responses stimulated rhythmic processes. In Walleczek J (ed.) Cambridge University Press, Cambridge UK, pp. 15–43
Kim D (2004). A spiking neuron model for synchronous flashing of fireflies. Biosystems 76: 7–20
Kiss IZ, Zhai Y, Hudson JL (2002). Emerging coherence in a population of chemical oscillators. Science 296: 1676–1678
Klevecz RR (1992). A precise circadian clock from chaotic cell cycle oscillations. In Lloyd D, Rossi EL (eds.) Ultradian Rhythms in Life Processes. Springer, London, pp. 41–70
Klevecz RR, Bolen J, Forrest G, Murray DB (2004). A genome wide oscillation in transcription gates the cell cycle. Proc Natl Acad Sci USA 101: 1200–1205
Kori H, Kuramoto Y (2001). Slow switching in globally coupled oscillators? Rubustness and occurrence through delayed coupling. Phys Rev E 63: 046–214
Krebs HA (1981). Reminiscences and Reflections. Clarendon Press, Oxford
Kuramoto Y (1975). Self-entrainment of a population of coupled non-linear oscillators, Araki H (ed). Lect Notes Theoret Phys 39: 420–422
Kuramoto Y (1984). Chemical Oscillations, Waves and Turbulence. Springer, Berlin
Kuramoto Y, Nishikawa I (1987). Dynamics of order parameters for globally coupled oscillators. J Stats Phys 49: 569–572
Landau L (1946). On the vibrations of electronic plasmas. J Phys USSR 10: 25–34
Lara-Aparicio M, Barriga-Montoya C, Fuentes-Pardo B (2006). A brief history of circadian rhythms: from Wigglesworth to Winfree. Sci Math Japn 2: 357–370
Lloyd D (1997). Circadian and ultradian clock-controlled rhythms in unicellular microorganisms. Adv Microbial Physiol Biochem 39: 291–338
Lloyd D (2007a) Rhythms, clocks and deterministic chaos in unicellular organisms. In Mancuso S and Shabala S (eds.) Rhythms in Plants. Springer, Berlin, pp. 267–294
Lloyd D (2007b) Oscillations in Yeasts. Landes Books, Austin
Lloyd D, Gilbert DA (1997). Temporal organization of the cell division cycle in unicellular microorganisms. Soc Gen Microbiol Symp 56: 271–278
Lloyd AL, Lloyd D (1993). Hypothesis: the central oscillator of the circadian clock is a controlled chaotic attractor. BioSystems 29: 77–85
Lloyd AL, Lloyd D (1995). Chaos: its significance and detection in biology. Biol Rhythm Res 26: 233–252
Lloyd D, Murray DB (2005). Ultradian metronome: timekeeper for the orchestration of cellular coherence. Trends Biochem Sci 30: 373–377
Lloyd D, Murray DB (2006). The temporal architecture of eukaryotic growth. FEBS Lett 580: 2830–2835
Lloyd D, Murray DB (2007). Redox rhythmicity: clocks at the core of temporal coherence. Bioessays 29: 465–473
Lloyd D, Volkov EI (1990). Quantized cell cycle times: interaction between a relaxation oscillator and ultradian clock pulses. Bio Systems 23: 305–310
Lloyd D, Volkov EI (1991). The ultradian clock: timekeeping for intracellular dynamics. In E Mosekilde (ed.) Complex Dynamics and Biological Evolution. Plenum, New York

Lloyd D, Rossi EL (1992). Ultradian Rhythms in Life Processes: an Inquiry into Fundamental Principles of Chronobiology and Psychobiology. Springer-Verlag, London
Lloyd D, Edwards SW, Fry JC (1982). Temperature-compensated oscillations in respiration and cellular protein content in synchronous cultures of *Accanthamoeba castellanii*. Proc Natl Acad Sci USA 79: 3785–3788
Lloyd D, Lloyd AL, Olsen LF (1992). The cell division cycle: a physiologically plausible dynamic model can exhibit chaotic solutions. BioSystems 27: 17–24
Lloyd D, Aon MA, Cortassa S (2001). Why homeodynamics not homeostasis? Sci World 1: 133–145
Lorenz EN (1963). Deterministic nonperiodic flow. J Atmos Sci 20: 130–141
Lorenz EN (1993). The Essence of Chaos. University College London Press, London
Lüttge U (2003). Circadian rhythmicity: is the "biological clock" hardware or software? Prog Bot 64: 277–319
Lüttge U, Hütt M-T (2004). High frequency or ultradian rhythms in plants. Prog Bot 65: 235–263
Lüttge U, Hütt M-T (2007). Spatio-temporal patterns and distributed computation – a formal link between CO_2 signalling, diffusion and stomatal regulation. Prog Bot 68: 242–260
Luzikov VN (1981). Control over assembly of the mitochondrial inner membrane: selection by a performance criterion. FEBS Lett 125: 131–133
Luzikov VN (2004). Quality control of proteins and organelles. Biochemistry (Moscow) 67: 171–183
Mancuso S, Shabala S (2007). Rhythms in Plants. Springer, Berlin
Masamizu Y, Ohtsuka T, Takashima Y, Nagahara H, Takenaka Y, Yoshikawa K, Okamura H, Kageyama R (2006). Real-time imaging of the somite segmentation clock: revelation of unstable oscillators in the individual presonitic mesoderm cells. Proc Natl Acad Sci USA 103: 1313–1318
Maroto M, Pourquié O (2001). A molecular clock involved in somite segmentation. Furr Top Dev Biol 51: 221–248
Maroto M, Dale JK, Dequéant ML, Petit AC, Pourquié O (2005). Synchronised cycling gene oscillations in presomitic mesoderm cells require cell-cell contact. Int J Dev Biol 49: 309–315
May RM (1976). Simple mathematical models with very complicated dynamics. Nature 261: 459–467
Mitsui A, Kumazawa S, Takahashi A, Ikemoto H, Cao S, Arai T (1986). Strategy by which nitrogen-fixing unicellular cyanobacteria grow photoantotrophically. Nature 323: 720–722
Moreno N, Colaço, Feijo JA (2007). The pollen tube oscillator: integrating biophysics and biochemistry. In Mancuso S and Shabala S (eds.) Rhythms in Plants. Springer, Berlin
Morré DJ, Church PJ, Pletcher T, Tang X, Wu LY, Morré DM (2002). Biochemical basis for the biological clock. Biochemistry 41: 11941–11945
Morré DJ, Healds M, Coleman J, Orezyk J, Jiang Z, Morré DM (2007). Structural observations of time dependent oscillatory behaviour of $Cu^{II}Cl_2$ solutions measured via extended x-ray absorption fine structure. J Inorg Biochem 101: 715–726
Morré DJ, Jiang Z, Marjonovic M, Orezyk J, Morré DM (2008). Response to electromagnetic fields of copper ^{II} containing ECTO-NOX proteins and Cu^{II} Cl_2 in solution related to equilibrium dynamics of ortholpara hydrogen spins isomers of water. J Inorg Biochem http://dx/doi.org.10.1016/j.jinorbio.2008.06.001
Murray DB, Lloyd D (2007). A tuneable attractor underlies yeast respiratory dynamics. BioSystems 90: 287–294
Murray DB, Klevecz RR, Lloyd D (2003). Generation and maintenance of synchrony in *saccharomyces cerevisiae* in continuous culture. Exp Cell Res 287: 10–15
Murray DB, Beckmann M, Kitano H (2007). The regulation of yeast oscillatory dynamics. Proc Natl Acad Sci USA 104: 2241–2246
Nityananda V, Balakrishnan R (2007). Synchrony during acoustic interactions in the bush cricket Mecopoda 'Chirpen' (*Tettigoniidae Orthoptera*) is generated by a combination of chirp by

chirp resetting and change in intrinsic chirp rate. J Comp Physiol A Neuroethol Sens Neunal Behav Physiol 193: 51–65
O'Rourke B (2000). Pathophysiological and protective roles of mitochondrial ion channels. J Physiol 529: 23–36
O'Rourke B, Ramza BM, Marban E (1994). Oscillations of membrane current and excitability driven by metabolic oscillations in heart cells. Science 265: 962–966
Ott E, Grebogic C, Yorke JA (1990). Controlling chaos. Phys Rev Lett 64: 1196–1199
Peel A, Akam M (2003). Evolution of segmentation: rolling back the clock. Curr Biol 13: R708–R710
Pittendrigh CS, Pikovsky A (1993). Temporal organization: reflections of a Darwinian clock-watcher. Annu Rev Physiol 55: 16–54
Pourquié O (2003). The segmentation clock: converting embryonic time into spatial pattern. Science 301: 328–330
Prigogine I, Nicolis G (1971). Biological order, structure and instabilities. Quart Rev Biophys 4: 107–148
Prytz G (2001). A biophysical study of oscillatory water regulation in plants measurements and Models. DSc Thesis, NTNU: Trondheim
Pyragas K (1992). Continuous control of chaos by self-controlling feed-back. Phys Lett 170: 421–428
Queiroz-Claret C, Valon C, Queiroz O (1988). Are spontaneous conformational interconversions a molecular basis for long-period oscillations in enzyme activity? Chronobiol Int 5: 301–309
Rascher V, Hütt M-T, Siebka K, Osmond B, Beck F, Lüttge U (2001). Spatiotemporal variation of metabolism in a plant circadian rhythm: the biological clock as an assembly of coupled individual oscillators. Proc Natl Acad Sci USA 98: 11801–11805
Reich JG, Sel'kov EE (1981). Energy Metabolism of the Cell. Academic Press, London
Rittenberg D (1948). Turnover of proteins. J Mount Sinai Hosp 14: 891–901
Romashko DN, Marban E, O'Rourke B (1998). Sub-cellular metabolic transients and mitochondrial redox waves in heart cells. Proc Natl Acad Sci USA 95: 1618–1623
Roussel MR, Lloyd D (2007). Observation of a chaotic multioscillatory metabolic attractor by real-time monitoring of a yeast continuous culture. FEBS 274: 1011–1018
Sagan D (1994). On the physics of Landau damping. Am J Phys 62: 450–462
Satroutdinov AD, Kuriyama H, Kobayashi H (1992). Oscillatory metabolism of *Saccharomyces cerevisiae* in continuous culture. FEMS Microbiol Lett 98: 261–268
Schneider ED, Sagan D (2005). Into the Cool: Energy Flow, Thermodynamics and Life. University of Chicago Press, Chicago.
Schoenheimer R (1942). The Dynamic State of Body Constituents. Harvard University Press, Cambridge, MA
Schuster HG (1988). Deterministic Chaos, An Introduction, 2nd edn. Physik-Verlag, Weinheim
Smith HM (1935). Synchronous flashing of fireflies. Science 82: 151–152
Strogatz SH (2000). From Kuramoto to Crawford: Exploring the onset of synchronization in populations of coupled oscillators. Physica D 143: 1–20
Strogatz SH (2003). Synch. The Emerging Science of Spontaneous Order. Hyperion, New York
Strogatz SH, Mirello RE (1991). Stability of incoherence in a population of coupled oscillators. Stat Phys 63: 613–635
Strogatz SH, Mirello RE, Matthews PC (1992). Coupled non-linear oscillators below the synchronization threshold: relaxation by generalized Landau damping. Phys Rev Lett 68: 2730–2733
Sweeney BM, Tuffli CF Jr, Rubin RH (1967). The circadian rhythm in photosynthesis in Acetabularia in the presence of actinomycin. D. J Gen Physiol 50: 647–659
Toth R, Taylor RE (2006). Loss of coherence in a population of diffusively coupled oscillators. J Chem Phys 125: 224708
Trimmer BA, Apritle JR, Dudzinski DM, Lagace CJ, Lewis SM, Michel T, Oazis S, Zagas RM (2001). Nitric oxide and the control of firefly flashing. Science 292: 2486–2488
Van de Pol B, Van der Mark (1928). The heartbeat considered as a relaxation oscillation and an electrical model of the heart. Phil Mag 6: 763–775

Visser GR, Reinten C, Coplam P, Gilbert DA, Hammond KD (1990). Oscillations in cell morphology and redox state. Biophys Chem 37: 383–394

Von Bertalanffy L (1952). Problems of Life. Harper, New York

Weber G (1976). Practical application and philosophy of optical spectroscopic probes. In Kasha M and Pullman (eds.) Horizons in Biochemistry and Biophysics, vol. 2. Addison-Wesley, Reading, MA, pp. 163–198

Wheatley DN (1989). Protein turnover in relation to growth and cell cycle stage of cultured mammalian cells. In Grisda S, Knecht E (eds.) Current Trends in Intracellular Protein Degradation Bilbao. Springer, Berlin, pp. 377–400

Wiener N (1961). Cybernetics, 2nd edn. MIT Press, Cambridge, MA

Wiley DA, Strogatz SH, Girvan M (2006). The size of the synch basin. Chaos. 16(1): 015103

Winfree AT (1967). Biological rhythms and the behaviour of populations of coupled oscillators. J. Theoret Biol 16: 15–42

Winfree AT (2001). The Geometry of Biological Time, 2nd ed. Springer, Berlin

Yates FE (1982). Outline of a physical theory of physiological systems. J Physiol Pharmacol 60: 217–248

Yates FE (ed.) (1987). Self-Organizing Systems: The Emergence of Order. Plenum, New York

Yates FE (1992). Fractal applications in biology:scaling time in biochemical networks. Methods Enzymol 210: 636–675

Yates FE, Yates LB (2008). Ultradian rhythms as the dynamic signature of life. In Lloyd D, Rossi ER (eds.) Ultradian Rhythms from Molecules to Mind: A New Vision of Life. Springer, Berlin

Yuan Z, Medina MA, Boiteux A, Müller SC, Hess B (1990). The role of fructose 2, 6-bisphosphate in glycollytic oscillations in extracts and cells of *Saccharomyces cerevisae*. Eur J Biochem 192: 791–795

Structure and Regulation of Plant Vacuolar H⁺-ATPase

T. Seidel

Contents

Abstract The vacuolar proton translocating ATPase (V-ATPase) is an essential protein complex present in all eukaryotes which functions as ATP-driven rotary motor. In higher eukaryotes, the V-ATPase consists of 13 different subunits. The V-ATPase assembly occurs at the stage of the ER and the presence of isogenes enables the formation of multiple V-ATPase-isoenzymes in dependence on tissue-specific and subcellular requirements. V-ATPases are essential for the cytosolic pH

T. Seidel
Department of Biochemistry and Physiology of Plants,
W5, University of Bielefeld, 33501 Bielefeld, Germany
e-mail: thorsten.seidel@uni-bielefeld.de

U. Lüttge et al. (eds.), *Progress in Botany 70*,
DOI:10.1007/978-3-540-68421-3, © Springer-Verlag Berlin Heidelberg 2009

homeostasis, the generation of a proton motive force and the pH-controlled termination of receptor–ligand interactions within secretory and endocytotic pathways. Especially in plants, V-ATPases are also involved in developmental events such as cell expansion and mechanisms of stress defense, e.g. vacuolar sequestration of toxic compounds. The regulation depends on the cellular redox state, nucleotide-availability and the developmental stage. This review focuses mainly on the structure of the V-ATPase: the subunit-composition, assembly, as well as regulatory structural alterations and protein–protein-interactions, of the V-ATPase are summarised.

1 Introduction

The vacuole fulfills multiple functions in plant cells. It represents a storage and a lytic compartment, and furthermore it is essential as both temporary and final waste deposit, e.g. of conjugated xenobiotics. These functions partly require contrasting lumenal milieus, resulting in the presence of at least two types of vacuoles in a single plant cell. One fundamental difference between both types of vacuoles is the lumenal pH-value. Whereas storage vacuoles are characterised by a neutral pH-value, lytic vacuoles are acidic. The dominant central vacuole in plant cells is initially of the lytic, acidic type and fuses later on with peripheral neutral vacuoles in some cells (Neuhaus 2000; Di Sansebastiano et al. 2001). In contrast, Paris et al. (1997) postulated the central vacuole formation to be an event involving the fusion of lytic and storage vacuoles, finally resulting in the breakdown and recycling of stored solutes. In *Mesembryanthemum crystallinum*, two different types of vacuoles develop in response to salinity. In this case, one serves as malate storage to maintain CAM-metabolism and the second is required for salt sequestration (Epimashko et al. 2004). Each type of vacuole requires specific transporters linked to its specific function. On the other hand, a set of primary active transporters is common for all vacuoles and generates the proton motive force and adjusts the lumenal pH-value. A primary active proton translocating pyrophosphatase (V-PPase) and an ATPase (V-ATPase) were identified at the tonoplast, so two distinct proton pumps coexist at the vacuolar membrane (Müller et al. 1996). The activity of V-PPase and V-ATPase can be distinguished by their specific response to inhibitors. The V-ATPase can be inhibited by concanamycin, bafilomycin or nitrate whereas the V-PPase is sensitive to aminomethylenebisphosphonate (Maeshima 2000). In lemon fruit, a nitrate-sensitive and nitrate-insensitive V-ATPase were identified at the tonoplast which differ in their H^+/ATP ratio (Müller et al. 2002). These isoenzymes may reflect the distinct types of vacuoles in plant cells. Recently, single V-ATPase subunits were identified at the tonoplast of tobacco and *Arabidopsis* by proteomic approaches. These data provide further evidence for the presence of isoenzymes at the tonoplast (Drobny et al. 2002; Carter et al. 2004; Shimaoka et al. 2004; Jaquinod et al. 2007). In accordance with their association with the secretory pathway, V-ATPases were also identified at the plasmamembrane, the ER, the Golgi and endosomal compartments (Robinson et al. 1996; Alexandersson et al. 2004; Dettmer et al. 2006; Seidel et al. 2008). Organelle-specific isoforms of

V-ATPase subunits indicate the involvement in organelle-specific functions in addition to the *de novo* biosynthesis along the secretory pathway.

The catalytic cycle of the V-ATPase involves a rotary mechanism similar to F-ATP-synthases. This mechanism distinguishes the complex ATPases from ABC-transporters like the plasmamembrane proton translocating ATPase (PM-ATPase) (Bowman 1983; Hirata et al. 2003). V-ATPase, F-ATP-synthase and the proton (or Na^+) translocating ATPase (A-ATPase) from archae bacteria evolved from a common ancestor. The H^+/ATP ratio determines whether the enzyme functions as ATP-synthase or proton pump (Arechaga and Jones 2001; Kettner et al. 2003): A coupling ratio of 2 favours proton transport, whereas a ratio of 4 enables ATP-synthesis (Cross and Müller 2004). Initially, the ATP-synthase probably derived from an ATPase by gene duplication of the catalytic head subunit, i.e. subunit β of F-ATPase and subunit A of A-type and V-type ATPase. A subsequent gene duplication and fusion of the proteolipid subunits resulted in the proton translocating ATPases (Cross and Taiz 1990). In some archae bacteria, the presence of two a-subunits corresponding to the A-ATPase subunit I allows the formation of two independent proton half-channel systems and subsequently the evolution to an ATP-synthase (Cross and Müller 2004). This mechanism was postulated for the archae bacteria *Caloramator fervidus* and *Pyrococcus furiosis* (Sapra et al. 2003; Cross and Müller 2004) and might represent the next step in evolution.

In general, V-ATPases are of importance for every eukaryotic cell. The silencing of genes encoding V-ATPase subunits results in lethality, e.g. of *Arabidopsis thaliana*, *Caenorhabditis elegans* and *Drosophila melanogaster*, and severely disturbs bone homeostasis in mice (Schumacher et al. 1999; Allen et al. 2000; Choi et al. 2003; Du et al. 2006; Lee et al. 2006). In yeast, V-ATPase mutants are characterised by a conditional lethality. The endocytotic uptake of acidic medium allows yeast to adjust the vacuolar pH in the absence of an active V-ATPase (Nelson and Nelson 1990). V-ATPase loss of function mutants can be screened by their inability to grow on neutral medium while they survive on acidic medium. In addition, these mutants are characterised by an increased sensitivity to metal ions (Nelson et al. 2000). Thus, yeast turned out to be the most suitable organism for mutational analysis of V-ATPase.

2 Structure and Subunit-Composition of V-ATPase

The knowledge on the V-ATPase-structure in plants is poor and even for yeast the available data are incomplete. The most complete structural overview derives from a species-independent point of view, keeping in mind their differences in cellular organisation and requirements. Numerous electron micrographs of the V-ATPase from *Saccharomyces cerevisiae*, *Bos taurus*, *Kalanchoe daigremontiana*, *Manduca sexta*, *Vigna radiata* and *Neurospora crassa* reveal the common bipartite structure of the V-ATPase indepedent of the source organism (Grüber et al. 2000; Domgall et al. 2002; Wilkens et al. 2004; Li and Zhang 2004; Venzke et al. 2005; Zhang et al. 2006). Even the amino acid sequences of various species are characterised by a high degree of similarities (Table 1).

Table 1 Protein-sequence identity of V-ATPases from *Arabidopsis thaliana* and *Saccharomyces cerevisiae* and A-ATPase from *Thermus thermophilus*

V_1	*Arabidopsis thaliana*	Locus	*Saccharomyces cerevisiae*	*Thermus thermophilus*
			Vma1	Subunit A*
	AtVHA-A	At1g78600	60%	51%
			Vma2	Subunit B*
	AtVHA-B1	At1g76030	73%	54%
	AtVHA-B2	At4g38510	73%	51%
	AtVHA-B3	At1g20260	70%	54%
			Vma5*	
	AtVHA-C	At1g12840	28%	–
			Vma8	Subunit D
	AtVHA-D	At3g58730	44%	26%
			Vma4	Subunit E*
	AtVHA-E1	At4g11150	32%	22%
	AtVHA-E2	At3g08560	30%	23%
	AtVHA-E3	At1g64200	32%	19%
			Vma7	Subunit F*
	AtVHA-F	At4g02260	51%	14%
			Vma10	Subunit G
	AtVHA-G1	At3g01390	30%	20%
	AtVHA-G2	At4g23710	29%	18%
	AtVHA-G3	At4g25950	30%	20%
			Vma13*	
	AtVHA-H	At3g42050	16%	–
V_0				
			Vph1p/Stv1p	Subunit I
	AtVHA-a1	At2g28520	33%/35%	17%
	AtVHA-a2	At2g21410	36%/33%	14%
	AtVHA-a3	At4g39080	37%/34%	12%
			Vma3	Subunit L
	AtVHA-c1	At4g34720	55%	32%
	AtVHA-c2	At1g19910	55%	32%
	AtVHA-c3	At4g38920	55%	32%
	AtVHA-c4	At1g75630	55%	32%
	AtVHA-c5	At2g16510	55%	32%
			Vma16	
	AtVHA-c"1	At4g32530	53%	–
	AtVHA-c"2	At2g25610	53%	–
			Vma6	Subunit C*
	AtVHA-d1	At3g28710	44%	15%
	AtVHA-d2	At3g28715	44%	15%
			Vma9	
	AtVHA-e1	At5g55290	17%	–
	AtVHA-e2	At4g26710	17%	–

Capital letters denominate subunits of the cytosolic, ATP-hydrolyzing V_1-sector whereas subunits of the transmembrane, proton translocating V_0-sector are denoted by *small letters*. Crystal structures of Vma5 (1u7l) and Vma13 (1ho8) as well as of subunits A (*Pyrococcus horikoshii*; 1vdz), B (*Methanosarcina mazei*; 2c61), C (1v9m), E (*Pyrococcus horikoshii*; 2dma) and F (2d00) from bacterial A-ATPase are available in the RCSB protein data base (subunits marked by an *asterisk*)

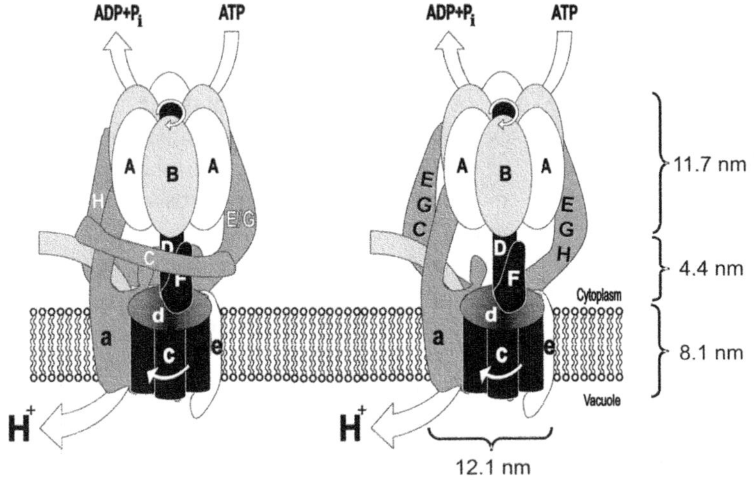

Fig. 1 Models of V-ATPase displaying alternative arrangements of the peripheral stalk Both models are based on the assumption of three stators. On the left hand, VHA-a, VHA-H and VHA-E/VHA-G form the stators, whereas on the right hand, VHA-a, VHA-E/VHA-G/VHA-C and VHA-E/VHA-G/VHA-H form the stators. V_1-subunits are encoded by *capital letters*, the V_0-subunits by *small letters*. The catalytic head subunits are shown in *white* (A) and *light grey* (B), rotor subunits (VHA-D, VHA-F, VHA-d, VHA-c) in *black* and peripheral stalk subunits (VHA-a, VHA-C, VHA-E, VHA-G, VHA-H) in *dark grey*. VHA-e is stained *white* as well. The given distances are based on the EM micrographs collected from the Kalanchoe V-ATPase (Domgall et al. 2002)

The apparent subsectors V_1 and V_0 represent two coupled engines which form the proton pump: the cytosol-exposed sector V_1 is the motor, which drives the rotation of the membrane integral proton-turbine V_0 (Grabe et al. 2000). V_1 consists of eight different subunits termed VHA-A to VHA-H and its mass represents approximately 70% of the holoenzyme. Three copies of each subunit VHA-A and VHA-B form a barrel-like hexameric structure of 11.7 nm length with a central cavity similar to the catalytic sector F_1 of the mitochondrial or chloroplastic ATP-synthases (Grüber et al. 2000, 2001; Domgall et al. 2002). The central cavity of the hexameric barrel serves as a bearing for the central stalk, which is formed by a heterodimer of VHA-D and VHA-F (Imamura et al. 2003; Maegawa et al. 2006). The other subunits, VHA-C, VHA-E, VHA-G and VHA-H, take part in peripheral stalks (Féthière et al. 2005), which prevent the contraproductive rotation of the hexameric head and stabilise the holoenzyme.

The precise number of peripheral stalks has not been elucidated up to now. One stator is formed by the V_0 subunit VHA-a, whereas 1–2 stators are formed by VHA-E and VHA-G. These may either interact with VHA-C and VHA-H and form the stators by EGC and EGH trimers or one stator is formed by the VHA-E/VHA-G subunits and another by VHA-H. In the latter case, VHA-C connects the stators (Fig. 1) (Domgall et al. 2002; Féthière et al. 2004, 2005, Zhang et al. 2006).

2.1 VHA-A

The 69kDa-subunit VHA-A is the dominant polypeptide of the V_1-sector. Together with VHA-B it forms the hexameric head and is responsible for ATP-hydrolysis. Each of the three copies of VHA-A is characterised by a distinct nucleotide-binding state at a defined moment, and the catalytic cycle proceeds in an alternating, but coordinated, manner in the three VHA-A copies (Kawamura et al. 2001). This enables the transduction of conformational changes in response to the nucleotide binding state into rotation of the central stalk (Nishi and Forgac 2002; Imamura et al. 2003). The VHA-A nucleotide binding state is further transmitted to VHA-a and VHA-E. Likewise, bafilomycin-binding to VHA-a results in conformational changes within VHA-A (Landolt-Marticorena et al. 1999; Kawamura et al. 2001). These interactions demonstrate the coordinated function and structural linkage of both subsectors. The structure of VHA-A can be divided into four domains (Fig. 2). The domains I, III and IV are conserved between V-ATPase and the F-ATP-synthase subunit β, whereas domain II is unique for V-ATPases (Shao et al. 2003). Accordingly, this 90 amino acid insertion is called the non-homologous region. It forms eight β-strands on the top of the hexameric barrel, which are visible as a knoblike extension in EM micrographs (Wilkens et al. 2004; Domgall et al. 2002). The homologous region is postulated to be important for coupling the V_0–V_1 and to enable the regulation via reversible dissociation (Shao and Forgac 2004). Two nucleotide binding P-loops are conserved in domains III and IV. The ATP-hydrolysis takes place in the first P-loop (Walker A motif), whereas the second P-loop (Walker B motif) is suggested to take part in regulating the coupling of VHA-A and the central stalk. The interaction between the central stalk and VHA-A is mediated by the DALPERE-motif (Maegawa et al. 2006). A conserved cysteine residue (Cys256 in *A. thaliana*) was identified in the catalytic P-loop region which results in redox-sensitivity of ATP-hydrolysis (Hager and Lanz 1989; Feng and Forgac 1992a, b). The additional conserved cysteine residues 279 and 535 were postulated to be involved in redox regulation by a disulfide transition mechanism. In this model, the active state is characterised by a disulfide formation between Cys279 and Cys535, whereas a disulfide bridge between Cys256 and Cys535 locks VHA-A in a closed conformation and hence inhibits the ATP-hydrolysis (Feng and Forgac 1994; Stevens and Forgac 1997). Alternatively, disulfide formation between VHA-A and Cys188 of VHA-B or modulation of a single cysteine residue by, e.g., glutathione might block the P-loop for nucleotide binding (Hager and Lanz 1989; Feng and Forgac 1992a; Taiz et al. 1994; Dietz et al. 1998). In chicken osteoclasts, the P-loop domain can be removed by alternative splicing (Hernando et al. 1995). However, a similar mechanism was not reported for any other organism. In *A. thaliana*, a T-DNA insertion into VHA-A results in severe alterations of the Golgi-morphology in male gametophytes. This leads to a failure in pollen formation and finally male lethality (Dettmer et al. 2005).

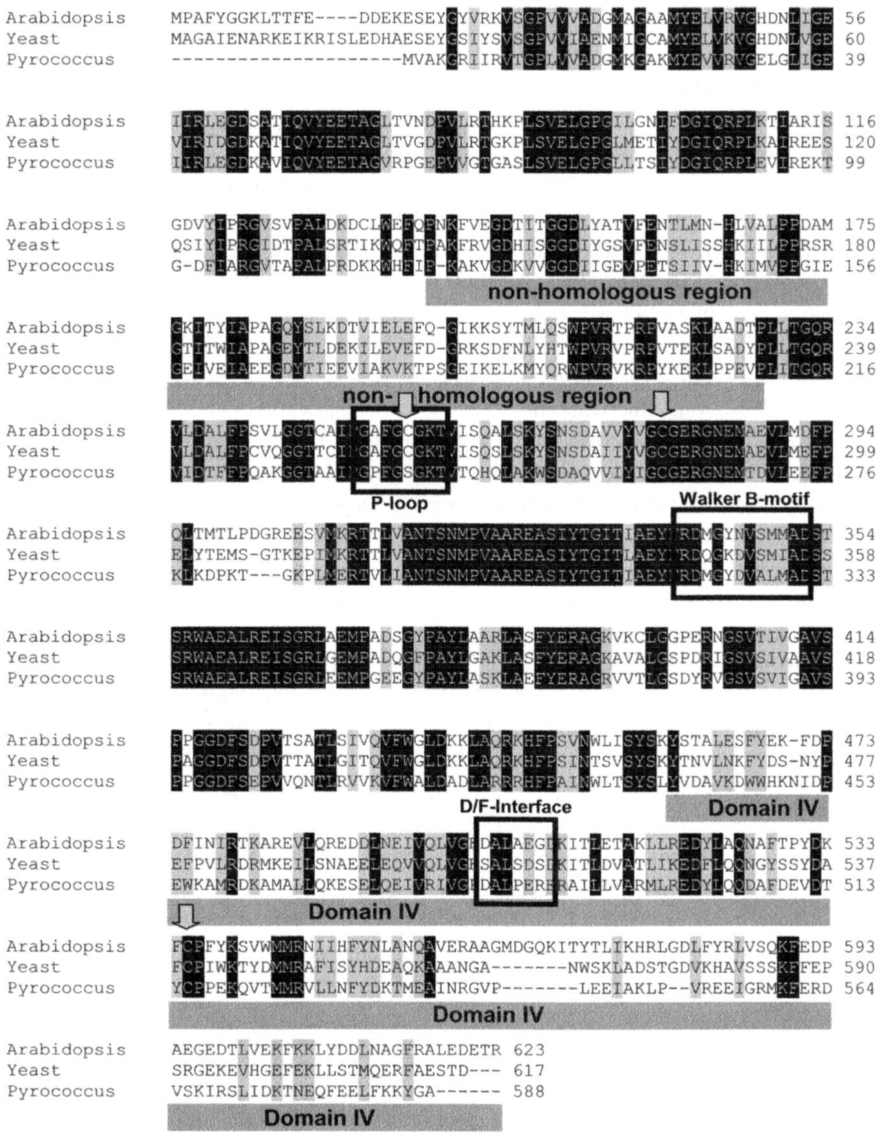

Fig. 2 Aminoacid-sequence alignment of VHA-A from *Arabidopsis thaliana*, *Saccharomyces. cerevisiae* and *Pyrococcus horikoshii*. Identical aminoacids are highlighted *black*, similar residues *light grey*. The *boxes* mark the conserved nucleotide-binding motifs (P-loop and Walker B-motif) and the central stalk interface (D/F-interface). The three cysteinyl residues, which are conserved in eukaryotes, are indicated by *arrows*. The four domains of VHA-A are displayed by *underlining* the non-homologous region and domain IV (according to Maegawa et al. 2006)

2.2 VHA-B

VHA-B is homologous to the F-ATP-synthase subunit α and takes part in nucleotide binding (Manolson et al. 1985). Although its structure resembles that of VHA-A and the presence of nucleotide-binding sites was proven, VHA-B lost its activity during evolution (Cross and Müller 2004). Whereas VHA-A is responsible for ATP-hydrolysis, VHA-B gained a regulatory function and by its N-terminal domain mediates the association of the V-ATPase with the actin-microfilaments of the cellular cytoskeleton (Holliday et al. 2000; Vasileya et al. 2000; Chen et al. 2004; Holliday et al. 2005). The activation state of V-ATPase is also altered by an interaction of VHA-B and the glycolytic aldolase (Lu et al. 2007). The V-ATPase of the facultative CAM-plant *Mesembryanthemum crystallinum* undergoes structural changes during salt-induced C_3-CAM-transition, which directly affect the structure of subunit VHA-B by proteolytic processing. Under these conditions, the polypeptides D_i and E_i appeared and remained associated with the holoenzyme (Bremberger et al. 1988; Bremberger and Lüttge 1991a, b; Ratajczak et al. 1994). These peptides cross-reacted with antibodies directed against VHA-B and hence represent fragments of VHA-B. The fragments also appeared and in parallel the amount of VHA-B decreased subsequent to an *in vitro* incubation of the holoenzyme with endogenous proteases or hydrogenperoxide (Krisch et al. 2000). The crystal structure of archael VHA-B from *Methanosarcina mazei* Gö1 identified three distinct domains. The central α-β-domain is formed by a nine-stranded β-sheet which is surrounded by seven α-helices. This arrangement matches the nucleotide binding site of the F-ATP-synthase subunit α, but the P-loop motif was identified 195 amino acids downstream of the typical position in F-ATP-synthases and V-ATPases (Schäfer et al. 2006). It is not clear whether these data can be transferred to predict functional domains of eukaryotic V-ATPases. Recently, the *Arabidopsis* VHA-B isoform 1 was demonstrated to form a glucose signalling complex with the nuclear hexokinase1 and the 19S regulatory particle of proteasome subunit (RPT5B), which directly affects target gene transcription in the nucleus (Cho et al. 2006). Besides the crosslinking between actin filaments by VHA-C (Vitavska et al. 2005), this report represents the second indication for additional functions of single V-ATPase subunits.

2.3 VHA-C

VHA-C is an elongated 40kDa protein of 12.5nm length (Tavakoli et al. 1999; Armbrüster et al. 2004). It consists of two globular domains, a globular "head" domain and a structurally similar, but on the level of primary structure rather different, "foot" domain, that are connected by an elongated neck-like bundle of four helices (Drory et al. 2004) (Fig. 3). In contrast to the earlier report of Armbrüster et al. (2004), the structure is distinct from VHA-H (Fig. 3), but VHA-C possesses several interaction domains similar to VHA-H (Drory et al. 2004; Jones et al. 2005). The lack of VHA-C or mutated VHA-C subunits did not affect assembly of V-ATPase in yeast, but these complexes turned out to be unstable *in vitro* and showed a reduced activity (Curtis and

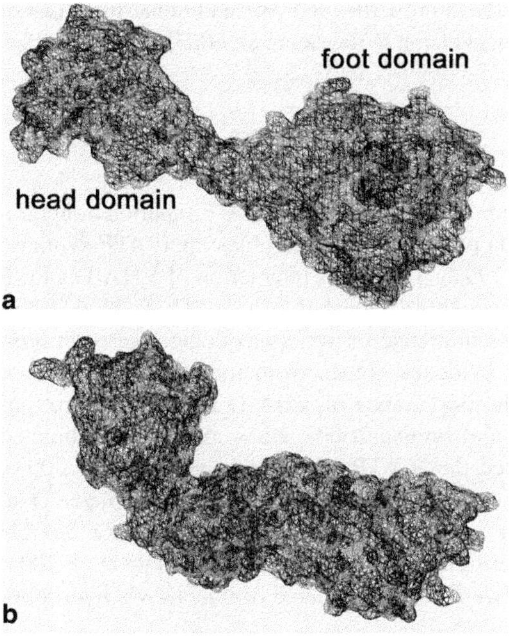

Fig. 3 Crystal structure of ScVHA-C (A) and ScVHA-H (B). VHA-C is characterised by a barbell-like structure (**A**, Iwata et al. 2004). Two domains are distinguished in VHA-H, which is characterised by an elongated structure (**B**, Sagermann et al. 2001) The structure of the yeast subunits Vma5 (pdb ID 1u7l) and Vma13 (pdb ID 1ho8) were obtained from the RCSB protein data bank and surfaces of the V-ATPase subunits were computed using the swiss prot pdb-viewer

Kane 2002; Smardon and Kane 2007). VHA-C mutants behave like RAVE-mutants in yeast, since VHA-C requires the RAVE-complex for assembling with the V-ATPase (Smardon and Kane 2007). In *A. thaliana*, lack of VHA-C resulted in defects in cell elongation and conditional defects in hypocotyl elongation (Schumacher et al. 1999). Additionally, the moderate over-expression of VHA-C inhibited ATP-hydrolysis and proton transport in yeast and higher VHA-C levels were lethal (Curtis et al. 2002). VHA-C represents a flexible stator subunit, which connects the transmembrane domain and the cytosolic stator subunits (Drory et al. 2004). Interactions between VHA-C and the peripheral stalk subunits VHA-E, VHA-G and VHA-a were demonstrated repeatedly in the past by crosslinking experiments (Xu et al. 1999; Inoue and Forgac 2005; Jones et al. 2005). Zhang et al. (2006) fitted the crystal structure of VHA-C into electron microscopic images of the yeast V-ATPase. The VHA-C-subunit appeared collar-shaped and surrounded the central stalk as connector between two postulated VHA-E/VHA-G-stators. Its ability to bind nucleotides indicates a regulatory role, e.g. in triggering reversible dissociation. Structural alterations in response to the nucleotide binding state might result in weakening of the stator and, hence, enable the detachment of the V_1-sector from the membrane in order to inhibit ATP-hydrolysis (Armbrüster et al. 2005; Grüber 2005). Several phosphorylation sites were identified within VHA-C, which are target of the WNK8-kinase and link V-ATPase regulation to the WNK-pathway in *Arabidopsis* (Hong-Hermesdorf et al. 2006). The assembled V-ATPase is anchored to the cytoskeleton via two actin binding sites of VHA-C in addition to the actin binding site of VHA-B (Vitavska et al. 2003). Furthermore, VHA-C is able to crosslink actin filaments and forms homodimers and oligomers in the absence of the V-ATPase complex. These properties indicate additional functions of VHA-C in the arrangement of the cytoskeleton (Vitavska et al. 2003, 2005).

2.4 VHA-D/F

It is now widely accepted that the subunits VHA-D and VHA-F function as central stalk subunits. VHA-D is a 29.1-kDa protein which resembles the F-ATP-synthase central stalk subunit γ (Kluge et al. 1999). The direct interaction of VHA-D and -F was confirmed by crosslinking experiments (Tomashek et al. 1997; Xu et al. 1999) and the rotation of their homologues in archae bacteria was demonstrated *in vitro* (Imamura et al. 2003). Additional evidence comes from mutational analysis of VHA-D in yeast, which revealed the importance of VHA-D for coupling proton-transport and ATP-hydrolysis (Xu and Forgac 2000). Also, an antibody directed towards VHA-D efficiently inhibited the V-ATPase (Aviezar-Hagai et al. 2003). Recently, the NMR-structure of the A-ATPase subunit F was obtained (Gayen et al. 2007). VHA-F is characterised by a two-domain structure. The more flexible C-terminal domain binds to the C-terminal domain of VHA-B (Coskun et al. 2004; Gayen et al. 2007). VHA-D and VHA-F were postulated to interact via hydrophobic interactions (Gayen et al. 2007).

2.5 VHA-E/VHA-G

VHA-E and VHA-G form the core-complex of the stator with a total length of 220 Å. The interaction of VHA-E and VHA-G is highly specific and the dimer cannot be formed *in vitro* (Fèthière et al. 2004). Over-expression, lack of VHA-G or mutations within VHA-G have a direct impact on the stability of VHA-E in yeast (Tomashek et al. 1997; Charsky et al. 2000; Ohira et al. 2006). The amino acids 19–38 are required for the electrostatic interaction with VHA-G, whereas the first 19 amino acids are essential for an interaction with VHA-C (Jones et al. 2005). Two heterodimers of VHA-E and VHA-G take part in the peripheral stalk formation and their C-termini are located at the V_0V_1-interface (Tomashek et al. 1997; Seidel et al. 2005; Ohira et al. 2006). The heterologous co-expression in *Escherichia coli* indicated the formation of one stator involving VHA-E, VHA-G and VHA-C and a second formed by VHA-E, VHA-G and VHA-H (Fèthière et al. 2005). However, VHA-E interacts with VHA-A and influences the catalytic activity directly (Owegi et al. 2006). On the other hand, VHA-E was demonstrated to form intramolecular disulfide bonds in response to oxidising conditions. This disulfide formation within VHA-E depends on the nucleotide binding state of VHA-A (Tavakoli et al. 2001), indicating a bidirectional communication between both subunits. VHA-E also interacts with the peripheral stalk subunit VHA-H and with regulatory elements, e.g. the glycolytic enzyme aldolase (Lu et al. 2002; Jones et al. 2005). The aldolase links the V-ATPase-activity to the cellular energy state and triggers the reversible dissociation of the complex in yeast. The detached subsector V_1 interacts with the so-called RAVE-complex via VHA-E in conjunction with reversible dissociation as well as *de novo*-biosynthesis of the complex (Smardon et al. 2002; Smardon and Kane 2007). Further evidence for the regulatory properties of VHA-E derived from overexpression of *Mesembryanthemum crystallinum* VHA-E in *A. thaliana* protoplasts. Cells expressing McVHA-E were characterised by a significant decrease of the vacuolar pH, a similar effect was observed for McVHA-H (Seidel et al. 2005). This result indicates either a stimulatory effect of excess VHA-E and VHA-H on V-ATPase activity or an antagonistic function, e.g. by binding to negative regulators. The lack of VHA-E isoform 1 in *A. thaliana* results in abnormal Golgi and vacuolar morphology and embryo lethality (Strompen et al. 2005).

2.6 VHA-H

The peripheral subunit VHA-H was the first crystallised V-ATPase subunit (Sagermann et al. 2001). It is characterised by two domains, which are connected by a short flexible linker (Liu et al. 2005). Both domains can be expressed independently and VHA-H-deletion mutants of yeast can be complemented by the unfused fragments (Liu et al. 2005). The overall structure is elongated and characterised by a multitude of putative protein–protein-interaction sites (Sagermann et al. 2001). Four Armadillo repeats within VHA-H were suggested to interact with the clathrin adaptor protein complex 2 (AP-2) and might link the V-ATPase to the endocytotic machinery in mammalia (Geyer et al. 2002). VHA-H interacts with multiple sites, i.e. the VHA-A, VHA-B and the first 78 amino acids of VHA-E, and also connects the stator to the membrane integral sector via VHA-a (Landolt-Marticorena et al. 2000; Lu et al. 2002). Domgall et al. (2002) fitted the crystal structure of VHA-H into the EM micrograph of the *Kalanchoë daigremontiana* V-ATPase and suggested VHA-H as additional third stator besides VHA-a and VHA-E/VHA-G. In a similar approach based on EM micrographs collected from bovine brain V-ATPases, Wilkens et al. (2004) located VHA-H in a converse manner at the V_0V_1-interface.

VHA-H is predicted to play a dual regulatory role, on the one hand by inhibiting the ATP-hydrolysis of detached V_1-sectors by bridging peripheral and central stalk, and on the other hand by activiting the assembled holoenzyme and modulating the coupling (Parra et al. 2000; Crider and Xie 2003; Liu et al. 2005).

2.7 VHA-a

The bidomain subunit VHA-a is the largest component of the V-ATPase complex. Its molecular mass is about 100 kDa, which is roughly equally distributed among the N-terminal and C-terminal domains. The C-terminal domain consists of nine transmembrane helices whereas the N-terminal half forms a cytosolic expansion, which is part of the stator and anchors the V_1-sector to the membrane

(Leng et al. 1999; Landolt-Marticorena et al. 2000; Harrison et al. 2003; Kluge et al. 2004; Dettmer et al. 2006). The C-terminal domain was suggested to function as lumenal pH-sensor, which transmits the information on vacuolar acidification to the cytoplasmic site (Marshansky 2007). The main function of VHA-a is linked to transmembrane proton transport. Cytosolic protons have access to the proton binding site of the V-ATPase via a half-channel formed by the C-terminal portion of VHA-a (Grabe et al. 2000; Kawasaki-Nishi et al. 2003b). The proton release depends on a positive charge located at an arginine residue (Arg735 in yeast) in the seventh helix of VHA-a. This charge alters the pK_a of the proton binding glutamate and the proton gets released (Grabe et al. 2000; Kawasaki-Nishi et al. 2001a; Harrison et al. 2003). The seventh helix was predicted to twist to enable channel gating and positioning of the barrier charge (Vos et al. 2007). Accordingly, the 100-kDa subunit VHA-a is arranged in close proximity to the proteolipid ring (Seidel et al. 2004). Two isoforms were identified in yeast, which differ in subcellular localisation, coupling and their ability of reversible dissociation (Manolson et al. 1994; Kawasaki-Nishi et al. 2001b). Mutational analysis in yeast revealed the importance of the N-terminal domain for targeting and reversible dissociation, and the influence of the C-terminal domain on the coupling between H^+-transport and ATP hydrolysis (Kawasaki-Nishi et al. 2001b). A similar subcellular differentiation was also observed for plant VHA-a isoforms (Kluge et al. 2004; Dettmer et al. 2006).

2.8 The Proteolipids VHA-c, VHA-c', VHA-c"

The central proteolipid ring of V-ATPase and F-ATP-synthases derived from a common ancestor. The main difference of proteolipid subunits in V-ATPases and F-ATP-synthases is the reduction of the number of proteolipid subunits following duplication of the transmembrane domains from F-ATP-synthase to V-ATPase (Powell et al. 2000; Cross and Müller 2004). In V-ATPases, the ring consists of 4–5 copies of the proteolipid subunit VHA-c, one copy of VHA-c', which is absent in higher eukaryotes, and one copy of VHA-c" (Powell et al. 2000; Graham et al. 2000; Sze et al. 2002). VHA-c and VHA-c' are composed of four transmembrane helices with lumenal-orientated C-termini (Flannery et al. 2004) and the proton binding site is located at the fourth helix (Fig. 4) (Harrison et al. 2003; Kawasaki-Nishi et al. 2003). The arrangement of the transmembrane regions of VHA-c" is subject of controversial discussions. In *Saccharomyces cerevisiae*, VHA-c" consists of five helices, but it is unclear if the first helix is a transmembrane or cytosolic domain (Powell et al. 2000; Nishi et al. 2001; Gibson et al. 2002; Aviezer-Hagai et al. 2003; Nishi et al. 2003; Flannery et al. 2004). Therefore, the prediction of the C-termini varies from cytosolic to lumenal location. A comparative analysis of VHA-c and VHA-c" revealed no structural differences between both subunits and a hybrid protein assembled with VHA-a. Results obtained by fluorescence energy transfer between fluorophore-fused VHA-c and VHA-c" subunits indicate a lumenal

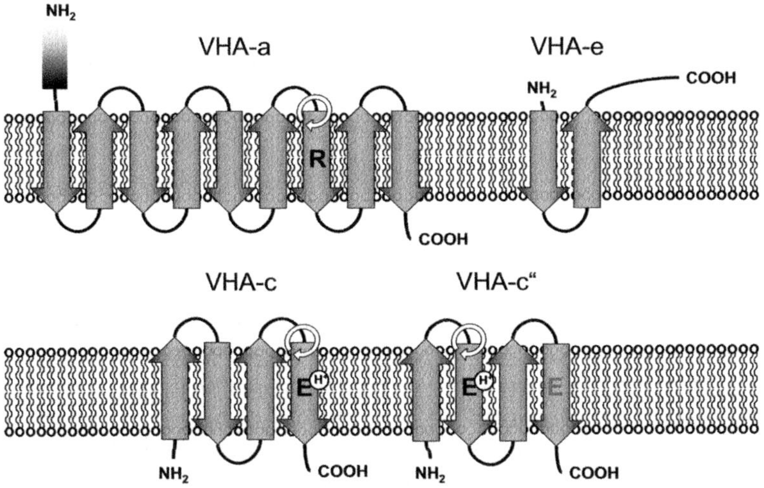

Fig. 4 Transmembrane topology of V_0-subunits. The C-terminal portion of VHA-a consists of nine transmembrane helices, which form the essential half channels. The stator charge surrounds a conserved arginine residue at the seventh helix. The proteolipids VHA-c and VHA-c" are characterised by similar transmembrane topology, but differ in the localisation of the essential glutamate residue. VHA-e consists of two transmembrane helices and a cytosolic extension

orientation of the VHA-c"-termini similar to VHA-c (Seidel et al. 2008). It has been shown that the first helix is not essential for V-ATPase function, in particular plant isoforms lack the first helix (Nishi et al. 2003; Aviezer-Hagai et al. 2003). Nevertheless, apart from the orientation of the termini the arrangement of the helices is similar between VHA-c and VHA-c". The transmembrane segments 1 and 3 are orientated towards the center of the ring and segments 2 and 4 face the membrane (Powell et al. 2000; Gibson et al. 2002). In contrast to VHA-c and VHA-c', VHA-c" contains two conserved glutamate residues located at the helices 2 and 4 (Fig. 4), but only the glutamate of helix 2 is essential and represents the proton binding site (Hirata et al. 1997; Powell et al. 2000; Wang et al. 2004). Both conserved glutamate residues were postulated to be equivalent to the proton binding sites of the F-synthase. They were conserved after the evolutionary gene fusion (Arechaga and Jones 2001). Similar to the seventh helix of VHA-a, the helices of the proteolipids which are involved in proton binding were also suggested to be bent to a certain degree (Wang et al. 2004). The stochiometry of the proteolipid ring is highly flexible and depends on the required transport capacity. The diameter of intramembrane particles, which correspond to the V_0-sector, increases after response to a high crassulacean acid metabolism (CAM) expression in *Kalanchoë blossfeldiana*. In parallel, the V-ATPase activity increased (Mariaux et al. 1997). A similar effect was observed for *Mesembryanthemum crystallinum* under salt-induced CAM (Rockel et al. 1994; Klink et al. 1997). Cells of the salt-sensitive rice cultivar BRRI29Dhan showed lower vacuolar pH values and this was related to a reduction of the proteolipid ring diameter. This tentatively indicates optimising of the H^+/ATP-ratio in favour of the generation of a higher proton gradient (Kader et al. 2006, 2007).

2.9 VHA-d

VHA-d is the only V_0 subunit, which lacks any transmembrane domain (Bauerle et al. 1998). Since the crystal structure of the A-ATPase subunit C is available, a role of this membrane associated protein in transmission of rotational force from the central stalk to the proteolipid ring was assumed (Iwata et al. 2004). This subunit C is homologous to the membrane-associated V-ATPase subunit VHA-d and a hybrid protein consisting of VHA-d and the C-terminus of the archael subunit C was functional in yeast (Owegi et al. 2006). VHA-d is also involved in coupling the proton transport and the ATP-hydrolysis (Nishi et al. 2003). Hence, the mutation F94A in the yeast subunit VHA-d affected solely the proton transport, but not the ATP-hydrolysis (Owegi et al. 2006). Alternatively, VHA-d might function as a peripheral stalk subunit in cooperation with VHA-a. In this scenario, its importance for coupling derives from linking VHA-a and the non-homologous region of VHA-A (Thaker et al. 2007). Three cysteine residues are conserved within VHA-d. Consequently, VHA-d was suggested to be involved in redox regulation (Merzendorfer et al. 1997). However, the folding of VHA-d involves a stabilising disulfide bridge, which increases the solubility of the protein (Thaker et al. 2007), and disulfide bridge formation was also observed in the crystal structure of the A-ATPase subunit C (Iwata et al. 2004)

2.10 VHA-e

For the first time, VHA-e was identified in V-ATPases from bovine chromaffin granule and the *Manduca sexta* midgut (Ludwig et al. 1998; Merzendorfer et al. 1999). So far, less is known about the 7.7 kDa subunit VHA-e. VHA-e has two transmembrane helices (Fig. 4) and is required for assembly of V-ATPase (Kluge et al. 2003; Sambade and Kane 2004; Compton et al. 2006). Recently, VHA-e was identified at the ER and endosomal compartments, but appeared absent from the tonoplast in plants (Drobny et al. 2002; Carter et al. 2004; Shimaoka et al. 2004; Jaquinod et al. 2007; Seidel et al. 2008). This data suggest VHA-e to be involved in the subcellular targeting or specific assembly of V-ATPase isoenzymes in *A. thaliana* rather than being important for proton transport (Seidel et al. 2008).

2.11 Additional Peptides (Predicted to be) Associated with the V-ATPase

The functional V-ATPase complex is not limited to the VHA-subunits *in vivo*. A multitude of additional peptides take part in complex formation, activation and regulation. For example, the V_1 subsector is linked to the cytoskeleton via VHA-B and VHA-C (Holliday et al. 2000, 2005; Vitavska et al. 2003, 2005; Chen et al. 2004). The VHA-complex further requires an interaction of the glycolytic aldolase and the subunits VHA-B and VHA-E in order to activate the enzyme (Lu et al. 2001, 2007; Konishi et al. 2004, 2005). The RAVE-complex and the VTC-protein family were demonstrated to stabilise the V-ATPase in yeast and both might be involved in assembly of the holoenzyme (Cohen et al. 1999; Müller et al. 2003; Sipos et al. 2004; Brace et al. 2006; Smardon and Kane 2007). Even though the vacuolar membrane integral protein Btn1p is of unknown function, it directly influences the coupling efficiency of the yeast V-ATPase (Padilla-López and Pearce 2006). The V-ATPase is involved in vesicle fusion via proton pore formation through two proteolipid rings on sandwiched membranes and thus is essential for vesicle trafficking (Mayer 2001; Peters et al. 2001; Müller et al. 2002, 2003; Bayer et al. 2003; Hiesinger et al. 2005). In this context, the V-ATPase was demonstrated to interact with the adaptor protein 2 of the clathrin-coated vesicles and the GTPases ARF6 and ARNO in mammalia (Hurtado-Lorenzo et al. 2006). It is not known up to now whether homologous peptides interact with the plant V-ATPases. Interestingly, many of these proteins and their associated processes are highly conserved among eukaryotes and, hence, it appears plausible that these interactions are also present in plants. The interaction of aldolase and V-ATPase was shown in rice, a homologue of the RAVE-subunits Rav1p and Rav2p was identified in the genome of *Arabidopsis* (Kane and Smardon 2003; Konishi et al. 2005), and the N-terminal actin-binding domain of VHA-B is highly conserved. Especially in plants, V-ATPases are activated in response to illumination. A tonoplast-bound kinase phosphorylates VHA-A and a 14–3–3 protein binds to phosphorylated VHA-A and finally activates the V-ATPase (Klychnikov et al. 2007). The observed interaction of V-PPase and V-ATPase in *Kalanchoë blossfeldiana* is a plant-specific feature, which may coordinate both vacuolar proton transporters by forming an enlarged proton pumping center (Fischer-Schliebs et al. 1997).

3 Assembly of V-ATPase

Two models exist for the assembly of the V-ATPase: (1) the assembly of the subsectors V_0 and V_1 may occur independently, alternatively (2) both sectors assemble in a concerted manner at the stage of the ER (Kane et al. 1999). Both pathways might occur *in vivo*, but the concerted pathway appears to be dominant (Kane et al. 1999). Since VHA-E is unstable without VHA-G, these subunits likely form the core of the peripheral stalk before association with the complex (Tomashek et al. 1997; Charsky et al. 2000; Féthière et al. 2004). Whereas an association of VHA-E/VHA-G/VHA-C and VHA-E/VHA-G/VHA-H was observed by heterologous co-expression in *E. coli* and a rabbit reticulocyte lysate system (Féthière et al. 2005; Jones et al. 2005), VHA-C and VHA-H were reported to require additional peptides like the RAVE-complex and VHA-a, respectively, to assemble with the complex in yeast (Landolt-Marticorena et al. 2000; Smardon and Kane 2007). Especially VHA-C was shown to require the RAVE-complex for a stable association between VHA-C and the V_1-sector. This mechanism might represent a posttranslational regulation, which allows the rapid acitivation of V-ATPases under conditions which require the active V-ATPase (Smardon and Kane 2007). VHA-a and V_1-subunits were demonstrated to associate in the range of minutes independent of the interaction of VHA-a and proteolipids, but the complex depends on the fully assembled V_0-sector to exit the ER (Kane et al. 1999). The formation of the V_0-sector depends on additional ER-associated assembly factors in yeast. The dimer Vma12p/Vma22p interacts transiently with the VHA-a homologue Vph1p and the membrane-integral assembly factor Vma21p interacts with the proteolipid ring via VHA-c' in order to assemble V_0 (Hill and Stevens 1994; Graham et al. 1998; Malkus et al. 2004). Additionally, the VTC-family was suggested to act as membrane chaperone in V-ATPase assembly (Cohen et al. 1999). In a different model, Müller et al. (2003) proposed that these peptides are involved in stabilisation of the V-ATPase and membrane fusion events rather than V-ATPase assembly. Besides the RAVE-complex, homologous assembly factors have not been identified in plants up to now and their relevance for V-ATPase assembly in plants will have to be investigated in future work. Species-specific differences become apparent from the fact that VHA-c' is not present in plants (Chavez et al. 2006) and thus cannot interact with a Vma21p-homologue. However, the ER-integral protein calnexin and the ER-lumenal chaperone BiP were co-purified with V_0 and V_1-subunits in oat (Li et al. 1998), and VHA-A assembled with the membrane before interacting with VHA-B (Frey and Randall 1998). Since VHA-A requires ATP to interact with VHA-B in A-ATPases (Imamura et al. 2006), ATP deficiency might trigger the additional function of VHA-B as co-regulator of transcription in the nucleus of *A. thaliana*. VHA-E was observed on its way along the secondary pathway, starting at the ER and finally identified at the tonoplast in *A. thaliana* (Seidel et al. 2005). These data support the functioning of the concerted assembly pathway in the plant ER. Similar to yeast, the sorting of the assembled holoenzyme depends on VHA-a-isoforms (Kawasaki-Nishi et al. 2001b; Dettmer et al 2006), and VHA-e-isoforms might also be involved in targeting (Seidel et al. 2008).

4 Catalytic Cycle of V-ATPase-*Two Coupled Enzymes*

The V-ATPase works as rotary pump: protons are transported by a central wheel and the rotation of this turbine is powered by an energy consuming, non-rotating movement of the motor-like catalytic head (Grabe et al. 2000). The enzyme is energised by nucleotides as substrate, preferably ATP. The order of the nucleotide-dependent activity is ATP>>GTP>>NTP (Ward and Sze 1992). Binding of a non-hydrolysable ATP-analogon like AMP-PNP arrests VHA-A in the nucleotide bound state. In addition, the hydrolysis requires cations as co-factor. Here, magnesium is favoured and leads to high activity, whereas manganese and calcium enable only a reduced level of activity (Ratajczak 2000).

The VHA-A conformation changes during the ATP-hydrolysis cycle, resulting in three different conformational states, i.e. (1) the MgATP-bound VHA-A, (2) the MgADP- and P_i-associated intermediate state, and (3) the open state after release of MgADP and P_i from VHA-A (Kawamura et al. 2001). Since the conformational changes alternate from one VHA-A to the next in an ordered manner and three copies of VHA-A are present in the complex, the catalytic cycle of VHA-A is transformed into a clockwise-oriented, stepwise rotation of the central stalk (Nishi and Forgac 2002; Imamura et al. 2003). The rotation of the central stalk is transmitted to the proteolipid ring via VHA-d, which serves as a bearing (Yokoyama et al. 2003; Iwata et al. 2004). The coupling ratio of ATP-hydrolysis and proton transport varies. VHA-H, VHA-C, VHA-a and VHA-d were demonstrated to influence the H^+/ATP-stochiometry by altering the structural state, and the proton motif force modulates the coupling ratio as thermodynamic counterforce of proton translocation (Landolt-Marticorena et al. 1999; Kawasaki-Nishi et al. 2001a, b; Curtis et al. 2002; Müller and Taiz 2002; Crider and Xie 2003; Kettner et al. 2003; Liu et al. 2005). The maximum proton gradient which can be established by the V-ATPase is $\Delta pH = 4$. This corresponds to a maximum membrane potential of $-240\,mV$ (Wieczorek et al. 2000; Müller and Taiz 2002). Lumenal nitrate as well as malate were demonstrated to improve the functional coupling of the V_0 and V_1 sector in lemon fruit (Müller and Taiz 2002). In the ground state, the proteolipid subunits are deprotonated. Cytosolic protons have access to an acidic glutamate of the proteolipid subunit via a half-channel formed by the transmembrane domain of VHA-a (Grabe et al. 2000). After approximately one clockwise turn the protonated proteolipid subunit is affected by electrostatic interactions with positively charged amino acids belonging to VHA-a. As mentioned above, this interference reduces the pK_a of VHA-c and thus the proton is released into the vacuolar lumen via another half channel formed by VHA-a. A hydrophobic barrier prevents the continuation of rotation of protonated proteolipid subunits without previous deprotonation (Grabe et al. 2000; Kawasaki-Nishi et al. 2003b). The half-channel gating and VHA-a-VHA-c-interactions were hypothesised to be controlled by rotation of the seventh helix of VHA-a and the proton-binding helix of the proteolipids (Wang et al. 2004; Vos et al. 2007). The proton transport capacity is defined by the number of proton binding sites within the proteolipid ring. Hence, the proton transport is thermodynamically facilitated by a decreased number of proton binding sites due to the increase of the distance of the proton binding sites. This impedes the diffusion of protons back to the cytosol by a counterclockwise rotation similar to F-ATP-synthases (Grabe et al. 2000): The required distance of proton diffusion back to the cytosol is about 6.2 nm in V-ATPase but only 3.1 nm in F-ATP-synthases. The distances are calculated based on the V_0-diameter of mungbean V-ATPases (Li and Zhang 2004) and 6 or 12 proton binding sites per proteolipid ring, respectively. Furthermore, the creation of a proton motive force favours the diffusion of protons across the membrane and back to the cytosol. This retrograde, contraproductive diffusion utilises the proteolipid ring by non-clockwise rotation (Grabe et al. 2000; Kettner et al. 2003). The increased distance of the proton binding sites of the proteolipid ring of the V-ATPase compared to the F-ATP-synthase minimises this effect since the release of a proton requires an increased step size during rotation. On the other hand, the resulting decrease of H^+/ATP stochiometry allows the creation of a high proton gradient (Powell et al. 2000). The ATP-hydrolysis activity of the plant V-ATPase under control conditions varies from 4.32 µmol released phosphate h^{-1} mg^{-1} protein in barley to 15.1 µmol h^{-1} mg^{-1} in mung bean (Dietz et al. 1998; Kawamura et al. 2000, 2001; Klobus and Janicka-Russak 2004). Exemplarily, the pea V-ATPase was further characterised by a $K_{M(ATP)} = 1.27 \pm 0.09$ mM and a $V_{max} = 36.2 \pm 1.2$ µmol h^{-1} mg^{-1} (Kawamura et al. 2000). In *Cucumis sativus*, the average activity increased from 10 to 30 µmol h^{-1} mg^{-1} in response to 200 mM salt treatment (Klobus and Janicka-Russak 2004). The yeast V-ATPase is characterised by a higher activity (29.7 µmol h^{-1} mg^{-1}) and the yeast enzyme produces a torque of 36 ± 4 pN nm. However, the torque was obtained by an actin filament labelled modified V-ATPase, which was characterised by a reduced activity of 6.3 µmol h^{-1} mg^{-1} even in the absence of a coupled filament (Hirata et al. 2003). For comparison, the A-ATPase produces a torque of 35 pN nm whereas the F-ATPase achieves 46 pN nm (Imamura et al. 2005).

5 Isoforms in Plants: Tissue- and Organelle-Specifity

Two V-ATPase with distinct catalytic properties were identified in lemon fruit (Müller et al. 1996, 2002). Plant cells are known to contain two functionally distinct vacuoles, a lytic and a storage vacuole. The lytic vacuole is characterised by an acidic lumen (Paris et al. 1997; Epimashko et al. 2004). Matsuoka et al. (1997) identified catalytically distinct V-ATPases at the tonoplast and the Golgi in tobacco. These observations provide evidence for the presence of V-ATPase subunit isoforms according to the organelle-specific requirements. Indeed, different expression of VHA-D and VHA-E isoforms affected the catalytic properties in pea, even though these isoforms were expressed in a tissue-specific manner (Kawamura et al. 2000).

In yeast, the V-ATPase subunits are encoded by single genes with the exception of VHA-a, which is represented by two isoforms, Vph1p and Stv1p (Manolson et al. 1994). In contrast, VHA-subunits are partially encoded by small gene families of 2–5 members in plants. The 13 V-ATPase subunits are encoded by 26 genes in *A. thaliana* (Sze et al. 2002) and 24 genes in *Oryza sativa* (Hanitzsch et al. 2007). Detailed information on specific function and catalytic properties of VHA-isoforms in plants is not available. Biochemical properties were analysed in various species, including *Arabidopsis*, rice, pea, carrot, tomato, common ice plant and barley. Generalising conclusions on the presence of isoenzymes and their catalytic properties are not yet available. However, subunits VHA-C, VHA-D and VHA-H are single copy genes in *Arabidopsis* as well as in rice, whereas VHA-A and VHA-F are encoded by two genes in rice but by single genes in *Arabidopsis* (Magnotta and Gogarten 2002; Sze et al. 2002; Hanitzsch et al. 2007). On the other hand, VHA-d as well as VHA-e are single copy genes in rice but encoded by two genes in *Arabidopsis* (Hanitzsch et al. 2007). Similar to rice, VHA-A isoforms were also identified in carrot; here, the expression was demonstrated to be organelle-specific whereas two VHA-A-isoforms differed in their tissue-specificity in tomato (Gogarten et al. 1992a, b; Bageshwar et al. 2005). VHA-B, VHA-E, VHA-G and VHA-a are encoded by three isogenes in *Arabidopsis*; especially VHA-E and VHA-G were subject of intensive analysis. VHA-E and VHA-G form a heterodimer as core complex of the peripheral stalk. It has been demonstrated that VHA-E2 and VHA-G3 are in a pollen-specific manner expressed in *A. thaliana*, suggesting some specificity of VHA-E and VHA-G isoforms in heterodimer formation. Some evidence for mainly tissue-specific expression of V_1 subunits and organelle-specific expression of V_0 subunits arises from *A. thaliana* (Hanitzsch et al. 2007). An organelle-specific expression was not observed for any V_1-subunit, whereas V_0-subunits like VHA-a-, VHA-c"- and VHA-e-isoforms differed in their subcellular localisation when expressed as GFP-fusion proteins (Dettmer et al. 2006; Seidel et al. 2008). VHA-a1 was localised at the trans-Golgi-network (TGN), but VHA-a2 and VHA-a3 were preferentially detected at the tonoplast (Dettmer et al. 2006). A difference in subcellular localisation of VHA-a was also shown in *Mesembryanthemum crystallinum* by immuno-localisation (Kluge et al. 2004). Both isoforms of VHA-c" and the second isoform of VHA-e were restricted to the ER whereas VHA-e1 was characterised by a punctated pattern tentatively identified as Golgi apparatus (Seidel et al. 2008). Both subunits were not identified at the tonoplast by proteomic approaches in *Arabidopsis* and in tobacco (Drobny et al. 2002; Carter et al. 2004; Shimaoka et al. 2004), confirming their absence from the tonoplast. Transcript analysis revealed the expression of all five VHA-c isoforms in roots as well as in leaves (Seidel et al. 2008). Promotor-activity of isoform 1 was shown to be responsive to light conditions in an organ-specific manner whereas the promotor-activity of isoform 3 was preferentially high in the root cap of *Arabidopsis* seedlings (Padmanaban et al. 2004). Due to the common localisation of both VHA-c" isoforms at the ER, their high identity on the amino acid level and the only marginal transcript level of the second isoform, these isoforms can likely replace each other (Seidel et al. 2008). Albeit less is known about VHA-d isoforms, the isoforms are identical except for one amino acid and seem to be redundant as well. Also, VHA-B isoform 3 appears to be redundant, since plants lacking this isoform were characterised by a phenotype-like wild type plants (Gaxiola et al 2007).

6 Regulation of Plant V-ATPase: Biochemical Modulation and Reversible Dissociation

The V-ATPase is subject of multiple biochemical modulations like phosphorylation, glutathionylation and oxidation of cysteinyl residues, target of protein–protein interactions and affected by adenylates, cold and its phospholipid- and ion-environment (Hager and Lanz 1989; Moriyama and Nelson 1989; Parry et al. 1989; Dietz et al. 1998; Landolt-Marticorena et al. 1999; Crider and Xie 2003; Konishi et al. 2005; Klychnikov et al. 2007; Qi and Forgac 2007). Even cytosolic chloride, as well as lumenal citrate and malate, have been demonstrated to stimulate V-ATPase activity whereas cytosolic nitrate impedes the ATP-hydrolysis (Randall and Sze 1986; Jochem and Lüttge 1987; Dschida and Bowman 1995; Müller and Taiz 2002; Fecht-Bartenbach et al. 2007). The effect of nitrate depends on chloride; both seem to be competitive interaction partners of one identical binding site (Randall and Sze 1986). The pH-gradient and hence the proton motive force influence the coupling efficiency and the H^+/ATP ratio. The ratio decreases with increasing ΔpH (Müller and Taiz 2002; Kettner et al. 2003). Finally, imposing an appropriate voltage at the tonoplast reverses the catalytic cycle of the V-ATPase in yeast and allows the complex to synthesise ATP from ADP and P_i (Kettner et al. 2003). On the level of the ATP-hydrolysing subcomplex V_1, activity depends on the adenylate ratio ADP/ATP (Dietz et al. 1998). In the absence of nucleotides, the V-ATPase peripheral stalk area appears diffuse and V_1 is tilted relative to V_0 in averaged EM micrographs (Domgall et al. 2002). Conformational changes within VHA-E, which involve disulfide formation as shown for the mung bean V-ATPase, are induced under these conditions (Kawamura et al. 2001). Besides the catalytic head subunits VHA-A and VHA-B, additional nucleotide binding sites have only been identified within the peripheral stalk subunit VHA-C up to now (Armbrüster et al. 2005). ADP inhibits V-ATPase activity (Dietz et al. 1998; Kettner et al. 2003). The nucleotide binding site of VHA-C may be the key player for ADP-inhibition, since this subunit is characterised by a higher affinity for ADP than ATP and ADP-binding induces structural alterations within VHA-C (Armbrüster et al. 2005). Furthermore, multiple phosphorylation sites have been identified especially in *A. thaliana* VHA-C and a single phosphorylation site in maize VHA-a. The phosphorylation of VHA-C occurs by a WNK8 kinase in *A. thaliana* (Liu et al. 2004; Hong-Hermesdorf et al. 2006). On the other hand, activation depends on phosphorylation of VHA-A by a blue light-responsive tonoplast-bound kinase and subsequent 14-3-3 protein binding, which links V-ATPase activity similar to the plasmamembrane P-type ATPase to light conditions and Ca^{2+}-signalling (Klychnikov et al. 2007). The barley V-ATPase is activated in response to gibberellic acid by the calcium-dependent kinase HvCDPK1, but the phosphorylation site remains unclear (McCubbin et al. 2004). The V-ATPase of *Cucumis sativa* might be stimulated by a Ca^{2+}/calmodulin-dependent kinase under salt stress (Klobus and Janicka-Russak 2004).

6.1 Regulation by Other Transporters and Channels

The primary H^+-pump V-ATPase energises secondary active transport processes. The Ca^{2+}-uptake in oat roots is sensitive to V-ATPase inhibitors like nitrate and stimulated by chloride, the Na^+/H^+-antiporter is also energised by the V-ATPase in *Mesembryanthemum crystallinum* (Schumaker et al. 1999; Barkla et al. 1995). In return, V-ATPase activity depends on dissipation of electrical potential by secondary active transporters like Ca^{2+}/H^+ exchanger or is regulated by influx of the counterion chloride through chloride channels (Nelson et al. 2000; Nishi and Forgac 2002; Cheng et al. 2003; Fecht-Bartenbach et al. 2007). However, a coordinated expression of the vacuolar chloride channel CLC-d and V-ATPase subunits was not observed in rice (Diedhiou and Golldack 2006). The presence of lumenal chloride minimises the electrogenic component and favours the generation of the concentration gradient of the proton motive force (Beyenbach and Wieczorek 2006). Hence, the resulting ΔpH across the tonoplast increases. The regulation occurs via kinase-mediated chloride-channel-gating in yeast and depends on Ca^{2+} in *Manduca sexta* (Wieczorek et al. 2000; Nishi and Forgac 2002). In the Golgi apparatus, the lumenal pH is directly linked to the chloride transport through the chloride channel AtCLC in *A. thaliana* (Fecht-Bartenbach et al. 2007). It should be noted that chilling stress also affects the V-ATPase activity via alterations of the membrane permeability to anions and protons (Yoshida 1991). In *Arabidopsis*, knockout mutants of the Ca^{2+}/H^+ antiporter CAX1 revealed a 50% decrease of Ca^{2+} transport into the vacuolar lumen accompanied by only 60% residual V-ATPase activity and 36% of Ca-ATPase activity (Cheng et al. 2003). The interplay of V-ATPase, Ca^{2+}/H^+-exchangers and Ca-ATPase is essential for the cellular Ca^{2+}-homeostasis and for Ca^{2+}-signalling (Martinoia et al. 2007). Since reduced Ca^{2+}-uptake into the vacuole should enhance proton uptake by dissipating the electrogenic counterforce, and since Ca^{2+}-signalling strongly depends on V-ATPase activity, the decrease of V-ATPase activity cannot be explained by simple electrogenic effects. The cooperated action of Ca^{2+}-uptake and V-ATPase activity has to take place on an unknown alternative regulatory level. In yeast, Ca^{2+}-releasing channels are coupled to the V_0-sector (Bayer et al. 2003). Further evidence for a coordinated action of transporters independent on electrogenic regulation derives from the observed interaction of V-PPase and V-ATPase in *Kalanchoë blossfeldiana* and the cooperation of the cation transporter Vnx1p and the V-ATPase in yeast (Fischer-Schliebs et al. 1997; Cagnac et al. 2007).

A new working hypothesis suggests the formation of transport supercomplexes at the tonoplast consisting of V-ATPase and other transporters. This would allow for a direct physical interaction and enable the coordinated regulation of multiple transporters at the tonoplast. In this model, the formation of locally limited proton gradients might be sufficient to drive secondary active transport without the need for massive acidification of the vacuolar lumen.

6.2 Reversible Dissociation

The yeast V-ATPase as well as the V-ATPase of the hornworm *Manduca sexta* are regulated by reversible dissociation of the membrane-associated V_1 sector from the membrane-integral V_0 sector (Kane 1995; Sumner et al. 1995). In yeast, the release of V_1 occurs in response to glucose deprivation or upon alteration of the carbon source to maintain the cytosolic ATP-level (Parra and Kane 1998). The glucose level is sensed by a glycolytic enzyme, the fructose-bisphosphate aldolase. In the presence of glucose the aldolase interacts with the peripheral stalk subunit VHA-E (Lu et al. 2001, 2007). Lack of glucose releases this interaction and VHA-C dissociates from the complex (Fig. 5). As a consequence, the association of the two subsectors V_1 and V_0 is weakened and V_1 detaches from the membrane (Kane and Smardon 2003). The membrane-associated V_0-subunit VHA-d was postulated to be essential for reversible dissociation (Iwata et al. 2004). Free cytosolic V_1 interacts with the RAVE-complex (Fig. 5), formed by the two proteins Rav1p and Rav2p as well as

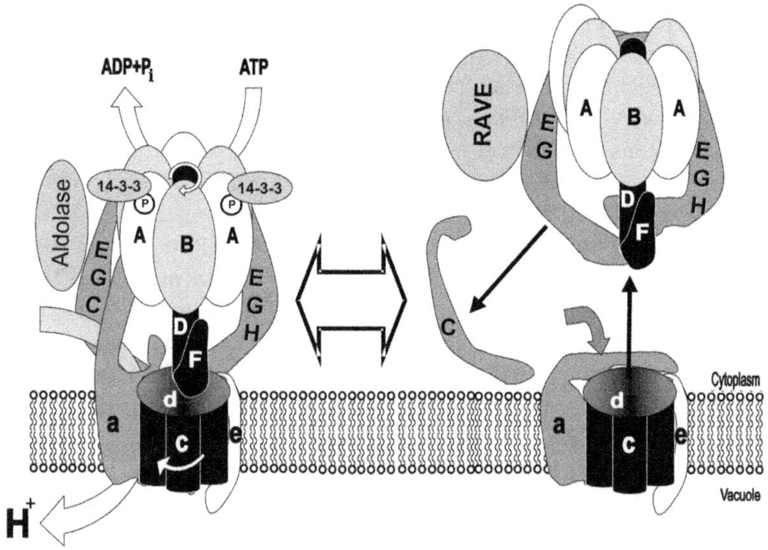

Fig. 5 Scheme of the reversible dissociation. The V-ATPase is regulated via detachment of the cytosolic V_1-sector in response to glucose starvation in yeast. The proton-transport is blocked by the N-terminal half of VHA-a and ATP-hydrolysis is inhibited by an interaction of VHA-H and the central stalk. VHA-C is released from the complex and the RAVE-complex associates with VHA-E. Recent data have shown that activation of V-ATPase involves binding of aldolase and 14-3-3-proteins

by the ubiquitin ligase Skp1p (Seol et al. 2001; Smardon et al. 2002). In order to avoid futile activity of separated subcomplexes, ATP-hydrolysis is inhibited by bound VHA-H and the N-terminal domain of VHA-a inhibits proton transport by blocking the proton pore (Parra et al. 2000; Kawasaki-Nishi et al. 2001a, b; Jefferies and Forgac 2008).

The observed interaction between actin filaments and VHA-B and VHA-C was postulated to play a role in rapid re-assembly of dissociated V-ATPase (Vitavska et al. 2003). Dissociation and re-association occur in the range of minutes after glucose deprivation or replete (Kane 1995). In plants, homologues of RAVE-subunits were identified and an interaction of aldolase and V-ATPase was described by Kane and Smardon (2003) and Konishi et al. (2004, 2005). In addition, the N-terminal domain of VHA-B, which carries the actin binding sites, is conserved among yeast and plant V-ATPases. Although the machinery of reversible dissociation is available, there is no evidence for the occurrence of reversible dissociation in plants. In contrast to yeast, the interaction of V-ATPase and aldolase is gibberellin-dependent and developmentally controlled in rice (Konishi et al. 2004, 2005). Furthermore, glucose deprivation results in rapid disassembly of more than 80% of tonoplast V-ATPase in yeast, but only a low effect of the ATP scavenger desoxyglucose was observed on the structure of the *A. thaliana* V-ATPase (Seidel and Dietz, unpublished data). The structural alterations in *A. thaliana* appear to be a nucleotide-dependent structural tilt of V_1 relative to V_0 as observed by Domgall et al. (2002) rather than reversible dissociation. The requirements of an autotrophic organism differs from heterotrophic organisms like yeast and *Manduca sexta*, perhaps explaining the minor impact of glucose availability on the plant V-ATPase.

6.3 Redox-Regulation of Plant V-ATPase

Redox regulation appears to be a highly conserved and dominant mechanism to regulate V-ATPase activity in dependence on alterations of the cellular environment. An inhibition by oxidising reagents like nitrate, $HgCl_2$, H_2O_2, N-ethylmaleimide (NEM), iodoacetamide (IAA) oxidised glutathione (GSSG), nitrosoglutathione (GSNO) and oxidised thioredoxin was observed in several studies with various species (Hager and Lanz 1989; Feng and Forgac 1992a, b; Dschida and Bowman 1995; Dietz et al. 1998; Forgac 1999; Tavakoli et al. 2001). The nitrate inhibition was prevented in the presence of reducing sulfide. A general observation was that ATP partially protected against inhibition (Dschida and Bowman 1995). Regeneration of oxidised V-ATPase was achieved by adding reducing agents like dithiothreitol (DTT), cysteine or reduced glutathione (GSH). In particular, the NEM-inhibition demonstrates the importance of cysteinyl residues for redox regulation of V-ATPases. Every V-ATPase subunit contains cysteinyl residues, but up to now mainly VHA-A and VHA-E were analysed with regard to redox regulation (Hager and Lanz 1989; Feng and Forgac 1992a, b; Tavakoli et al. 2001). Although VHA-d was also suggested as redox regulator of the V-ATPase, which might modulate the coupling between ATP hydrolysis and proton transport in *Manduca sexta* (Merzendorfer et al. 1997).

For the bovine V-ATPase, a mechanism of disulfide interconversion within VHA-A was suggested to regulate V-ATPase activity (Feng and Forgac 1992b). Under reducing conditions, the disulfide formation between Cys277 and Cys532 represents the active state, whereas V-ATPase is inhibited after establishing a disulfide bridge between Cys532 and Cys254. The cysteinyl residue 254 resides within the nucleotide binding P-loop of VHA-A and, thus, is thought to prevent ATP-binding by disulfide formation (Feng and Forgac 1994). Furthermore, Cys254 was shown to be the target of NEM-inhibition (Feng and Forgac 1992a). Based on crystal structures of the F-ATP-synthase subunit β, Stevens and Forgac (1997) estimated a distance of 5–6 Å between Cys254 and Cys532 and postulated a 20 Å movement of the C-terminus away from the P-loop in the absence of ATP as regular step of the catalytic cycle similar to F-ATP-synthases. In this scenario, a disulfide formation between Cys254 and Cys532 would lock the closed conformation and thereby inhibit the V-ATPase. However, mobility shifts under reducing and oxidising conditions, which indicate formation, release or reorganisation of disulfide bridges, were not seen in the case of VHA-A from *A. thaliana* (M. Hanitzsch, T. Seidel, K.J. Dietz, unpublished data). It appears more plausible that redox regulation in *A. thaliana* takes place by modulation of the single cysteine residue Cys256, including processes such as nitrosylation and glutathionylation. This conclusion is supported by the fact that the substitution of the VHA-A cysteine residue 261 (corresponding to Cys256 in *A. thaliana)* by serine totally abolished the NEM-sensitivity of ATP-hydrolysis in yeast (Taiz et al. 1994). The recently published crystal structure of the A-ATPase subunit A revealed a separation of the cysteines in the range of 12.5–29.15 Å, which appears incompatible for disulfide-formation (Maegawa et al. 2006). Interestingly, Cys256 is less conserved among V- and A-ATPases. Thermophilic and anaerobic archae bacteria lack this cysteinyl residue, whereas Cys279 and Cys535 are conserved (Fig. 2). It should be mentioned that a Cys located in the P-loop is a common feature of many ATPases and thus might be involved in a general mechanism of redox control (Yatsuhashi et al. 2002; Messens et al. 2004; Buhrmann et al. 2005). On the other hand, VHA-A and VHA-B can be cross-linked through endogeneous cysteine residues in yeast, and Cys188 of VHA-B resembles the P-loop-located cysteine residue of VHA-A in *Neurospora crassa* (Adachi et al. 1990; Taiz et al. 1994; Tomashek et al. 1996).

Only two cysteines are conserved in VHA-E of plants, which form a disulfide bridge *in vitro* under oxidising conditions (Tavakoli et al. 2001). These cysteines were suggested to regulate V-ATPase directly in response to cellular redox conditions. However, the redox potential (−330 mV) is rather low so that VHA-E is more likely oxidised than reduced under cellular redox conditions (C. Methner, T. Sander, T. Seidel, K.J. Dietz, unpublished data). Hence, VHA-E should be inhibitory under most conditions and an efficient electron donor is required to maintain the reduced state by permanently feeding electrons to VHA-E. The finding, that VHA-E disulfide bridge formation is linked to the nucleotide binding state of VHA-A indicates its role rather in nucleotide-depending regulation than in redox regulation (Kawamura et al. 2001). VHA-E seems not to be a key player in redox regulation in contrast to VHA-A. In the near future, attention has to be paid to the cysteinyl residues of the transmembrane sector V_0 and their impact on proton transport and coupling.

7 Conclusions

In recent years, the structure and particularly the regulation of the V-ATPase turned out to be more complex than had been assumed. The assembly of the V_0- and V_1-sectors occurs in a coordinated fashion and requires chaperones and assembly factors. On the structural level, a multitude of additional peptides interact with the holoenzyme in order to regulate activity and coupling or to anchor the enzyme to the cytoskeleton. Coordinated activities of vacuolar transporters with the V-ATPase further indicate the existence of transport supercomplexes at the tonoplast. These complexes may tentatively allow the mutual coordination of primary active and secondary active transporters besides electrogenic effects. Finally, the regulation of plant V-ATPase depends on the cytosolic redox conditions and energy state, but a regulatory mechanism via reversible dissociation seems to be absent in plants.

References

Adachi I, Puopolo K, Marquez-Sterling N, Arai H, Forgac M (1990) Dissociation, cross-linking, and glykosylation of the coated vesicle proton pump. J Biol Chem 265:967–973

Alexandersson E, Saalbach G, Larsson C, Kjellbom P (2004) *Arabidopsis* plasma membrane proteomics identifies components of transport, signal transduction and membrane trafficking. Plant Cell Physiol. 45:1543–1556

Allen GJ, Chu SP, Schumacher K, Shimazaki CT, Vafeados D, Kemper A, Hawke SD, Tallman G, Tsien RY, Harper JF, Chory J, Schroeder JI (2000) Alteration of stimulus-specific guard cell calcium oscillations and stomatal closing in Arabidopsis det3 mutant. Science 289:2338–2342

Arechaga I, Jones PC (2001) The rotor in the membrane of the ATPsynthase and its relatives. FEBS Lett 494:1–5

Armbrüster A, Svergun DI, Coskun Ü, Juliano S, Bailer SM, Grüber G (2004) Structural analysis of the stalk subunit Vma5p of the yeast V-ATPase in solution. FEBS Lett 570:119–125

Armbrüster A, Hohn C, Hermesdorf A, Schumacher K, Börsch M, Grüber G (2005) Evidence for major structural changes in subunit C of the vacuolar ATPase due to nucleotide binding. FEBS Lett 579:1961–1967

Aviezer-Hagai K, Padler-Karavani V, Nelson N (2003) Biochemical support for the V-ATPase rotary mechanism: antibody against HA-tagged Vma7p or Vma16p but not Vma10p inhibits activity. J Exp Biol 206:3227–3237

Bageshwar UK, Taneja-Bageshwar S, Moharram HM, Binzel ML (2005) Two isoforms of the A subunit of the vacuolar H^+-ATPase in *Lycopersicon esculentum*: highly similar proteins but divergent patterns of tissue localization. Planta 220:632–643

Barkla BJ, Zingarelli L, Blumwald E, Smith J (1995) Tonoplast Na^+/H^+ antiport activity and its energetization by the vacuolar H^+-ATPase in the halophytic plant *Mesembryanthemum crystallinum* L. Plant Physiol 109:549–556

Bauerle C, Magembe C, Briskin DP (1998) Characterization of a red beet protein homologous to the essential 36-kilodalton subunit of the yeast V-type ATPase. Plant Physiol 117:859–867

Bayer MJ, Reese C, Buhler S, Peters C, Mayer A (2003) Vacuole membrane fusion: V_0 functions after *trans*-SNARE pairing and is coupled to the Ca^{2+}-releasing channel. J Cell Biol 162: 211–222

Beyenbach KW, Wieczorek H (2006) The V-type H+ ATPase: molecular structure and function, physiological roles and regulation. J Exp Biol 209:577–589

Bowman EJ (1983) Comparison of the vacuolar membrane ATPase of *Neurospora crassa* with the mitochondrial and plasma membrane ATPases. J Biol Chem 258:15238–15244

Brace EJ, Parkinson LP, Fuller RS (2006) Skp1p regulates Soi3p/Rav1p association with endosomal membranes but is not required for vacuolar ATPase assembly. Eukaryot Cell 5: 2104–2113

Bremberger, Lüttge U (1991a) Dynamics of tonoplast proton pumps and other tonoplast proteins of Mesembryanthemum-crystallinum L. during the induction of crassulacean acid metabolism. Planta 188:575–580

Bremberger, Lüttge U (1991b) Tonoplast ATPase of *Mesembryanthemum-crystallinum* – partial amino-acid-sequence of subunits induced during the transition from C3-photosynthesis to crassulacean acid metabolism. Comptes Rendus de L Academie des Sciences Serie III-Sciences De La Vie-Life Sciences 315:119–125

Bremberger C, Haschke HP, Lüttge U (1988) Separation and purification of the tonoplast atpase and pyrophosphatase from plants with constitutive and inducible crassulacean acid metabolism. Planta 175:465–470

Buhrmann G, Parker B, Sohn J, Rudolph J, Mattos C (2005) Structural mechanism of oxidative regulation of the phosphatase Cdc25B via an intramolecular disulfide bond. Biochem 44:5307–5316

Cagnac O, Leterrier M, Yeager M, Blumwald E (2007) Identification and characterization of Vnx1p, a novel type of vacuolar monovalent cation/H^+ antiporter of *Saccharomyces cerevisiae*. J Biol Chem 282:24284–24293

Carter C, Pan S, Zouhar J, Avila EL, Girke T, Raikhel NV (2004) The vegetative vacuole proteome of *Arabidopsis thaliana* reveals predicted and unexpected proteins. Plant Cell 16:3285–3303

Charsky CM, Schumann NJ, Kane PM (2000) Mutational analysis of subunit G (Vma10p) of the yeast vacuolar H^+-ATPase. J Biol Chem 275:37232–37239

Chavez C, Bowman EJ, Reidling JC, Haw KH, Bowman BJ (2006) Analysis of strains with mutations in six genes encoding subunits of the V-ATPase: eukaryotes differ in the composition of the V_0 sector of the enzyme. J Biol Chem 281:27052–27062

Chen SH, Bubb MR, Yarmola EG, Zuo J, Jiang J, Lee BS, Lu M, Gluck SL, Hurst IR, Holliday LS (2004) Vacuolar H^+-ATPase binding to microfilaments: regulation in response to phosphatidylinositol 3-kinase activity and detailed characterization of the actin-binding site in subunit B. J Biol Chem 279:7988–7998

Cheng NH, Pittman JK, Barkla BJ, Shigaki T, Hirschi KD (2003) The Arabidopsis cax1 mutant exhibits impaired ion homeostasis, development, and hormonal responses and reveals interplay among vacuolar transporters. Plant Cell 22:104–107

Cho YH, Yoo SD, Sheen J (2006) Regulatory functions of nuclear hexokinase1 complex in glucose signalling. Cell 127:579–589

Choi KY, Ji YJ, Dhakal BK, Yu JR, Cho C, Song WK, Ahnn J (2003) Vacuolar-type H + -ATPase E subunit is required for embryogenesis and yolk transfer in *Caenorhabditis elegans*. Gene 311:13–23

Cohen A, Perzov N, Nelson H, Nelson N (1999) A novel family of yeast chaperones involved in the distribution of V-ATPase and other membrane proteins. J Biol Chem 274:26885–26893

Compton MA, Graham LA, Stevens TH (2006) Vma9p (subunit e) is an integral membrane V_0 subunit of the yeast V-ATPase. J Biol Chem 281:15312–15319

Coskun U, Radermacher M, Müller V, Ruiz T, Grüber G (2004) Three-dimensional organization of the archaeal A1-ATPase from *Methanosarcina mazei* Gö1. J Biol Chem 279:22759–22764

Crider BP, Xie X-S (2003) Characterization of the functional coupling of bovine brain V-ATPase Effect of divalent cations, phospholipids, and subunit H (SFD). J Biol Chem 278:44281–44288
Cross RL, Müller V (2004) The evolution of A-, F-, and V-type ATP synthases and ATPases: reversals in function and changes in the H+/ATP coupling ratio. FEBS Lett 576:1–4
Cross RL, Taiz L (1990) Gene duplication as a means for altering H + /ATP ratios during the evolution of FoF1 ATPases and synthases. FEBS Lett 259:227–229
Curtis KK, Kane PM (2002) Novel vacuolar H+-ATPase complexes resulting from overproduction of Vma5p and Vma13p. J Biol Chem 277:2716–2724
Curtis KK, Francis SA, Oluwatosin Y, Kane PM (2002) Mutational analysis of the subunit C (Vma5p) of the yeast vacuolar H + -ATPase. J Biol Chem 277:8979–8988
Dettmer J, Schubert D, Calvo-Weimar O, Stierhof YD, Schmidt R, Schumacher K (2005) Essential role of the V-ATPase in male gametophyte development. Plant J 41:117–124
Dettmer J, Hong-Hermesdorf A, Stierhof YD, Schumacher K (2006) Vacuolar H+-ATPase activity is required for endocytic and secretory trafficking in *Arabidopsis*. Plant Cell 18: 715–730
Di Sansebastiano GP, Paris N, Marc-Martin S, Neuhaus JM (2001) Regeneration of a lytic central vacuole and of neutral peripheral vacuoles can be visualized by green fluorescent proteins targeted to either type of vacuoles. Plant Physiol 126:78–86
Diedhiou C, Golldack D (2006) Salt-dependent regulation of chloride channel transcripts in rice. Plant Science 170:793–800
Dietz KJ, Heber U, Mimura T (1998) Modulation of the vacuolar H+-ATPase by adenylates as basis for the transient CO_2-dependent acidification of the leaf vacuole upon illumination. Biochim Biophys Acta 1373:87–92
Domgall I, Venzke D, Lüttge U, Ratajczak R, Böttcher B (2002) Three dimensional map of a plant V-ATPase based on electron microscopy. J Biol Chem 277:13115–13121
Drobny M, Schnölzer M, Fiedler S, Lüttge U, Fischer-Schliebs E, Christian AL, Ratajczak R (2002) Phenotypic subunit composition of the tobacco (*Nicotiana tabacum* L) vacuolar-type H+-translocating ATPase. Biochim Biophys Acta 1564:243–55
Drory O, Frolow F, Nelson N (2004) Crystal structure of yeast V-ATPase subunit C reveals its stator function. EMBO Rep 5:1148–1152
Dschida WJ, Bowman BJ (1995) The vacuolar ATPase: sulfite stabilization and the mechanism of nitrate inactivation. J Biol Chem 270:1557–1563
Du J, Kean L, Allan AK, Southall TD, Davies SA, McInerny CJ, Dow JA (2006) The Sza mutations of the B subunit of the Drosophila vacuolar H+-ATPase identify conserved residues essential for function in fly and yeast. J Cell Sci 119:2542–2551
Epimashko S, Meckel T, Fischer-Schliebs E, Lüttge U, Thiel G (2004) Two functionally different vacuoles for static and dynamic purposes in one plant mesophyll leaf cell. Plant J 37:294–300
Fecht-Bartenbach J, Bogner M, Krebs M, Stierhof YD, Schumacher K (2007) Function of the AtCLC-d in the trans-Golgi network. Plant J 50:466–474
Feng Y, Forgac M (1992a) Cysteine 254 of the 73-kDa A subunit is responsible for inhibition of the coated vesicle (H+)-ATPase upon modification by sulfhydryl reagents. J Biol Chem 267:5817–5822
Feng Y, Forgac M (1992b) A novel mechanism for regulation of vacuolar acidification. J Biol Chem 267:19769–19772
Feng Y, Forgac M (1994) Inhibition of vacuolar H+-ATPase by disulfide bond formation between Cys254 and Cys 532 in subunit A. J Biol Chem 269:13224–13230
Féthière J, Venzke D, Diepholz M, Seybert A, Geerlof A, Gentzel M, Wilm M, Böttcher B (2004) Building the stator of the yeast vacuolar ATPase. J Biol Chem 279:40670–40676
Féthière J, Venzke D, Madden DR, Böttcher B (2005) Peripheral stator of the yeast V-ATPase: stochiometry and specificity of interaction between the EG complex and subunits C and H. Biochem 44:15906–15914
Fischer-Schliebs E, Mariaux JB, Lüttge U (1997) Stimulation of H+-transport activity of vacuolar H+-ATPase by activation of H+-PPase in *Kalanchoë blossfeldiana*. Biol Plant 39:169–177

Flannery AR, Graham LA, Stevens TH (2004) Topological characterization of the c, c', and c" subunits of the vacuolar ATPase from the yeast *Saccharomyces cerevisiae*. J Biol Chem 279:39856–39862
Forgac M (1999) Structure and properties of the clathrin-coated vesicle and yeast vacuolar ATPase. J Bioenerg Biomembr 31:57–65
Frey RK, Randall SK (1998) Initial steps in the assembly of the vacuole-type H^+-ATPase. Plant Physiol 118:137–147
Gaxiola RA, Palmgreen MG, Schumacher K (2007) Plant proton pumps. FEBS Lett 581:2204–2214
Geyer M, Yu H, Mandic R, Linnemann T, Zheng YH, Fackler OT, Peterlin BM (2002) Subunit H of the V-ATPase binds to the medium chain of adaptor protein complex 2 and connects Nef to the endocytic machinery. J Biol Chem 277:28521–28529
Gayen S, Vivekanandan S, Biukovic G, Grüber G, Yoon HS (2007) NMR solution structure of subunit F of the methanogenic A1AO adenosine triphosphate synthase and its interaction with the nucleotide-binding subunit B. Biochemistry 46:11684–11694
Gibson LC, Cadwallader G, Finbow ME (2002) Evidence that there are two copies of subunit c" in V_0 complexes in the vacuolar H^+-ATPase. Biochem J 366:911–919
Gogarten JP, Fichmann J, Braun Y, Morgan L, Styles P, Taiz SL, Delapp K, Taiz L (1992a) The use of antisense mRNA to inhibit the tonoplast H^+ATPase in carrot. Plant Cell 4:851–864
Gogarten JP, Starke T, Kibak H, Fishman J, Taiz L (1992b) Evolution and isoforms of V-ATPase subunits. J Exp Biol 172:137–147
Grabe M, Wang H, Oster G (2000) The mechanochemistry of V-ATPase proton pumps. Biophys J 78:2798–2813
Graham LA, Hill KJ, Stevens TH (1998) Assembly of the vacuolar H^+-ATPase occurs in the endoplasmic reticulum and requires a Vma12p/Vma22p assembly complex. J Cell Biol 142:39–49
Graham LA, Powell B, Stevens TH (2000) Composition and assembly of the yeast vecuolar H^+-ATPase complex. J Exp Biol 203:61–70
Grüber G (2005) Structural features and nucleotide-binding capability of the C subunit are integral to the regulation of the eukaryotic V_1V_0 ATPases. Biochem Soc Trans 33:883–885
Grüber G, Radermacher M, Ruiz T, Godovac-Zimmermann J, Canas B, Kleine-Kohlbrecher D, Huss M, Harvey WR, Wieczorek H (2000) Three dimensional structure and subunit topology of the V_1 ATPase from *Manduca sexta* midgut. Biochemistry 39:8609–8616
Grüber G, Wieczorek H, Harvey WR, Müller V (2001) Structure-function relationship of A-, F- and V-ATPases. J Exp Biol 204:2597–2605
Hager A, Lanz C (1989) Essential sulfhydryl groups in the catalytic center of the tonoplast H^+-ATPase from coleoptiles of *Zea mays* L as demonstrated by the biotin-streptavidin-peroxidase system. Planta 180:116–122
Hanitzsch M, Schnitzer D, Seidel T, Golldack D, Dietz KJ (2007) Transcript level regulation of the vacuolar H(+)-ATPase subunit isoforms VHA-a, VHA-E and VHA-G in *Arabidopsis thaliana*. Mol Membr Biol 24:507–508
Harrison M, Durose L, Song CF, Barrat E, Trinick J, Jones R, Findlay JBC (2003) Structure and function of the vacuolar H^+-ATPase: moving from low-resolution models to high resolution structures. J Bioenerg Biomembr 35:337–345
Hernando N, Bartkiewicz M, Collin-Osdoby P, Osdoby P, Baron P (1995) Alternative splicing generates a second isoform of the catalytic A subunit of the vacuolar H(+)-ATPase. Proc Natl Acad Sci USA 92:6087–6091
Hiesinger PR, Fayyazuddin A, Mehta SQ, Rosenmund T, Schulze KL, Zhai RG, Verstreken P, Cao Y, Zhou Y, Kunz J, Bellen HJ (2005) The V-ATPase V_0 subunit a1 is required for a late step in synaptic vesicle exocytosis in Drosophila. Cell 121:607–620
Hill KJ, Stevens TH (1994) Vma21p is a yeast membrane protein retained in the endoplasmic reticulum by a di-lysine motif and is required for the assembly of the vacuolar $H^{(+)}$-ATPase complex. Mol Biol Cell 5:1039–1050
Hirata T, Iwamoto-Kihara A, Sun-Wada G-H, Okajima T, Wada Y, Futai M (2003) Subunit rotation of vacuolar-type proton pumping ATPase: relative rotation of the G and c subunits. J Biol Chem 278:23714–23719

Holliday LS, Lu M, Lee BS, Nelson RD, Solivan S, Zhang L, Gluck SL (2000) The amino-terminal domain of the B subunit of vacuolar H^+-ATPase contains a filamentous actin binding site. J Biol Chem 275:32331–32337

Holliday LS, Bubb MR, Jiang J, Hurst IR, Zuo J (2005) Interactions between vacuolar H + - ATPases and microfilaments in osteoclasts. J Bioenerg Biomembr 37:419–423

Hong-Hermesdorf A, Brüx A, Grüber A, Grüber G, Schumacher K (2006) A WNK-kinase binds and phosphorylates V-ATP subunit C. FEBS Lett 580:932–939

Hurtado-Lorenzo A, Skinner M, El Annan J, Futai M, Sun-Wada GH, Bourgoin S, Casanova J, Wildeman A, Bechoua S, Ausiello DA, Brown D, Marshansky V (2006) V-ATPase interacts with ARNO and Arf6 in early endosomes and regulates the protein degradative pathway. Nat Cell Biol 8:124–136

Imamura H, Nakano M, Noji H, Muneyuki E, Ohkuma S, Yoshida M, Yokoyama K (2003) Evidence for rotation of V_1-ATPase. Proc Natl Acad Sci USA 100:2312–2315

Imamura H, Takeda M, Funamoto S, Shimabukuro K, Yoshida M, Yokoyama K (2005) Rotation scheme of V_1-motor is different from that of F_1-motor. Proc Natl Acad Sci USA 102: 17929–17933

Imamura H, Funamoto S, Yoshida M, Yokoyama K (2006) Reconstitution *in vitro* of V_1 complex of *Thermus thermophilus* V-ATPase revealed that ATP binding to the A subunit is crucial for V_1 formation. J Biol Chem 281:38582–38591

Inoue T, Forgac M (2005) Cysteine-mediated cross-linking indicates that subunit C of the V-ATPase is in close proximity to subunits E and G of the V_1 domain and subunit a of the V_0 domain. J Biol Chem 280:27896–27903

Iwata M, Imamura H, Stambouli E, Ikeda C, Tamakoshi M, Nagata K, Makyio H, Hankamer B, Barber J, Yoshida M, Yokoyama K, Iwata S (2004) Crystal structure of a central stalk subunit C and reversible association/dissociation of vacuole-type ATPase. Proc Natl Acad Sci USA 101:59–64

Jaquinod M, Villiers F, Kieffer-Jaquinod S, Huqouvieux V, Bruley C, Garin J, Bourquiqnon J (2007) A proteomics dissection of *Arabidopsis thaliana* vacuoles isolated from cell culture. Mol Cell Proteomics 6:394–412

Jefferies KC, Forgac M (2008) Subunit H of the vacuolar (H+) ATPase inhibits ATP hydrolysis by the free V1 domain by interaction with the rotary subunit F. J Biol Chem 283:4512–4519

Jochem P, Lüttge U (1987) Proton transporting enzymes at the tonoplast of leaf cells of the CAM plant *Kalanchoe daigremontiana*. Plant Physiol 129:251–268

Jones RP, Durose LJ, Findlay JB, Harrison MA (2005) Defined sites of interaction between subunits E (Vma4p), C (Vma5p), and G (Vma10p) within the stator structure of the vacuolar H^+-ATPase. Biochem 44:3933–3941

Kader MA, Seidel T, Golldack D, Lindberg S (2006) Expression of OsHKT1, OsHKT2, and OsVHA are differentially regulated under NaCl stress in salt-sensitive and salt-tolerant rice (*Oryza sativa* L) cultivars. J Exp Bot 57:4257–4268

Kader MA, Lindberg S, Seidel T, Golldack D, Yemelyamov V (2007) Sodium sensing induces different changes in free cytosolic calcium concentration and pH in salt-tolerant and –sensitive rice cultivars. Physiol Plant 130: 99–111

Kane PM (1995) Disassembly and reassembly of the yeast H(+)-ATPase *in vivo*. J Biol Chem 270:17025–17032

Kane PM, Smardon AM (2003) Assembly and regulation of the yeast vacuolar H^+-ATPase. J Bioenerg Biomembr 35:313–321

Kane PM, Tarsio M, Liu J (1999) Early steps in assembly of the yeast vacuolar H^+-ATPase. J Biol Chem 274:17275–17283

Kawamura Y, Arakawa K, Maeshima M, Yoshida S (2000) Tissue specificity of E subunit isoforms of plant vacuolar H^+-ATPase and existence of isotype enzymes. J Biol Chem 275:6515–6522

Kawamura Y, Arakawa K, Maeshima M, Yoshida S (2001) ATP analogue binding to the A subunit induces conformational changes in the E subunit that involves a disulfide bond formation in plant V-ATPase. Eur J Biochem 268:2801–2809

Kawasaki-Nishi S, Nishi T, Forgac M (2001a) Arg-735 of the 100-kDa subunit a of the yeast V-ATPase is essential for proton translocation. Proc Natl Acad Sci USA 98:12397–12402

Kawasaki-Nishi S, Bowers K, Nishi T, Forgac M, Stevens TH (2001b) The amino-terminal domain of the vacuolar proton-translocating ATPase a subunit controls targeting and *in vivo* dissociation, and the carboxyl-terminal domain affects coupling of proton transport and ATP-hydrolysis. J Biol Chem 276:47411–47420

Kawasaki-Nishi S, Nishi T, Forgac M (2003a) Interacting helical surfaces of the transmembrane segments of subunits a and c' of the yeast V-ATPase defined by disulfide-mediated cross-linking. J Biol Chem 278:41908–41913

Kawasaki-Nishi S, Nishi T, Forgac M (2003b) Proton translocation driven by ATP-hydrolysis in V-ATPases. FEBS Lett 545:76–85

Kettner C, Bertl A, Obermeyer G, Slayman C, Bihler H (2003) Electrophysiological analysis of the yeast V-type proton pump: variable coupling ratio and proton shunt. Biophys J 85:3730–3738

Klink R, Haschke HP, Kramer D, Lüttge U (1997) Membrane particles, proteins and ATPase activity of tonoplast vesicles of *Mesembryanthemum crystallinum* in the C_3 and CAM-state. Botanica Acta 103:24–31

Klobus G, Janicka-Russak M (2004) Modulation by cytosolic components of proton pump activities in plasma membrane and tonoplast from *Cucumis sativus* roots during salt stress. Physiol Plant 121: 84–92

Kluge C, Golldack D, Dietz KJ (1999) Subunit D of the vacuolar H^+-ATPase of *Arabidopsis thaliana*. Biochim Biophys Acta 1419:105–110

Kluge C, Lahr J, Hanitzsch M, Bolte S, Golldack D, Dietz KJ (2003) New insight into the structure and regulation of the plant vacuolar V-ATPase. J Bioenerg Biomembr 35:377–388

Kluge C, Seidel T, Bolte S, Sharma SS, Hanitzsch M, Satiat-Jeunemaitre B, Ross J, Sauer M, Golldack D, Dietz KJ (2004) Subcellular distribution of the V-ATPase complex in plant cells, and *in vivo* localisation of the 100kDa subunit VHA-a within the complex. BMC Cell Biol 5:29

Klychnikov OI, Li KW, Lill H, de Boer AH (2007) The V-ATPase from etiolated barley (*Hordeum vulgare* L) shoots is activated by blue light and interacts with 14–3–3 proteins. J Exp Bot 58:1013–1023

Konishi H, Yamane H, Maeshima M, Komatsu S (2004) Characterization of fructose-bisphosphate aldolase regulated by gibberellin in roots of rice seedlings. Plant Mol Biol 56:839–848

Konishi H, Maeshima M, Komatsu S (2005) Characterization of vacuolar membrane proteins changed in rice root treated with gibberellin. J Proteome Res 4:1775–1780

Krisch R, Rakowski K, Ratajczak R (2000) Processing of V-ATPase subunit B of *Mesembryanthemum crystallinum* L. is mediated *in vitro* by a protease and/or reactive oxygen species. Biol Chem 381:583–592

Landolt-Marticorena C, Kahr WH, Zawarinski P, Correa J, Manolson MF (1999) Substrate and inhibitor- induced conformational changes in the yeast V-ATPase provide evidence for communication between the catalytic and proton-translocating sectors. J Biol Chem 274:26057–26064

Landolt-Marticorena C, Williams KM, Correa J, Chen W, Manolson MF (2000) Evidence that the NH_2 terminus of Vph1p, an integral subunit of the vector V_0 sector of the yeast V-ATPase, interacts directly with the Vma1p and Vma13p subunits of the V_1 sector. J Biol Chem 275:15449–15457

Lee SH, Rho J, Jeong D, Sul JY, Kim T, Kim N, Kang JS, Miyamoto T, Suda T, Lee SK, Pignolo RJ, Koczon-Jaremko B, Lorenzo J, Choi Y (2006) V-ATPase V_0 subunit d2-deficient mice exhibit impaired osteoclast fusion and increased bone formation. Nat Med 12:1403–1409

Leng XH, Nishi T, Forgac M (1999) Transmembrane topology of the 100-kDa a subunit (Vph1p) of the yeast vacuolar proton-translocating ATPase. J Biol Chem 274:14655–14661

Li Z, Zhang X (2004) Electron-microscopic structure of the V-ATPase from mung bean. Planta 219:948–954

Li X, Su RT, Hsu HT, Sze H (1998) The molecular chaperone calnexin associates with the vacuolar H+-ATPase from oat seedlings. Plant Cell 10: 119–130
Liu GS, Chen S, Chen J, Wang XC (2004) Identification of the phosphorylation site of the V-ATPase subunit a in maize roots. Acta Bot Sinica 46:428–435
Liu M, Tarsio M, Charsky CM, Kane PM (2005) Structural and functional separation of the N- and C-terminal domains of the yeast V-ATPase subunit H. J Biol Chem 280:36978–36985
Lu M, Holliday LS, Zhang L, Dunn Jr WA, Gluck SL (2001) Interaction between aldolase and vacuolar H+-ATPase. J Biol Chem 276:30407–30413
Lu M, Vergara S, Zhang L, Holliday LS, Aris J, Gluck SL (2002) The amino-terminal domain of the E subunit of vacuolar H+-ATPase (V-ATPase) interacts with the H subunit and is required for V-ATPase function. J Biol Chem 277:38409–38415
Lu M, Ammar D, Ives H, Albrecht F, Gluck SL (2007) Physical interaction between aldolase and vacuolar H+-ATPase is essential for the assembly and activity of the proton pump. J Biol Chem 282: 24495–24503
Ludwig J, Kerscher S, Brandt U, Pfeiffer K, Getlawi F, Apps DK, Schagger H (1998) Identification and characterization of a novel 92-kDa membrane sector-associated protein of vacuolar proton-ATPase from chromaffin granules. J Biol Chem 273:10939–10947
Maegawa Y, Morita H, Iyaguchi D, Yao M, Watanabe N, Tanaka I (2006) Structure of the catalytic nucleotide-binding subunit A of A-type ATP synthase from *Pyrococcus horikoshii* reveals a novel domain related to the peripheral stalk. Acta Crystallogr D Biol Crystallogr 62:483–488
Maeshima M (2000) Vacuolar H+-pyrophosphatase. Biochim Biophys Acta 1465:37–51
Magnotta SM, Gogarten JP (2002) Multi site polyadenylation and transcriptional response to stress of a vacuolar type H+-ATPase subunit A gene in *Arabidopsis thaliana*. BMC Plant Biol 2:3
Malkus P, Graham LA, Stevens TH, Schekman R (2004) Role of the Vma21p in assembly and transport of the yeast vacuolar ATPase. Mol Biol Cell 15:5075–5091
Manolson MF, Rea PA, Poole RJ (1985) Identification of 3-O-(4-benzoyl)benzoyladenosine 5' - triphosphate- and N,N'-dicyclohexylcarbodiimide-binding subunits of a higher plant H+-translocating tonoplast ATPase. J Biol Chem 260:12273–12279
Manolson MF, Wu B, Proteau D, Taillon BE, Roberts BT, Hoyt MA, Jones EW (1994) STV1 gene encodes functional homologue of 95-kDa yeast vacuolar H(+)-ATPase subunit Vph1p. J Biol Chem 269:14064–14074
Mariaux JB, Fischer-Schliebs E, Luttge U, Ratajczak R (1997) Dynamics of activity and structure of the tonoplast vacuolar-type H+-ATPase in plants with differing CAM expression and in a C-3 plant under salt stress. Protoplasma 196:181–189
Marshansky V (2007) The-V-ATPase a2-subunit as a putative endosomal pH-sensor. Biochem Soc Trans 35:1092–1099
Martinoia E, Maeshima M, Neuhaus HE (2007) Vacuolar transporters and their essential role in plant metabolism. J Exp Bot 58:83–102
Matsuoka K, Higuchi T, Maeshima M, Nakamura K (1997) A vacuolar-type H+-ATPase in a nonvacuolar organelle is required for the sorting of soluble vacuolar protein precursors in tobacco cells. Plant Cell 9:533–546
Mayer A (2001) What drives membrane fusion in eukaryotes? Trends Biochem Sci 26(12):717–723
McCubbin AG, Ritchie SM, Swanson SJ, Gilroy S (2004) The calcium-dependent protein kinase HvCDPK1 mediates the gibberellic acid response of the barley aleurone through regulation of vacuolar function. Plant J 39:206–218
Merzendorfer H, Graf R, Huss M, Harvey WR, Wieczorek H (1997) Regulation of proton-translocating V-ATPases. J Exp Biol 200:225–235
Merzendorfer H, Huss M, Schmid R, Harvey WR, Wieczorek H (1999) A novel insect V-ATPase subunit M97 is glycosylated extensively. J Biol Chem 274:17372–17378
Messens J, van Molle I, Vanhaesebrouck P, Limbourg M, van Belle K, Wahni K, Martins JC, Loris R, Wyns L (2004) How thioredoxin can reduce a buried disulphide bond. J Mol Biol 339: 527–537
Moriyama Y, Nelson N (1989) Cold inactivation of vacuolar proton-ATPases. J Biol Chem 264: 3577–3582

Müller ML, Taiz L (2002) Regulation of the lemon-fruit V-ATPase by variable stoichiometry and organic acids. J Membr Biol 185:209–220

Müller M, Irkens-Kiesecker U, Rubinstein B, Taiz L (1996) On the mechanism of hyperacidification in lemon. Comparison of the vacuolar H(+)-ATPase activities of fruits and epicotyls. J Biol Chem 271:1916–1924

Müller O, Bayer MJ, Peters C, Andersen JS, Mann M, Mayer A (2002) The Vtc proteins in vacuole fusion: coupling NSF activity to V(0) trans-complex formation. EMBO J 21:259–269

Müller O, Neumann H, Bayer MJ, Mayer A (2003) Role of the Vtc proteins in V-ATPase stability and membrane trafficking. J Cell Sci 116:1107–1115

Nelson H, Nelson N (1990) Disruption of genes encoding subunits of yeast vacuolar H(+)-ATPase causes conditional lethality. Proc Natl Acad Sci USA 87:3503–3507.

Nelson N, Perzov N, Cohen A, Hagai K, Padler V, Nelson H (2000) The cellular biology of proton-motive force generated by V-ATPases. J Exp Biol 203:89–95

Neuhaus JM (2000) GFP as a marker for vacuoles in plants. Ann Plant Rev 5: 254–269

Nishi T, Forgac M (2002) The vacuolar (H^+)-ATPases – nature's most versatile proton pumps. Nat Rev Mol Cell Biol 3:94–103

Nishi T, Kawasaki-Nishi S, Forgac M (2001) Expression and localization of the mouse homologue of the yeast V-ATPase 21-kDa subunit "c" (Vma16p). J Biol Chem 276:34122–34130

Nishi T, Kawasaki-Nishi S, Forgac M (2003) The first putative transmembrane segment of subunit "c" (Vma16p) of the yeast V-ATPase is not necessary for function. J Biol Chem 278: 5821–5827

Ohira M, Smardon AM, Charsky CM, Liu J, Tarsio M, Kane PM (2006) The E and G subunits of the yeast V-ATPase interact tightly and are both present at more than one copy per V_1 complex. J Biol Chem 281:22752–22760

Owegi MA, Pappas DL, Finch MW Jr, Bilbo SA, Resendiz CA, Jaquemin LJ, Warrier A, Trombley JD, McCulloch KM, Margalef KL, Mertz MJ, Storms JM, Damin CA, Parra KJ (2006) Identification of a domain in the V_0 subunit d that is critical for coupling of the yeast vacuolar proton-translocating ATPase. J Biol Chem 281:30001–30014

Padilla-López S, Pearce DA (2006) *Saccharomyces cerevisiae* lacking Btn1p modulate vacuolar ATPase activity to regulate pH imbalance in the vacuole. J Biol Chem 14:10273–10280

Padmanaban S, Lin X, Perera I, Kawamura Y, Sze H (2004) Differential expression of vacuolar H^+-ATPase subunit c genes in tissues active in membrane trafficking and their roles in plant growth as revealed by RNAi. Plant Physiol 134:1514–1526

Paris N, Rogers SW, Jiang L, Kirsch T, Beevers L, Phillips TE, Rogers JC (1997) Molecular cloning and further characterization of a probable plant vacuolar sorting receptor. Plant Physiol 115:29–39

Parra KJ, Kane PM (1998) Reversible association between the V_1 and V_0 domains of yeast vacuolar H^+-ATPase is an unconventional glucose-induced effect. Mol Cell Biol 18:7064–7074

Parra KJ, Keenan KL, Kane PM (2000) The H subunit (Vma13p) of the yeast V-ATPase inhibits the ATPase activity of cytosolic V_1-complexes. J Biol Chem 275:21761–21767

Parry RV, Turner JC, Rea PA (1989) High purity preparations of higher plant vacuolar H+-ATPase reveal additional subunits. Revised subunit composition. J Biol Chem 264:20025–20032

Peters C, Bayer MJ, Bühler S, Andersen JS, Mann M, Mayer A (2001) Trans-complex formation by proteolipid channels in the terminal phase of membrane fusion. Nature 409:581–588

Powell B, Graham LA, Stevens TH (2000) Molecular characterization of the yeast vacuolar H^+-ATPase proton pore. J Biol Chem 275:23654–23660

Qi J, Forgac M (2007) Cellular environment is important in controlling V-ATPase dissociation and its dependency on activity. J Biol Chem 282:24743–24751

Randall SK, Sze H (1986) Properties of the partially purified tonoplast H+-pumping ATPase from oat roots. J Biol Chem 261:1364–1371

Ratajczak R (2000) Structure, function and regulation of the plant vacuolar H^+-translocating ATPase. Biochim Biophys Acta 1465:17–36

Ratajczak R, Richter J, Lüttge U (1994) Adaptation of the tonoplast V-type H+-ATPase of *Mesembryanthemum crystallinum* to salt stress C3-CAM transition and plant age. Plant Cell Environm 17:1101–1112
Robinson DG, Haschke HP, Hinz G, Hoh B, Maeshima M, Marty F (1996) Immunological detection of tonoplast polypeptides in the plasma membrane of pea cotyledons. Planta 198:95–103
Rockel B, Ratajczak R, Becker A, Lüttge U (1994) Changed densities and diameters of intramembrane tonoplast particles of *Mesembryanthemum crystallinum* in correlation with NaCl-induced CAM. J Plant Physiol 143:318–324
Sagermann M, Stevens TH, Matthews BW (2001) Crystal structure of the regulatory subunit H of the V-type ATPase of *Saccharomyces cerevisiae*. Proc Natl Acad Sci USA 98:7134–7139
Sambade M, Kane PM (2004) The yeast vacuolar proton-translocating ATPase contains a subunit homologous to the *Manduca sexta* and bovine e subunits that is essential for function. J Biol Chem 279:17361–17365
Sapra R, Bagramyan K, Adams MW (2003) A simple energy-conserving system: proton reduction coupled to proton translocation. Proc Natl Acad Sci USA 100:7545–7550
Schäfer IB, Bailer SM, Düser MG, Börsch M, Bernal RA, Stock D, Grüber G (2006) Crystal structure of the archaeal A_1A_0 ATP synthase subunit B from *Methanosarcina mazei* Gö1: implications of nucleotide-binding differences in the major A1Ao subunits A and B. J Mol Biol 358:725–740
Schumacher K, Vafeados D, McCarthy M, Sze H, Wilkins T, Chory J (1999) The Arabidopsis det3 mutant reveals a central role for the vacuolar H+-ATPase in plant growth and development. Genes Dev 13:3259–3270
Seidel T, Schnitzer D, Golldack D, Sauer M, Dietz KJ (2008) Organelle-specific isoenzymes of plant V-ATPase as revealed by *in vivo*-FRET analysis. BMC Cell Biol. 9:28
Seidel T, Kluge C, Hanitzsch M, Ross J, Sauer M, Dietz KJ, Golldack D (2004) Colocalization and FRET-analysis of subunits c and a of the vacuolar H+-ATPase in living plant cells. J Biotech 112(1–2): 165–175
Seidel T, Golldack D, Dietz KJ (2005) Mapping of C-termini of V-ATPase subunits by *in vivo*-FRET measurements. FEBS Lett 579:4374–4382
Seidel T, Schnitzer D, Golldack D, Sauer M, Dietz KJ (2008) Organelle-specific isoenzymes of plant V-ATPae as revealed by *in vivo*-FRET analysis. BMC Cell Biol 9:28
Seol JH, Shevchenko A, Shevchenko A, Deshaies RJ (2001) Skp1 forms multiple protein complexes, including RAVE, a regulator of V-ATPase assembly. Nat Cell Biol 3:384–391
Shao E, Forgac M (2004) Involvement of the nonhomologues region of subunit A of the yeast V-ATPase in coupling and *in vivo* dissociation. J Biol Chem 279:48663–48670
Shao E, Nishi T, Kawasaki-Nishi S, Forgac M (2003) Mutational analysis of the non-homologous region of subunit A of the yeast V-ATPase. J Biol Chem 278:12985–12991
Shimaoka T, Ohnishi M, Sazuka T, Mitsuhashi N, Hara-Nishimura I, Shimazaki K, Maeshima M, Yokota A, Tomizawa K, Mimura T: isolation of intact vacuoles and proteomic analysis of tonoplast from suspension-cultured cells of *Arabidopsis thaliana*. Plant Cell Physiol 45:672–683
Sipos G, Brickner JH, Brace EJ, Chen L, Rambourg A, Kepes F, Fuller RS (2004) Soi3p/Rav1p functions at the early endosome to regulate endocytic trafficking to the vacuole and localization of trans-Golgi network transmembrane proteins. Mol Biol Cell 15:3196–3209
Smardon AM, Kane PM (2007) RAVE is essential for the efficient assembly of the C-subunit with the vacuolar H+-ATPase. J Biol Chem 282:26185–26194
Smardon AM, Tarsio M, Kane PM (2002) The RAVE complex is essential for stable assembly of the yeast V-ATPase. J Biol Chem 277:13831–13839
Stevens TH, Forgac M (1997) Structure, function and regulation of the vacuolar (H+)-ATPase. Annu Rev Cell Dev Biol 13:779–808
Strompen G, Dettmer J, Stierhof YD, Schuhmacher K, Jürgens G, Mayer U (2005) Arabidopsis vacuolar H+-ATPase subunit E isoform 1 is required for Golgi organization and vacuolar function in embryogenesis. Plant J 41:125–132

Sumner JP, Dow JA, Earley FG, Klein U, Jager D, Wieczorek H (1995) Regulation of plasma membrane V-ATPase activity by dissociation of peripheral subunits. J Biol Chem 270:5649–5653

Taiz L, Nelson H, Maggert K, Morgan L, Yatabe B, Taiz SL, Rubinstein B, Nelson N (1994) Functional analysis of conserved cysteine residues in the catalytic subunit of the yeast vacuolar H(+)-ATPase. Biochim Biophy Acta 1194:329–334

Tavakoli N, Eckerskorn C, Golldack D, Dietz KJ (1999) Subunit C of the vacuolar H^+-ATPase of *Hordeum vulgare*. FEBS Lett 456:68–72

Tavakoli N, Kluge C, Golldack D, Mimura T, Dietz KJ (2001) Reversible redox control of plant vacuolar H^+-ATPase is related to disulfide bridge formation in subunit E as well as subunit A. Plant J 28:51–59

Thaker YR, Roesle M, Grüber G (2007) The boxing glove shape of subunit d of the yeast V-ATPase in solution and the importance of disulfide formation for folding of this protein. J Bioenerg Biomembr 39:275–289

Tomashek JJ, Sonnenburg JL, Artimovich JM, Klionsky DJ (1996) Resolution of subunit interactions and cytoplasmic subcomplexes of the yeast vacuolar proton-translocating ATPase. J Biol Chem 271:10397–10404

Tomashek JJ, Graham LA, Hutchins MU, Stevens TH, Klionsky DJ (1997) V_1-situated stalk subunits of the yeast vacuolar proton-translocating ATPase. J Biol Chem 272:26787–26793

Vasileya E, Liu Q, MacLeod KJ, Baleja JD, Forgac M (2000) Cysteine scanning mutagenesis of the noncatalytic nucleotide binbding site of the yeast V-ATPase. J Biol Chem 275: 255–260

Venzke D, Domgall I, Köcher T, Féthière J, Fischer S, Böttcher B (2005) Elucidation of the stator organization in the V-ATPase of *Neurospora crassa*. J Mol Biol 349:659–669

Vitavska O, Wieczorek H, Merzendorfer H (2003) A novel role for subunit C in mediating binding of the H^+-ATPase to the actin cytoskeleton. J Biol Chem 278:18499–18505

Vitavska O, Merzendorfer H, Wieczorek H (2005) The V-ATPase subunit C binds to polymeric F-actin as well as to monomeric G-actin and induces cross-linking of actin filaments. J Biol Chem 280:1070–1076

Vos WL, Vermeer LS, Hemminga MA (2007) Conformation of a peptide encompassing the proton translocation channel of vacuolar H(+)-ATPase. Biophys J 92:138–146

Wang Y, Inoue T, Forgac M (2004) TM2 but not TM4 of subunit c" interacts with TM7 of subunit a of the yeast V-ATPase as defined by disulfide-mediated cross-linking. J Biol Chem 279:44628–44638

Ward JM, Sze H (1992) Proton Transport Activity of the Purified Vacuolar H-ATPase from Oats: Direct Stimulation by Cl. Plant Physiol 99:925–931

Wieczorek H, Grüber G, Harvey WR, Huss M, Merzendorfer H, Zeiske W (2000) Structure and regulation of insect plasma membrane H(+)V-ATPase. J Exp Biol 203:127–135

Wilkens S, Inoue T, Forgac M (2004) Three-dimensional structure of the vacuolar ATPase from electron microscopy. J Biol Chem 279:41942–41949

Xu T, Forgac M (2000) Subunit D (Vma8p) of the yeast vacuolar H^+-ATPase plays a role in coupling of proton transport and ATP-hydrolysis. J Biol Chem 275:22075–22081

Xu T, Vasiliyeva E, Forgac M (1999) Subunit interactions in the clathrin-coated vesicle vacuolar H^+-ATPase complex. J Biol Chem 274:28909–28915

Yatsuhashi T, Hisatome I, Kurata Y, Sasaki N, Ogura K, Kato M, Kinugasa R, Matsubara K, Yamawaki M, Yamamoto Y, Tanaka Y, Ogino K, Igawa O, Makita N, Shigemasa C (2002) L-cysteine prevents oxidation-induced block of the cardiac Na+ channel via interaction with heart-specific cysteinyl residues in the P-loop region. Circ J 66:846–850

Yokoyama K, Nakano M, Imamura H, Yoshida M, Tamakoshi M (2003) Rotation of the proteolipidring in the V-ATPase. J Biol Chem 278:24255–24258

Yoshida S (1991) Chilling-Induced Inactivation and Its Recovery of Tonoplast H-ATPase in Mung Bean Cell Suspension Cultures. Plant Physiol. 95: 456-460

Zhang Z, Inoue T, Forgac M, Wilkens S (2006) Localization of subunit C (Vma5p) in the yeast vacuolar ATPase by immuno electron microscopy. FEBS Lett 580: 2006–2010

The Role and Regulation of Sugar Transporters in Plants with Crassulacean Acid Metabolism

E. Antony and A.M. Borland(✉)

Contents

Abstract The photosynthetic specialisation of Crassulacean acid metabolism (CAM) operates via a diel separation of carboxylation processes mediated by phosphoenolpyruvate carboxylase (PEPC) and Rubisco which optimise photosynthetic performance and carbon gain in potentially limiting environments. Carbohydrates are a key limiting factor for nocturnal CO_2 uptake in CAM plants, and the partitioning of sugars towards providing substrate for dark carboxylation must be balanced with the requirements of other competing sinks that include growth and export. Sugar transporters have been recognised previously as key targets for regulatory roles in the long-distance and subcellular distribution and partitioning of assimilates in plants. Thus, in CAM plants, sugar transporters may represent a strategic checkpoint for regulating the partitioning of photosynthetically fixed carbon between provision of substrate for nocturnal carboxylation and export for growth. In this review, we examine the major pathways underpinning sugar flux in CAM plants and describe the requirements for sugar transport across the chloroplastic and vacuolar membranes as well as for long-distance transport in the phloem. We review current knowledge on the structure, function and regulation of plant sugar transporters and highlight candidate sugar transporters that could play cardinal roles in regulating carbohydrate partitioning between the potentially conflicting fates of storage for CAM and for growth and productivity.

A.M. Borland
Institute for Research on Environment and Sustainability, Devonshire Building, Newcastle University, NE1 7RU, UK
e-mail: a.m.borland@ncl.ac.uk

U. Lüttge et al. (eds.), *Progress in Botany 70*,

DOI:10.1007/978-3-540-68421-3, © Springer-Verlag Berlin Heidelberg 2009

1 Introduction

As major, primary end products of photosynthesis, sugars play key roles in plant growth, metabolism and signalling processes. In plants with the photosynthetic specialisation of Crassulacean acid metabolism (CAM), sugars have a central role as providers of phosphoenolpyruvate (PEP), the substrate for nocturnal CO_2 uptake. Depending upon the species of CAM plant, PEP may be derived from the glycolytic breakdown at night of sugars/sugar phosphates that are exported from the chloroplast or from sugars that are exported from the vacuole (Holtum et al. 2005). Up to 20% of leaf dry weight can be tied up as carbohydrate reserves for CAM yet, despite these significant costs in terms of carbon investment, in many cases the potential for high productivity is not compromised. Agronomically important CAM species, including pineapple (*Ananas comosus*) and several of the Agaves, can show productivities rivalling that of the C_4 plant sugar cane (Bartholomew and Kadzimin 1977; Nobel 1996). Clearly, the partitioning of sugars to dark carboxylation in CAM plants must be balanced with the requirements of other competing sinks that include export for growth. Sugar transporters have been recognised previously as key targets for regulatory roles in the long-distance and subcellular distribution and partitioning of assimilates in plants (Williams et al. 2000; Lalonde et al. 2004). Thus, it can be hypothesised that, in CAM plants, sugar transporters represent a strategic checkpoint for regulating the partitioning of photosynthetically fixed carbon between provision of substrate for nocturnal carboxylation and export for growth. In this review, we examine the major pathways underpinning sugar flux in CAM plants and describe the requirements for sugar transport across various intracellular membranes. We review current knowledge on the structure, function and regulation of plant sugar transporters and highlight candidate sugar transporters that could play cardinal roles in regulating carbohydrate partitioning between the potentially conflicting fates of storage for CAM and for growth and productivity.

2 Carbohydrate Metabolism and Provision of Sugars in CAM Plants

In essence, CAM is expressed on a background of C_3 photosynthesis via the deployment of additional carboxylation processes that operate at night and decarboxylation processes that operate during the day (Fig. 1). At night, when evapotranspiration rates are low, atmospheric CO_2 and/or respiratory CO_2 are fixed in the cytosol by the enzyme phosphoenolpyruvate carboxylase (PEPC). The 3-C substrate, phosphoenolpyruvate (PEP), is produced via the glycolytic breakdown of carbohydrate formed during the previous day. The final 4-C product, malic acid, is stored in a large central vacuole. During the day, malate exits the vacuole and decarboxylation may occur through the single or combined action of three decarboxylases (depending on plant species): NADP-malic enzyme (NADP-ME), NAD-ME and phosphoenolpyruvate carboxykinase (PEPCK). In addition to the 3-C products PEP or

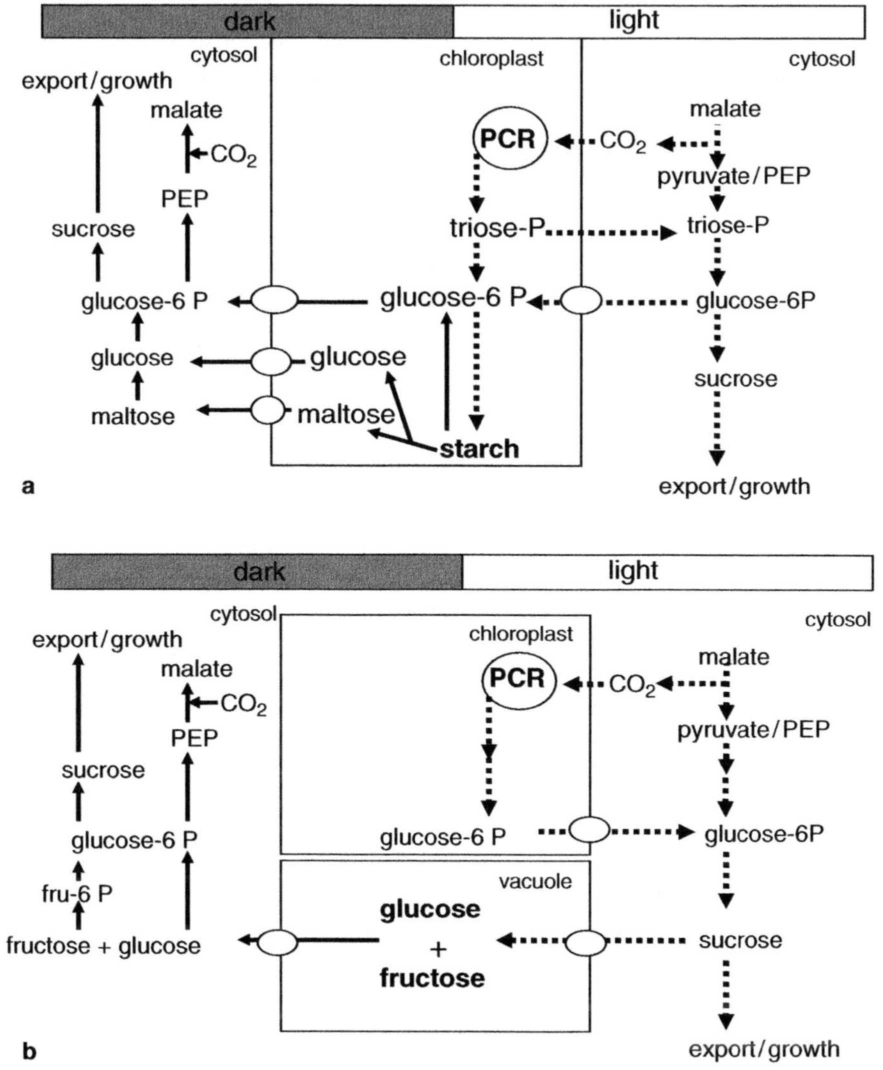

Fig. 1 Proposed flow of carbon and intercellular sugar transport processes for CAM plants using **a** starch or **b** soluble sugars as the major source of substrate for nocturnal carboxylation. Day-time fluxes are shown by *dotted lines* and night-time fluxes by *solid lines*. Sugar and sugar-P transporters implicated in the intercellular exchanges are represented by the *circles* located on the chloroplast and vacuole membranes

pyruvate, CO_2 is released at a high internal partial pressure (pCO_2) which is often sufficient to result in stomatal closure and thus conserve water. Indeed, CAM results in water-use efficiencies that exceed those of C_4 plants up to five-fold and C_3 plants by up to ten-fold (Borland et al. 2000). The high pCO_2 generated behind closed stomata via decarboxylation promotes Rubisco-mediated carboxylation and suppresses photorespiration. During this decarboxylation phase, carbohydrate is recovered via gluconeogenesis which imposes a high energetic cost (when compared to C_3 plants) but ensures the production of substrate for subsequent nocturnal carboxylation and partitioning for growth. The carbohydrates that will provide substrates for the nocturnal reactions are transported either into the chloroplast and stored as starch or transported into the vacuole and stored as sucrose and/or hexose (Christopher and Holtum 1996). Towards the latter part of the photoperiod, once malate reserves have been depleted and pCO_2 declines, the stomata may re-open and atmospheric CO_2 may be taken up and fixed directly via Rubisco for several hours, a period designated Phase IV (sensu Osmond 1978). It is believed that sugars synthesised during Phase IV are predominantly exported for growth (Borland et al. 1994; Borland and Dodd 2002). In terms of net carbon flux, the result of these temporally separated carboxylation and decarboxylation processes is a reciprocal cycling of malate and carbohydrate which can represent up to 20% of leaf dry weight.

The taxonomic diversity of CAM plants is reflected by diversity in carbohydrate species that constitute the major storage pools of photosynthetically fixed carbon within the leaf (Christopher and Holtum 1996). CAM species that predominantly accumulate starch in the chloroplast include the well-studied or "model" species such as the inducible CAM species *Mesembryanthemum crystallinum* and the constitutive CAM species *Kalanchoë daigremenotiana* and *K. fedstchenkoi* (Dodd et al. 2002). Species that predominantly accumulate soluble sugars within the vacuole include the constitutive CAM species, *Ananas comosus* (pineapple), and the C_3-CAM intermediate, *Clusia minor* (Black et al. 1996). The intercellular sugar transport requirements for starch-storing and soluble sugar-storing CAM plants are summarised and compared in Fig. 1. The degradation of chloroplastic starch at night can proceed via a hydrolytic pathway that releases maltose and glucose for export or via a phosphorylytic pathway that releases glucose-6 P for export (Fig. 1a). During the day in starch-storing CAM plants, the gluconeogenic processing of cytosolic PEP produces triose-P and glucose-6 P for import to the chloroplast (Fig. 1a). For CAM plants where soluble sugars are the major storage carbohydrate and the source of PEP for nocturnal carboxylation, vacuolar sugar transporters play a key role in the diel operation of the CAM cycle. It has been reported that isolated vacuoles of *A. comosus* contain mainly glucose and fructose (Kenyon et al. 1985; Christopher and Holtum 1998), whilst whole-leaf extracts contain substantial amounts of sucrose (Kenyon et al. 1985). Given the very high extractable activities of acid invertase reported for pineapple leaves and indeed for *Clusia rosea*, a CAM species that also accumulates soluble sugars (Black et al. 1996), it can be proposed that sucrose is synthesised in the cytoplasm during the day, transported across the tonoplast into the vacuole, and hydrolysed in the vacuolar lumen by acid invertase (Smith and Bryce 1992; and see Fig. 1b). Such data indicate that sucrose and hexose transporters localised to the tonoplast could play strategic roles in regulating carbon fluxes in soluble-sugar-storing CAM species.

The diagrams presented in Fig. 1 are likely to be an over-simplification of sugar fluxes *in vivo* since some soluble-sugar-accumulating CAM species like *A. comosus* and *Clusia* also accumulate starch over the photoperiod and some or all of this starch may be used to provide substrate for nocturnal carboxylation (Table 1). The nocturnal production of PEP in different CAM species can be estimated from measurements of

Table 1 Nocturnal hexose requirements for provision of PEP as substrate for measured organic acid accumulation (where 1 hexose ≡ 2 mal or 1 cit) in comparison with hexoses potentially provided by the measured nocturnal breakdown of soluble sugars or starch. Hexoses degraded in excess of those required for organic acid synthesis are potentially available for growth/export

Species	Nocturnal hexose requirements	Hexoses from soluble sugar degradation	Hexoses from starch degradation	Hexoses for growth/export
Mesembryanthemum crystallinum	30	5	30	5
Ananas comosus	55	45	55	45
Clusia lanceolata	30	32	40	42
Clusia fluminensis	98	82	20	4

the amounts of organic acids accumulated at night with 1 mol malate ≈ 1 mol PEP or 1 mol citrate ≈ 2 mol PEP (NB. *A. comosus* and *Clusia species* can accumulate both malate and citrate overnight; Lüttge 1988; Borland and Griffiths 1989). In comparison with the measured depletion of soluble sugars and starch at night (1 mol hexose ≈ 2 mol PEP), the data in Table 1 indicate that *C. lanceolata* could use soluble sugars to provide all the PEP required for dark carboxylation, *C. fluminensis* requires both chloroplastic and vacuolar carbohydrates to furnish PEP production whilst *A. comosus* could, in principle, use starch to provide all or some of the substrate for nocturnal carboxylation. Any carbohydrate depleted over and above the nocturnal PEP requirements could, in principle, be available for growth and export at night, so all the CAM species considered in Table 1 require additional sugar transporters that can facilitate export out of the cell and into the phloem for subsequent translocation to sinks such as developing leaves, reproductive tissues and the roots. In the following sections, we review current knowledge on sugar transporters localised to the chloroplast and to the vacuolar membrane as well as sugar transporters involved in phloem transport, and we discuss the possible functions of these transporters in the operation of CAM. We also consider how these categories of sugar transporters might be regulated to achieve a balance between the dynamic shifts in supply and demand for carbohydrate that occur over the 24 h CAM cycle.

3 Chloroplastic Sugar Transporters and CAM

The hydrolytic degradation of starch in the chloroplast results in the production of maltose and glucose for export from the chloroplast whilst starch degradation via the phosphorylytic route releases glucose-6 P for export (Fig. 1a). In leaves of *Arabidopsis thaliana*, the hydrolytic pathway has been shown to be of prime importance for the nocturnal conversion of starch to sucrose with maltose being the major export product from chloroplasts degrading starch at night (Niittyla et al. 2004). In contrast, the phosphorolytic pathway of starch degradation has been proposed to supply carbon for metabolism inside the chloroplast of *A. thaliana* with starch phosphorylase providing glucose-6 P as substrate for the oxidative pentose phosphate pathway, particularly under conditions of stress and when photorespiration is elevated (Zeeman et al. 2004; Weise et al. 2006). The induction of CAM in *M. crystallinum* by exposure to salinity is accompanied by increased activities of a range of starch degrading enzymes that are implicated in both the hydrolytic and phosphorolytic pathways (Paul et al. 1993). However, the switch to CAM in *M. crystallinum* appears to result in a change in transport activities across the chloroplast envelope with chloroplasts isolated from C_3 plants reported to export mainly maltose whilst chloroplasts isolated from plants in the CAM mode export mainly glucose-6 P (Neuhaus and Schulte 1996). Glucose-6 P transport across the CAM chloroplasts of *M. crystallinum* has been shown to be inhibited competitively by P_i and 3-phosphogycerate (3-PGA) suggesting that glucose-6 P is transported by one or more plastidic Pi translocators in this species (Kore-eda and Kanai 1997). Two isogenes that encode glucose-6 P/Pi translocators have been isolated from *M. crystallinum* (*Mc*GPT1 and *Mc*GPT2) and phylogenetic analyses indicate that the predicted proteins cluster together on a branch of the plastidic Pi translocator tree that includes a GPT translocator from pea as well as two isogenes of GPT from *A. thaliana* (Kore-eda et al. 2005). The switch to CAM in *M. crystallinum* is accompanied by an increase in transcript abundance of both *Mc*GPT1 and *Mc*GPT2 which also show circadian patterns of abundance, although *Mc*GPT2 transcripts exhibited the more pronounced diurnal change (Kore-eda et al. 2005). A functional division of labour between *Mc*GPT1 and *Mc*GPT2 has been suggested with *Mc*GPT2 proposed as a CAM-specific isoform for chloroplastic glucose-6 P transport whilst *Mc*GPT1 facilitates CAM as well as heterotrophic metabolism in non-green plastids (Kore-eda et al. 2005).

The functional role of the glucose-6 P transporters in the CAM cycle is yet to be definitively demonstrated, but recent theoretical deductions have proposed both import and export roles for these chloroplast transporters. During the day-time period of decarboxylation, the products of PEP metabolism could potentially enter the chloroplast through the triose-P and glucose-6 P translocators. Complexity arises with the need to export some carbon from the chloroplast during the day to support sucrose synthesis in the cytosol (Kluge 1969). Thus, it has been suggested that phosphorylated precursors of starch formation are imported from cytosol to chloroplast via the glucose-6 P transporter and that chloroplastic precursors for sucrose formation in the cytosol are exported from the chloroplast via the triose-P translocator (Häusler et al. 2000; Holtum et al. 2005). At night, starch degradation is thought to proceed via the phosphorylytic pathway and the glucose-6 P translocator is proposed to facilitate export of glucose-6 P to the cytosol where it might be predicted to allosterically activate PEPC (Osmond and Holtum 1982). The glycolytic conversion of glucose-6 P in the cytosol would provide ATP by substrate-level phosphorylation and thus help to energise the nocturnal accumulation of malate in the vacuole (Holtum et al. 2005). The mechanisms that regulate glucose-6 P partitioning between the production of malate or sucrose at night are unknown, but it has been proposed that, as the night progresses, starch degradation may shift from the phosphorylytic route to the hydrolytic route. This assertion is based on observations in CAM-performing *M. crystallinum* of a declining concentration of glucose-6

P as the night progresses and an increase in the activity of chloroplastic phosphofructokinase which could increase flux to phosphorylated trioses with the potential for export via the triose phosphate translocator (Häusler et al. 2000). It is currently unknown how a qualitative and/or quantitative shift in the export of sugars/metabolites from the chloroplast might be regulated, although it is worthy of note that transcript abundance of the CAM-induced *Mc*GPT2 peaks early in the dark period and transcripts decline throughout the remainder of the night (Kore-eda et al. 2005). There is also some evidence that GPTs are subject to post-translational regulation (Kore-eda and Kanai 1997). Measurements are needed of glucose-6 P export from the chloroplasts of *M. crystallinum* throughout the night in order to ascertain if transcriptional and/or post-translational control of *Mc*GPT2 might regulate qualitative shifts in chloroplastic export and, by analogy, regulate the subsequent shift in the partitioning of exported products between nocturnal carboxylation and export for growth.

It is interesting to review the above suggestions for chloroplast transport processes in starch-accumulating CAM plants alongside recent findings for *A. thaliana* which indicate that maltose is the major export product and glucose a minor export product from chloroplasts of C_3 plants that degrade starch at night (Zeeman et al. 2004, 2007). Interrogation of an expressed sequence tag (EST) database for *M. crystallinum* containing 25,000 ESTs has failed to identify any candidate chloroplastic maltose transporters (James Hartwell, personal communication), which might suggest that maltose export from the chloroplast in *M. crystallinum* is quantitatively less important than that in *A. thaliana*. Whilst chloroplastic glucose transporters have been identified from *M. crystallinum* (Häusler et al. 2000; E. Antony, T. Taybi and A. Borland, unpublished observations), there is no change in transcript abundance of these transporters with CAM induction and, unlike many other CAM-associated genes and transporters, the transcript abundance of the chloroplastic glucose transporters do not show diurnal or circadian patterns of abundance (E. Antony and A. Borland, unpublished observations). As more sequence information from other CAM species becomes available it will be interesting to examine the hypothesis that the CAM cycle operates using a different route of starch degradation to that found in C_3 species.

4 Vacuolar Sugar Transporters and CAM

The model presented in Fig. 1b, proposes that for soluble sugar-storing CAM plants, sucrose is synthesised in the cytoplasm during the day, transported across the tonoplast into the vacuole, and hydrolysed in the vacuolar lumen by acid invertase (Smith and Bryce 1992). To avoid futile cycling, the hexoses thus produced would need to be stored in the vacuole and not released to the cytoplasm until the following dark period, when they would be metabolised by glycolysis to provide the C_3 substrate (PEP) for nocturnal malate synthesis. This model is supported by the finding that the tonoplast of pineapple possesses a sucrose-transport system with kinetics appropriate to catalyse sucrose fluxes of the required magnitude. McRae et al. (2002) demonstrated that sucrose uptake by isolated tonoplast-enriched vesicles prepared from pineapple leaves exhibits concentration-dependent saturation kinetics, ATP-independence and trans-stimulation by internal sucrose, characteristics that are consistent with the operation of a carrier type of sucrose transporter. The apparent Km for sucrose uptake was 50 mM and transport was not competitively affected by glucose or fructose. Recently, a specific hexose-transport system has also been observed in isolated pineapple tonoplast vesicles (D. Haines and J.A.M. Holtum, personal communication).

The role of vacuolar sucrose and hexose transporters in regulating assimilate partitioning in plants has received relatively little attention due to limited information on the molecular characterisation of these proteins. In all plant species analysed to date, small families of sugar transporters have been found with nine sucrose transporter-like sequences reported for *A. thaliana* (Sauer et al. 2004). Sucrose transporters possess 12 transmembrane α-helices and are comprised of two units of six connected by a central loop which varies in length (Lemoine 2000). A recent phylogenetic analysis of confirmed and predicted sucrose transporters revealed four distinct groups in plants; group 1 contains only monocot sequences, group 2 contains only dicot sequences, group 3 contains low-affinity sucrose transporters with alleged sucrose-sensing properties, whilst group 4 sucrose transporters were originally proposed as plasma membrane-localised high-capacity sucrose transporters that are expressed in sieve elements of the phloem (Sauer 2007). To date, the only vacuolar sucrose transporters to be identified at the molecular level (*Hv*SUT2 from barley and *At*SUC4 from *Arabidopsis*) belong to group 4 of the sucrose transporter phylogenetic tree (Sauer 2007). Both of these proteins catalyse the co-transport of H^+/sucrose when expressed in yeast, implying that the function of *Hv*SUT2 and *At*SUC4 is to facilitate sucrose export from the vacuole. To date, sucrose transporter(s) that catalyse the import of sucrose to the vacuole, analogous to that proposed for the operation of CAM in sugar-accumulating CAM plants (McRae et al. 2002 and Fig. 1b), have not yet been unequivocally identified at the molecular level in any plant species.

A sucrose transporter that shows homology to group 3 of the sucrose transporters has recently been identified from pineapple (designated nomenclature *Ac*SUT1, accession number EF460878; Antony et al. 2008). Group 3 sucrose transporters have been categorised as low affinity/high capacity transporters, characteristics that are consistent with the reported sucrose transport kinetics of isolated tonoplast vesicles of pineapple (McRae et al. 2002). In common with the other group 3 sucrose transporters, *Ac*SUT1 possesses an extended cytoplasmic loop between transmembrane helices 6 and 7. This extended cytoplasmic loop of the group 3 sucrose transporters appears to have no significant function in controlling substrate affinity or transport (Schulze et al. 2000), and a proposed function in sucrose-sensing, based on the similarity of extended domains of yeast sensors RGT2 and SNF3 (Kuhn et al. 1997), has yet to be proven. The pineapple sucrose transporter *Ac*SUT1 was found to localise to pre-vacuolar compartments in tobacco leaf epidermal cells and is the first sucrose transporter shown to be localised to this sub-cellular compartment (Antony et al. 2008). In plant cells, pre-vacuolar compartments (PVC) mediate transport of proteins from the trans-Golgi network to the vacuole (Tse et al. 2004; Miao et al. 2006). The localisation of *Ac*SUT1 in PVC could facilitate vesicle-mediated transport of sucrose into the apoplast for subsequent loading into the phloem (Echeverria 2000) or could be involved in regulation of turnover of membrane sucrose transporters. To date, the functions and localisation of group 3 sucrose transporters (of which there is one in *Arabidopsis*) have been studied only in eudicots, in which, due to elevated expression in sink tissues, they have been implicated in phloem unloading (Sauer 2007). The higher transcript abundance of *Ac*SUT1 in pineapple source leaves compared to sink tissues might imply a different function for this transporter compared to that of the *Arabidopsis* homologue (Antony et al. 2008). It remains to be established if group 3 sucrose transporters have different localisation patterns and functions in dicots and monocots, or indeed in C_3 and CAM plants.

Vacuolar hexose transporters are likely to belong to the monosaccharide transporter (–like) (MST) gene family of which there are 53 members within the *Arabidopsis* genome (Lalonde et al. 2004). The MST transporters possess 12 transmembrane domains and 10 or more of these proteins have been characterised as plasma membrane-localised monosaccharide importers (Lalonde et al. 2004). In a recent phylogenetic review of the *A. thaliana* MST family, Büttner (2007) grouped the genes into seven distinct sub-families which have one common origin. Recently, the first tonoplast monosaccharide transporters (*At*TMT) were directly identified from *Arabidopsis*; these form a subfamily of three members with an extended middle loop between the predicted transmembrane helices six and seven (Wormit et al. 2006). We have recently identified a homologue of the *Arabidopsis* tonoplast monosaccharide transporter (*At*TMT) sub-group from an *A. comosus* fruit and root EST database that is also expressed in leaves (designated *Ac*MST2). The AtTMT transporters are believed to operate via a proton-coupled antiport mechanism and to allow the active transport and accumulation of hexoses (glucose and fructose) in the vacuole, particularly in response to stimuli (i.e. cold, drought, salinity) that promote sugar accumulation in *Arabidopsis* (Wormit et al. 2006). The function of such a vacuolar hexose importer in the operation of CAM is not clear. However, we have recently identified another MST sequence from pineapple (*Ac*MST1, Accession number EF460876) that localised to the tonoplast in tobacco leaf epidermis and that shows homology to a further and distinct subfamily of MST genes in *A. thaliana* designated AtERD6-like (Antony et al. 2008). The *Arabidopsis* sequences with greatest similarity to *Ac*MST1 (At1g75220 and At1g19450) have also previously been proposed as candidate vacuolar sugar transporters on the basis of a tonoplast proteomics approach (Endler et al. 2006), although localisation to the tonoplast was not verified for these proteins. Although the 19 genes within the AtERD6-like family are the least investigated sub-group within the MST family, the results described above imply that some of these genes could encode tonoplast-localised sugar transporters. The AtERD6-like genes are named after the ERD6 (early-responsive to dehydration) gene that codes for a putative sugar transporter (Kiyosue et al. 1998), and other gene members of this group are induced/up-regulated in response to senescence, drought and wounding (Büttner 2007). Transport activity has not been demonstrated for any of the AtERD6-like genes but, as found for *Ac*MST1, the members of this group show considerable sequence similarity to the mammalian GLUT family of proteins that transport glucose via facilitated diffusion (Lalonde et al. 2004; Büttner 2007). To date, no energy-independent monosaccharide transporters have been identified at the molecular level in plants, and it has been suggested that AtERD6 homologues (like *Ac*MST1) could play a role in the transport of sugars out of the vacuole (Büttner 2007). The model proposed for vacuolar sugar transport in the leaves of *A. comosus* (McRae et al. 2002) implies the existence of a tonoplast-localised hexose transporter to permit efflux of glucose and fructose at night to provide substrates for dark CO_2 uptake (Fig. 1b). It remains to be established if *Ac*MST1 fulfills this role in the leaves of *A. comosus*. Indeed, the mechanism that restricts efflux of vacuolar hexose to the dark period and avoids futile cycling during the light period is still unknown. There is no evidence for diurnal or circadian control of transcript abundance of this pineapple tonoplast hexose transporter nor was there any difference in transcript abundance of *Ac*MST1 between two pineapple cultivars that differ in the magnitude of CAM (Antony et al. 2008). Moreover, transcript abundance of *Ac*MST1 was found to be higher in pineapple fruit tissue compared to leaves (Antony et al. 2008). The fruit tissue does not perform CAM or turn over soluble sugars on a day/night basis, but contains twice as much hexose on a fresh weight basis as the leaves. Further work is required to characterise the transport activity of *Ac*MST1 and to compare the physiological characteristics and energetic requirements of hexose transport across tonoplast vesicles in the leaves and fruits of this economically important tropical fruit crop.

The monosaccharide transporter (-like) family also contains proteins that facilitate the sub-cellular distribution and long-distance transport of sugar alcohols and polyols. A further tonoplast-localised MST sequence has been isolated from pineapple (designated *Ac*INT1, Accession number EF460877, Antony et al. 2008) that shows homology to the families of *myo*inositol transporters in *Arabidopsis (At*INT*)* and *Mesembryanthemum crystallinum (M*ITR*)*. *Myo*inositol and its phosphorylated derivatives play important roles as osmolytes and in cell wall biosynthesis, ascorbate biosynthesis, mineral storage, cellular energy currency and in several signalling pathways. Unlike the other vacuolar sugar transporters that have been isolated from pineapple and are described above, *Ac*INT1 shows diurnal patterns of transcript abundance that are similar to those found for the homologue of this vacuolar transporter (*MItr3*) in CAM-performing *M. crystallinum* (J. Cushman, personal communication). Moreover, there was significantly higher transcript abundance of *Ac*INT1 in a high-CAM expressing cultivar of pineapple compared to a cultivar with less CAM (Antony et al. 2008). Given that inositol is implicated in diverse aspects of metabolism, the significance of a day/night shift in transcript abundance of a vacuolar inositol transporter for the operation of CAM remains to be established. Conceivably, this might be involved in transport of inositols to help maintain a balance of cytoplasmic and vacuolar osmotic pressures in the course of the day/night cycle.

5 Partitioning of Sugars for Growth in CAM Plants

The intra-cellular sugar transport requirements for nocturnal generation of PEP and day-time recovery of carbohydrate for subsequent nocturnal reactions must be integrated with the metabolism of sugars for immediate leaf growth, as well as with the export of sugars out of the photosynthetic "source" cells to sink tissues, such as developing leaves, roots and reproductive tissues, to maintain growth of the whole plant. The long-distance transport of sucrose in the phloem is likely to play a key role in coordinating the supply and demand for sugars between intracellular and extra-cellular sugar transport processes. In order to decipher how phloem-localised sugar transporters might be regulated in CAM plants, it is important to consider how carbohydrates are partitioned between storage and growth/export processes over the day/night cycle. The relative amounts of sugars that are utilised in growth/export at night versus the day can be quantified by constructing net 24 h carbon budgets that measure: (1) the source of photosynthetically fixed carbon during the day, i.e. direct C_3 photosynthesis of atmospheric CO_2 or C_4 (i.e. carbon derived from breakdown of malate over the photoperiod where 1 mol malate = 4 mol C); (2) the partitioning of this carbon between starch or soluble sugars (1 mol glucose equivalent = 6 mol C) with the excess partitioned to day-time growth/export; and (3) the partitioning of carbohydrates at night between the generation of substrates for CAM, respiratory CO_2 or dark growth/export (Borland 1996). Using this approach, it was shown that salt-induced CAM in *M. crystallinum* was accompanied by a decrease in net growth and export from source leaves. The relative proportions of day/night growth/export was dependent upon leaf ontogeny with the CAM-performing primary leaves utilising sugars for growth and export only at night whilst the CAM secondary leaves utilised sugars for growth and export predominantly during the day (Borland and Dodd 2002). Such ontogenetic differences in day/night growth/export processes have been attributed to the different roles that primary and secondary leaves play in the annual life cycle of *M. crystallinum* (Borland and Dodd 2002) and indicate that the long-distance transport of sugars may be subject to day/night regulation that is superimposed upon a longer-term developmental modulation of sugar transport activity.

The phasing of leaf expansion growth over the diel CAM cycle has important implications for the mechanisms that regulate assimilate partitioning over a 24 h period. Using high-resolution digital image sequence processing (DISP) to examine diel patterns of leaf growth, Gouws et al. (2005) showed that leaf growth accelerated at night and decelerated during the day in C_3-performing *M. crystallinum*. This pattern has been reported for other C_3 species (Walter and Schurr 2005). In contrast, leaves of the CAM species *Kalanchoë beharensis* and cladodes of *Opuntia engelmanii* and *O. oricola* showed accelerated growth in the middle part of the day (phase III of CAM) and little or no growth at night (Gouws et al. 2005). The same authors proposed that the markedly different diel growth patterns in the CAM species can be explained in terms of both the distinctive carbon supply and turgor relations of CAM plants, implying that sugars are made available for growth predominantly during phases III and IV of CAM. During phase III, 75% of the malic acid that is decarboxylated must be sequestered as starch in the chloroplast or soluble sugars in the vacuole to provide PEP for the subsequent dark period. The remaining 25% of malic acid (equivalent to gross nocturnal CO_2 uptake) is thus available for growth and export. The mechanisms that regulate partitioning between storage reserves and growth in phase III are unknown. For starch accumulators (Fig. 1a) cycling of carbon between the chloroplast and cytosol is implied and the cytosolic glucose-6P pool would appear to be a branch point supplying carbon for both starch and sucrose synthesis (Fig. 1a). For soluble sugar accumulators, cytosolic sucrose would appear to be the branch point for storage or growth/export (Fig. 1b). The existence of discrete pools of metabolites within the cytosol that are either destined for storage or for export has been suggested by measurements of the carbon isotopic signatures ($\delta^{13}C$) of starch and sugar pools which have been found to differ by up to 4‰ (Deléens et al. 1979; Borland et al. 1994; Borland and Dodd 2002). The organisation of glycolytic enzymes into discrete functional complexes within the cytosol could, in theory be achieved by the formation of associations with organelles and microtubules (Giegé et al. 2003). However, it remains a distinct possibility that the variations in $\delta^{13}C$ in leaf carbohydrate fractions could be due to cell to cell variations in the level of CAM expression within the thick succulent leaves that are characteristic of CAM species. In general, it would appear that, for most CAM plants, the use of sugars for growth and export is curtailed at night relative to that during the day. Further research is needed to establish if sugar transporters localised on the plasma membrane and those associated with the phloem play a central role in regulating the day/night shift in the export of sugars from CAM leaves.

6 Regulation of Sugar Transport Processes

To date, most studies on the regulation of plant sugar transporters have focused on plasma membrane-localised sucrose transporters that are implicated in phloem transport. Heterotrophic cells in roots, reproductive structures, storage and developing organs rely on sugars that are exported from the photosynthetic 'source' leaves. The long-distance transport of sucrose in the phloem can thus be perceived as a systemic plant growth regulator that coordinates the supply and demand for carbon, thereby maintaining cellular homeostasis and growth. Sugar transport represents a process that is highly regulated at multiple levels and in response to changing sucrose concentration (Vaughn et al. 2002; Ransom-Hodgkins et al. 2003). For sugar transporters that function as H^+-symporters, regulation of activity by the proton-motive force is clearly important, and factors that modulate the activity of the H^+ ATPase are likely to impact upon sugar transport kinetics. Examples of direct transcriptional regulation of sucrose and hexose transporters include reports on changes in transcript abundance in leaves undergoing the source-sink transition, and there is also evidence that changes in environmental conditions (light, water, salinity) can alter the expression of certain sugar transporters (Lalonde et al. 2004).

Post-transcriptional control of sugar transporters can operate via changes in mRNA stability (turnover) whilst mRNA translation and post-translational control have also been described for some sugar transporters. Examples include the diurnal regulation of sucrose transporters (Kuhn et al. 1997; Hirose et al. 1997), phosphorylation-dephosphorylation (Roblin et al. 1998) and possible modification of activity by the lipid environment (Delrot et al. 2003). There is also evidence that changing sugar levels can modulate the expression and activity of certain sugar transporters at transcriptional and post-transcriptional levels (Chiou and Bush 1998; Ransom-Hodgkins et al. 2003). The responses to sugars are often fully reversible, indicating the existence of dynamic sucrose and hexose-sensing pathways that control sugar transport activities (Atanassova et al. 2003).

Given the distinctive diel patterns of leaf growth and carbon supply described for CAM species (see Sect. 5), it might be hypothesised that diel regulation of phloem-localised sugar transporters may be altered with a shift from C_3 to CAM. To date, only one EST for a putative sucrose transporter has been identified from the facultative species *M. crystallinum. McSut3* was found to show 98–99% similarity (Blast sequence comparison) with sucrose transporters from *Phaseolus vulgaris* and *Beta vulgaris* (Antony 2007) which belong to group 2 of the sucrose transporter family as recently re-classified by Sauer (2007). The close homology with the Group 2 sucrose transporters which have a high affinity for sucrose suggest a putative role for *Mc*Sut3 in phloem loading in *M. crystallinum*. Despite the reported shift in day/night export of sugars from leaves of *M. crystallinum* as CAM is induced (Borland and Dodd 2002), there was no appreciable change in diel patterns of transcript abundance of *Mc*Sut3 in C_3 or CAM leaves of *M. crystallinum* (Antony 2007). The transcript abundance of *Mc*Sut3, however, was depressed when leaves were fed with sucrose (up to 250 mM) whilst glucose had no effect (Antony 2007). Such results are consistent with those reported by Chiou and Bush (1998) for a homologue of *Mc*Sut3 in *B. vulgaris* where it was proposed that sucrose transport activity in the phloem can be modulated in response to changing sucrose concentrations in the leaf. It remains to be established how the control of sugar transporters by their own substrate at both transcriptional and post-transcriptional levels might be integrated within the distinctive day/night patterns of phloem translocation in CAM plants.

Although the mechanisms that regulate the activity of vacuolar sugar transporters have not been studied in any great detail, there is evidence for stimulation of transcript abundance of the *Arabidopsis* vacuolar monosaccharide transporter (*At*TMT) by glucose (Wormit et al. 2006). There is no evidence for day/night regulation of vacuolar sugar transporters in *Arabidopsis* or for the vacuolar hexose and sucrose transporters described for pineapple (Antony et al. 2008). However, the existence of phosphorylation motifs in the sequences of the vacuolar transporters identified from pineapple (Antony 2007) suggests the possibility of post-translational regulation of sugar fluxes across the vacuolar membrane. Further work is required to determine how such control mechanisms could be integrated with sugar signalling pathways to regulate vacuolar import of sucrose during phase III of CAM whilst restricting hexose export from the vacuole until the night (phase I).

7 Conclusions: Integration and Regulation of Sugar Transport Processes over the CAM Cycle

The mechanisms that synchronise the supply and demand for carbon over the diel CAM cycle must be subject to modulation by internal and external factors so that CAM plants can adjust the relative proportions of C_3 and C_4 carboxylation in response to changing environmental conditions. It has been proposed that two fundamental layers of control preside over the metabolic components of CAM (Borland and Taybi 2004). Control by a circadian clock sets up the diel phases of CAM and achieves appropriate synchronisation of metabolic and transport processes by phasing the transcription of particular genes to specific times in the day/night cycle (Hartwell 2005). The clock can also act at post-transcriptional and post-translational levels by, for example, regulating the transcription of the kinase that activates PEPC and dark CO_2 uptake (Nimmo 2000). Activation/repression of sugar transport processes by circadianly regulated kinases is a further possible aspect of control of the CAM cycle that should be explored. Circadian control of CAM is overlain by metabolite control, acting at both the transcriptional and post-transcriptional levels, which facilitates photosynthetic plasticity by entraining output from the clock to fluctuations in the environment. By serving as an internal gauge of external environmental conditions, the reciprocating pools of organic acids and carbohydrates are central to the metabolic control of CAM. In relation to resource allocation, the ability to sense altered sugar concentrations provides a mechanism by which carbohydrate metabolism in the source leaf may be tailored to meet sink demand. Membrane-localised transporters and sensors for sugars and metabolic intermediates thus occupy strategic checkpoints for integrating the circadian and metabolic signals that synchronise and modulate the diel phases in response to the environment. The tonoplast in particular plays a key role in orchestrating metabolism over the day/night cycle by controlling the uptake and release of organic acids to the cytosol which have been shown to over-ride the circadian regulation of PEPC phosphorylation/activation (Borland et al. 1999; Taybi et al. 2004). In soluble-sugar-storing CAM plants, a situation can be perceived where the uptake and release of sucrose and glucose/fructose from the vacuole could impinge on sugar-mediated regulation of the expression of genes implicated in photosynthesis, primary metabolism and sugar transport. As more sequence information becomes available for other CAM species, it seems likely that more sugar transporters will be identified in plants with this mode of carbon metabolism. Furthermore, recent progress in the genetic transformation of the CAM species *K. fedstchenkoi* (James Hartwell, personal communication) will provide a timely platform for testing the functional significance of candidate sugar transporters in regulating the distinctive day/night shifts in carbon partitioning that define this photosynthetic specialisation.

Acknowledgements We gratefully acknowledge funding provided via a Commonwealth Scholarship awarded to E.A. and by the Natural Environment Research Council UK.

References

Antony E (2007) Cloning localisation and expression analysis of sugar transporters in CAM plants. PhD Thesis, Newcastle University

Antony E, Taybi T, Courbot M, Mugford ST, Smith JAC, Borland AM (2008) Cloning, localisation and expression analysis of vacuolar sugar transporters in the CAM plant *Ananas comosus*. J Exp Bot 59: 1895–1908

Atanassova R, Leterrier M, Gaillard C, Agasse A, Sagot E, Coutos-Thevenot P, Delrot S (2003) Sugar-regulated expression of a putative hexose transport gene in grape. Plant Physiol 131: 326–334

Bartholomew D, Kadzimin S (1977) Pineapple. In: Alvim P de T, TT Kozolowski TT (eds) Ecophysiology of Tropical Crops. Academic Press, New York, pp. 113–156

Black CC, Chen JQ, Doong RL, Angelov MN, Sung SJS (1996) Alternative carbohydrate reserves used in the daily cycle of Crassulacean acid metabolism. In: Winter K, Smith, JAC (eds) Crassulacean Acid Metabolism; Biochemistry, Ecophysiology and Evolution. Springer, Berlin, pp. 31–43

Borland AM (1996) A model for the partitioning of photosynthetically fixed carbon during the C_3-CAM transition in *Sedum telephium*. New Phytol 134: 433–444

Borland AM, Dodd AN (2002) Carbohydrate partitioning in crassulacean acid metabolism plants: reconciling potential conflicts of interest. Fun Plant Biol 29: 707–716

Borland AM, Griffiths H (1989) The regulation of citric acid accumulation and carbon recycling during CAM in *Ananas comosus*. J Exp Bot 40: 53–60

Borland AM, Taybi T (2004) Synchronization of metabolic processes in plants with Crassulacean acid metabolism. J Exp Botany 55: 1255–1265

Borland AM, Griffiths H, Broadmeadow MSJ, Fordham MC, Maxwell C (1994) Carbon-isotope composition of biochemical fractions and the regulation of carbon balance in leaves of the C_3-Crassulacean acid metabolism intermediate *Clusia minor* L growing in Trinidad. Plant Physiol 106: 493–501

Borland AM, Hartwell J, Jenkins GI, Wilkins MB, Nimmo HG (1999) Metabolite control overrides circadian regulation of phosphoenolpyruvate carboxylase kinase and CO2 fixation in Crassulacean acid metabolism. Plant Physiol 121: 889–896

Borland AM, Maxwell K, Griffiths H (2000) Ecophysiology of the CAM pathway. In: Leegood RC, Sharkey TD, von Caemmerer S (eds) Photosynthesis: Physiology and Metabolism. Kluwer, The Netherlands, pp. 583–605.

Büttner M (2007) The monosaccharide transporter(-like) gene family in *Arabidopsis*. FEBS Letts 581: 2318–2324

Chiou T-J, Bush DR (1998) Sucrose is a signal molecule in assimilate partitioning. PNAS 95: 4784–4788

Christopher JT, Holtum JAM (1996) Patterns of carbon partitioning in leaves of Crassulacean acid metabolism species during deacidification. Plant Physiol 112: 393–399

Christopher JT, Holtum JAM (1998) Carbohydrate partitioning in the leaves of Bromeliaceae performing C_3 photosynthesis or Crassulacean acid metabolism. Aus J Plant Physiol 25: 371–376

Deléens E, Garnierdardart J, Queiroz O (1979) Carbon isotope composition of intermediates of the starch-malate sequence and level of Crassulacean acid metabolism in leaves of *Kalanchoë blossfeldiana* Tom Thumb. Planta 146: 441–449

Delrot S, Atanassova R, Maurousset L (2003) Regulation of sugar, amino acid and peptide plant membrane transporters. Biochim Biophys Acta 1465: 281–306

Dodd AN, Borland AM, Haslam RP, Griffiths H, Maxwell K (2002) Crassulacean acid metabolism: plastic, fantastic. J Exp Bot 53: 569–580

Echeverria E (2000) Vesicle-mediated solute transport between the vacuole and the plasma membrane. Plant Physiol 123: 1217–1226

Endler A, Meyer S, Schelbert S, Schneider T, Weschke W, Peters SW, Keller F, Baginsky S, Martinoia E, Schmidt UG (2006) Identification of a vacuolar sucrose transporter in barley and *Arabidopsis thaliana* mesophyll cells by a tonoplast proteomic approach. Plant Physiol 141: 196–207

Giegé P, Heazlewood JL, Roessner-Tunali U, Millar AH, Fernie AR, Leaver CJ, Sweetlove LJ (2003) Enzymes of glycolysis are functionally associated with the mitochondrion in *Arabidopsis* cells. Plant Cell 15: 2140–2151

Gouws LM, Osmond CB, Schurr U, Walter A (2005) Distinctive diel growth cycles in leaves and cladodes of CAM plants: differences from C_3 plants and putative interactions with substrate availability, turgor and cytoplasmic pH. Fun Plant Biol 32: 421–428

Häusler RE, Baur B, Scharte J, Teichmann T, Eicks M, Fischer KL, Flugge UI, Schubert S, Weber A, Fischer K (2000) Plastidic metabolite transporters and their physiological functions in the inducible crassulacean acid metabolism plant *Mesembryanthemum crystallinum*. Plant J 24: 285–296

Hartwell J (2005) The co-ordination of central plant metabolism by the circadian clock. Biochem Soc Trans 33: 945–948

Hirose T, Imaizumi N, Scofield GN, Furbank RT, Ohsugi R (1997) cDNA cloning and tissue specific expression of a gene for sucrose transporter from rice (*Oryza sativa L.*) Plant Cell Physiol 38: 1389–1396

Holtum JAM, Smith JAC, Neuhaus HE (2005) Intracellular transport and pathways of carbon flow in plants with crassulacean acid metabolism. Fun Plant Biol 32: 429–450

Kenyon WH, Severson RF, Black CC (1985) Maintenance carbon cycle in Crassulacean acid metabolism plant leaves: source and compartmentation of carbon for nocturnal malate synthesis. Plant Physiol 77: 183–189

Kiyosue T, Abe H, Yamaguchi-Shinozaki K, Shinozaki K (1998) ERD6, a cDNA clone for an early dehydration-induced gene of Arabidopsis, encodes a putative sugar transporter. Biochim Biophys Acta 1370: 187–191

Kluge M (1969) Veränderliche Markierungsmuster bei $^{14}CO_2$ Fütterung von *Bryophyllum tubiflorum* zu verschiedenen Zeipunkten der Hell-Dunkel-Periode. I. Die Fütterung unter Belichtung Planta 88: 113–129

Kore-eda S, Kanai R (1997) Induction of glucose 6-phosphate transport activity in chloroplasts of *Mesembryanthemum crystallinum* by the C3-CAM transition Plant Cell Physiol 38: 895–901

Kore-eda S, Noake C, Ohishi M, Ohnishi J-I, Cushman JC (2005) Transcriptional regulation of organellar metabolite transporters during induction of crassulacean acid metabolism in *Mesembryanthemum crystallinum*. Fun Plant Biol 32: 451–466

Kuhn C, Franceschi VR, Schulz A, Lemoine R, Frommer WB (1997) Macromolecular trafficking indicated by localization and turnover of sucrose transporters in enucleate sieve elements. Science 275: 1298–1300

Lalonde S, Wipf D, Frommer WB (2004) Transport mechanisms for organic forms of carbon and nitrogen between source and sink. Ann Rev Plant Biol 55: 341–372

Lemoine R (2000) Sucrose transporters in plants: update on function and structure. Biochim Biophys Acta 1465: 246–262

Lüttge U (1988) Day–night changes of citric acid levels in CAM: phenomenon and ecophysiological significance. Plant Cell Environ 11: 445–451

McRae SR, Christopher JT, Smith JAC, Holtum JAM (2002) Sucrose transport across the vacuolar membrane of *Ananas comosus*. Fun Plant Biol 29: 717–726

Miao Y, Yan PK, Kim H, Hwang I, Jiang L (2006) Localization of green fluorescent protein fusions with the seven arabidopsis vacuolar sorting receptors to prevacuolar compartments in tobacco BY-2 cells. Plant Physiol 142: 945–962

Neuhaus HE, Schulte N (1996) Starch degradation in chloroplasts isolated from C_3 or CAM (crassulacean acid metabolism)-induced *Mesembryanthemum crystallinum* L. Biochem J 318: 945–953

Niittyla T, Messerli G, Trevisan M, Chen J, Smith AM, Zeeman SC (2004) A previously unknown maltose transporter essential for starch degradation in leaves. Science 303: 87–89

Nimmo HG (2000) The regulation of phosphoenolpyruvate carboxylase in CAM plants. Trends Plant Sci 5: 75–80

Nobel PS (1996) High productivity of certain agronomic CAM species. In: Winter K, Smith JAC (eds) Crassulacean Acid Metabolism. Biochemistry, Ecophysiology and Evolution. Springer, Berlin, pp. 255–265

Osmond CB (1978) Crassulacean acid metabolism – curiosity in context. Ann Rev Plant Phys Plant Mol Biol 29: 379–414

Osmond CB, Holtum JAM (1982) Crassulacean acid metabolism. In: Hatch MD, Boardman NK (eds) The Biochemistry of Plants, Vol 8. Academic Press, New York, pp. 283–370

Paul MJ, Loos K, Stitt M, Ziegler P (1993) Starch-degrading enzymes during the induction of CAM in *Mesembryanthemum crystallinum*. Plant Cell Environ 16: 531–538

Ransom-Hodgkins WD, Vaughn MW, Bush DR (2003) Protein phosphorylation plays a key role in sucrose-mediated transcriptional regulation of a phloem-specific proton-sucrose symporter. Planta 217: 483–489

Roblin G, Sakr S, Bonmort J, Delrot S (1998) Regulation of a plant plasma membrane sucrose transporter by phosphorylation. FEBS Lett 424: 165–168

Sauer N (2007) Molecular physiology of higher plant sucrose transporters. FEBS Lett 581: 2309–2317

Sauer N, Ludwig A, Knoblauch A, Rothe P, Gahrtz M, Klebl F (2004) AtSUC8 and AtSUC9 encode functional sucrose transporters, but the closely related AtSUC6 and AtSUC7 genes encode aberrant proteins in different *Arabidopsis* ecotypes. Plant J 40: 120–130

Schulze W, Weise A, Frommer WB, Ward JM (2000) Function of the cytosolic N-terminus of sucrose transporter AtSUT2 in substrate affinity. FEBS Lett 485: 189–194

Smith JAC, Bryce JH (1992) Metabolic compartmentation and transport in CAM plants. In: Tobin A (ed) Plant Organelles: Compartmentation of Metabolism in Photosynthetic Cells. Cambridge University Press, Cambridge, pp. 141–167

Taybi T, Nimmo HG, Borland AM (2004) Expression of phosphoenolpyruvate carboxylase and phosphoenolpyruvate carboxylase kinase genes: implications for genotypic capacity and phenotypic plasticity in the expression of Crassulacean acid metabolism. Plant Physiol 135: 587–598

Tse YC, Mo BX, Hillmer S, Zhao M, Lo SW, Robinson DG, Jiang LW (2004) Identification of multi-vesicular bodies as prevacuolar compartments in *Nicotiana tabacum* BY-2 cells. Plant Cell 16: 672–693

Vaughn MW, Harrington GN, Bush DR (2002) Sucrose-mediated transcriptional regulation of sucrose symporter activity in the phloem. PNAS 99: 10876–10880

Walter A, Schurr U. (2005) Dynamics of leaf and root growth: endogenous control versus environmental impact. Ann Bot 95: 891–900

Weise SE, Schrader SM, Kleinbeck KR, Sharkey TD (2006) Carbon balance and circadian regulation of hydrolytic and phosphorolytic breakdown of transitory starch. Plant Physiol 141: 879–886

Williams LE, Lemoine R, Sauer N (2000) Sugar transporters in higher plants – a diversity of roles and complex regulation. Trend Plant Sci 5: 283–290

Wormit A, Trentmann O, Feifer I, Lohr C, Tjaden J, Meyer S, Schmidt U, Martinoia E, Neuhaus HE (2006) Molecular identification and physiological characterization of a novel monosaccharide transporter from *Arabidopsis* involved in vacuolar sugar transport. Plant Cell 18: 3476–3490

Zeeman SC, Smith SM, Smith AM (2004) The breakdown of starch in leaves. New Phytol 163: 247–261

Zeeman SC, Delatte T, Messerli G, Umhang M, Stettler M, Mettler T, Streb S, Reinhold H, Kotting O (2007) Starch breakdown: recent discoveries suggest distinct pathways and novel mechanisms. Fun Plant Biol 34: 465–473

Ecology

Epiphytic Plants in a Changing World-Global: Change Effects on Vascular and Non-Vascular Epiphytes

G. Zotz(✉) and M.Y. Bader

Contents

Abstract Epiphytes have been called particularly vulnerable to climate change because of their existence at the interface of vegetation and atmosphere. We review the available evidence for this notion and put our analysis into the larger context of human-induced changes in general. Besides climate change, land use changes adversely affect epiphytes, while other factors, e.g. biotic exchange, are of lesser importance in this life form. Both land use change and climate change primarily affect hygrophilic taxa, while drought-resistant species may sometimes even benefit. Vascular and non-vascular epiphytes in tropical cloud forests will seriously suffer from decreased moisture input. In contrast, varying precipitation in more seasonal lowland forests seems to affect vascular species rather little, but a possible negative

G. Zotz
Functional Ecology of Plants, Institute of Biology and Environmental Sciences, University of Oldenburg, P.O. Box 2503, 26111 Oldenburg, Germany
e-mail: gerhard.zotz@uni-oldenburg.de

U. Lüttge et al. (eds.), *Progress in Botany 70*,
DOI:10.1007/978-3-540-68421-3,

impact of rising temperatures on plant performance is unexplored. For co-occurring lichens and bryophytes, however, rising temperatures could have disastrous effects, as suggested by model calculations. In the temperate zones, global warming should allow range extensions towards the poles for vascular epiphytes and lead to new assemblages among non-vascular epiphytes. In spite of this mixed picture, epiphytes as a group may indeed be "particularly" threatened by global change, because the habitats characterised by exceptional species richness, e.g. tropical cloud forests, are those most seriously affected.

1 Introduction

The global climate is changing rapidly due to anthropogenic activities (Solomon et al. 2007) and there is a growing concern that global biodiversity will be severely affected (Lovejoy and Hannah 2005). Naturally, this preoccupation is particularly pronounced when it comes to hyperdiverse ecosystems such as tropical rain forests. Although initially the scope of most publications was restricted mainly to the tree component of these forests (Malhi and Phillips 2005), there is now a growing number of studies addressing the impact of global change on other life forms, such as climbing plants and epiphytes, with particular attention to possible feedbacks on ecosystem functioning. For example, increases in liana abundance may directly increase the dynamics of the entire forest (Phillips et al. 2002), and a possible dieback of epiphytes may lower the efficiency with which cloud forests use mist as a moisture source, which may endanger the entire ecosystem (Benzing 1998; Still et al. 1999).

This review will focus on the response of epiphytic plants, both vascular and non-vascular (including lichens, excluding epiphylls), to anthropogenic changes. Although our scope includes the entire globe, there is a clear focus on tropical systems, which reflects the much higher importance of epiphytes at lower latitudes. Vascular epiphytes are almost completely restricted to the tropics, with noteworthy temperate diversity "hotspots" in New Zealand, Chile and the Himalayas (Zotz 2005). Non-vascular epiphytes, on the other hand, can be found from the tropics to the polar treeline, but their diversity and importance relative to terrestrial species is also highest in the tropics (S. Robbert Gradstein, André Aptroot and Robert Lücking, personal communication).

1.1 *Why Single Out Epiphytes?*

All plant life is likely to be affected by global climate change, e.g. changes in precipitation regimes (Solomon et al. 2007), more frequent climatic extremes (Timmermann et al. 1999), or reduced cloud water in tropical montane forests (Still et al. 1999). However, since epiphytic diversity and abundance responds more strongly to water availability than any other life form, such as trees or terrestrial herbs (Gentry and Dodson 1987), epiphytes are likely to be more dramatically affected. Indeed, because of their tight coupling to atmospheric inputs, epiphytes have been called "particularly vulnerable" to global change (Benzing 1998; Lugo and Scatena 1992).

Adverse effects on epiphytes cannot be viewed in isolation. In many tropical but also temperate forests, epiphytes represent a substantial proportion of the entire biomass (Hofstede et al. 1993), play a vital role in forest hydrology (Pócs 1980; Pypker et al. 2006; Veneklaas et al. 1990), influence forest nutrient fluxes (Clark et al. 2005; Forman 1975; Knops et al. 1996), and also facilitate animal life in tree canopies (Richardson et al. 2000). Thus, any negative impact will almost immediately affect other components of the forests.

Unfortunately, the global database which documents already observed global climate change on natural biological systems is extremely biased towards mid- and high latitudes (Parmesan 2006). Thus, in spite of the above-mentioned alarming reviews, possibly already on-going climate change impacts on epiphytic flora in tropical forests remain largely undocumented, and our understanding of their future prospects is still sketchy.

As a basis for future work, this paper summarises our current knowledge and identifies research needs by putting the likely impact of global climate change on epiphytic flora in the larger context of anthropogenic changes in general. Sala et al. (2000) have identified five major drivers of change for global biodiversity: land use changes, climate change, rising atmospheric CO_2 levels, nitrogen deposition and acid rain, and biotic exchange (invasive species). They argued that, globally, the relative effect of land use changes is currently larger than that of any other factor, this dominance being particularly pronounced in tropical biomes. However, the relevance of a particular driver may change in time. For example, Malcolm et al. (2006) predict that in the next century the number of extinctions in tropical biodiversity hotspots caused by climate change may exceed those caused by deforestation. Thus, the relative importance of climatic changes versus other drivers as discussed in this paper is not static and any evaluation has to be revised through time.

2 Vascular Epiphytes

2.1 General Ecology and Distribution

Epiphytism is defined by the non-parasitic use of other plants, usually trees, as structural support (Benzing 1990). Consequently, epiphytes are ecologically quite a heterogeneous group, ranging from extreme xerophytes at sun-exposed sites in the upper forest canopy to hygrophilous shade plants. There are more than 20,000 species of vascular epiphytes worldwide, equivalent to almost 10% of all vascular plant species (Benzing 1990). This diversity is very unevenly distributed, both taxonomically and geographically. First, just a few families (e.g. Orchidaceae, Araceae and Bromeliaceae) represent the large majority of all species. Second, there are steep latitudinal and altitudinal gradients both in species richness and abundance (Cardelús et al. 2006; Zotz 2005). The ecosystems richest in vascular epiphytes are wet tropical montane forests (Foster 2001). In such forests, which are characterised by high precipitation and regular cloud immersion, epiphytes may represent the most species-rich life form encompassing up to 50% of the local vascular flora.

The exceptional promotion of this life form with increased precipitation highlights the overwhelming importance of the factor water in epiphyte ecology (Gentry and Dodson 1987; Zotz and Hietz 2001). This importance is further emphasised by the high proportion of epiphytic plants using the water-saving crassulacean acid metabolism (CAM). Originally, this photosynthetic pathway was believed to be associated mainly with succulents from arid environments, but we now assume that the majority of CAM species are epiphytic (Winter and Smith 1996). The proportion of CAM species is inversely related to precipitation and to the distance to the forest floor, but CAM species are not excluded from wet forests or the shaded understorey (Pierce et al. 2002).

In conclusion, the diversity of habitat types in which epiphytes are found, their varied taxonomy and their diverse physiology should cause a rather mixed response to anthropogenic changes by members of this plant group.

2.2 Land Use Changes

Removing trees in the course of selective logging activities has obvious consequences for structurally dependent flora. However, the damage inflicted on local epiphyte assemblages by such activities is larger than the number of removed trees implies, because selective logging preferentially involves large trees (Peres et al. 2006), which normally host an over-proportional fraction of the local epiphyte community.

Replacing primary with secondary forest may have dramatic effects on epiphytic flora (Krömer and Gradstein 2003), although some studies found little or no change in species numbers (Williams-Linera et al. 1995). Even if total species numbers change little, conversion of primary forests normally produces “winners” and “losers” among epiphytic flora. Considering the importance of moisture for this plant group, there is a non-surprising decrease or even entire loss of hygrophilous taxa such as many Hymenophyllaceae and other drought-sensitive ferns in secondary vegetation (Barthlott et al. 2001). On the other hand, there is the frequent albeit not universal observation (Krömer and Gradstein 2003) of increases in bromeliads in secondary forests in the Neotropics, sometimes in both species number and abundance (Barthlott et al. 2001), but more often only in the abundance of certain species (Cascante-Marin et al. 2006; Hietz et al. 2006). This increase can be impressive: the latter two reports document three- and four-fold increases in total epiphyte densities. The reasons for the promotion of epiphytic bromeliads in secondary vegetation are subject to speculation. Cascante-Marin et al. (2006) suggest differences in propagule dispersal, while Hietz et al. (2006) invoke physiological advantages of light-loving taxa, particularly atmospheric species. Indeed, shifts from C3 tank bromeliads to light-loving and drought-resistant atmospheric CAM species have also been reported (Wolf 2005). Conversion of primary forests may affect epiphytes not only immediately via microclimatic changes, but also by altering biotic interactions. For example, potential pollinators like bees, butterflies and bats all show different affinity to primary and secondary forests (Barlow et al. 2007).

Forest clearance also produces more or less fragmented landscapes, which are likely to affect epiphytes adversely in the remaining, seemingly intact forests areas by (1) changes in local microclimate and (2) changes in forest dynamics. The notion of a "vegetation breeze" refers to moist air being pulled away from forests into adjoining pastures and clearings (Laurance et al. 2002). Here, this humid air condenses into clouds, and dry air is then recycled back over the forest. Considering the importance of moisture for epiphytes, a direct regional effect on epiphytic flora seems likely, but this possibility has not yet been studied. It has been shown, however, that deforestation in the lowlands may directly influence cloud formation and reduce moisture input in adjacent mountains (Lawton et al. 2001).

Drought at the edge of forest fragments has also been shown to increase tree dynamics for depths of at least 1,000 m into a patch of forest. Laurance et al. (2000) documented a particularly pronounced increase in mortality among large trees and suggested as possible mechanisms vulnerability to drought, increased wind throw close to the forest edge, but also liana infection. Besides an edge-related increase in tree mortality, the wood-debris production also increased substantially (Nascimento and Laurance 2004). However, even in remote forests areas there are indications of increased dynamics, possibly due to climate change (Lewis et al. 2004). Any increase in forest dynamics, from the scale of twigs or branches (=wood-debris production) to shorter tree longevity, has immediate consequences for dependent flora. It has been shown repeatedly that, besides drought, substrate instability is the major cause of mortality among epiphytes (Zotz and Schmidt 2006). Thus, epiphytes may already be negatively affected by global change even in remote or protected areas (Zotz and Schmidt 2006), but the complete lack of pertinent data does not allow us to evaluate this possibility unambiguously.

In conclusion, current anthropogenic perturbations of forests affect epiphytic plants in a multi-faceted manner, from localised physical destruction to more or less subtle effects on their growth and survival via increased forest dynamics or altered climatic conditions.

2.3 *Climatic Change*

2.3.1 Warming

Together with precipitation, temperature is a prime determinant of altitudinal and latitudinal plant distributions (Archibold 1995). If global average land surface temperatures rise, as expected, by ca. 3°C in tropical latitudes and ca. 6°C at higher latitudes by the end of the century (Solomon et al. 2007), a cascade of biological responses should follow. First, plant metabolism will be immediately affected, but the effects on photosynthesis, respiration and growth are hard to predict, as plants acclimate in a species-specific manner (Campbell et al. 2007). Second, phenological patterns may also change, because temperature triggers many natural history traits, and, finally, we should observe shifts in species ranges either poleward or up in elevation as species move to occupy areas within their metabolic tolerances.

Of note, the lower predicted temperature increase in the tropics may still produce a biological effect of similar or even stronger magnitude than that at higher latitudes. This is suggested by the results of comparative studies with tropical and temperate trees that demonstrated much lower acclimation potentials for tropical taxa (Cunningham and Read 2003). Unfortunately, few physiological and ecological data are available for vascular epiphytes to make specific predictions of their likely response. For example, although temperature effects have been studied in a number of bromeliads or orchids, most reports had a horticultural context, i.e. used cultivated plant material under conditions relevant for mass cultivation (Lopez and Runkle 2005). There are studies, however, which are directly relevant to natural conditions. For example, Vaz et al. (2004) investigated temperature effects on *in vitro* growth and flowering in *Psygmorchis pusilla* (Orchidaceae). Growth and reproductive investment was largest at 27°C, representing the average temperature at natural growing sites. An increase in growth temperature of just 3°C reduced final plant biomass by >30% and the length of the floral spike by >50%. Other studies focussed on the effect of temperature on germination. Downs (1964), e.g. showed that the temperature responses in a number of bromeliads were more or less consistent with current temperatures prevailing in situ, e.g. the lowland species *Aechmea nudicaulis* did not germinate at temperatures below ca. 15°C or above 30°C. Clearly, such a species could substantially shift its altitudinal range upwards in future decades, if germination conditions were indeed the life history bottleneck.

For geographic distributions of plants, temperature extremes are probably at least as important as averages. During the last decades there has already been a widespread reduction in the number of frost days in mid-latitude regions, and the frequency of frost is likely to decrease in tropical mountains as well (Diaz and Graham 1996). Although some vascular epiphytes can tolerate at least moderate frost (Jenkins 1999), freezing constitutes a major limitation to the occurrence of epiphytes and explains, e.g. to a large part the northern limits of vascular epiphytes in the southern USA. Thus, we may observe not only altitudinal range changes, but also a poleward expansion of epiphytic taxa with a further increase in minimum temperatures. For example, we expect the northern limit of *Tillandsia usneoides*, which currently extends north up to Virginia, to shift further north. A case in point may also be the appearance of abundant epiphytic *Polypodium vulgare* in the Rhine valley in Germany in recent decades (Frahm 2002).

Changes in phenological patterns, which are amply documented for terrestrial plants in the temperate zone (Fitter and Fitter 2002), have been evoked at least once for a tropical epiphyte in Puerto Rico: Lasso and Ackerman (2003) linked changes in flowering phenology in *Vriesea sintenisii* with an increase in daily minimum temperatures during the last decades.

2.3.2 Changes in Rainfall and Mist

Although rising temperatures will almost certainly affect epiphytic vegetation to some degree, changes in moisture input are expected to have a more dramatic effect. Globally, annual precipitation is expected to increase with rising temperatures (Solomon et al. 2007), but there are large regional differences. For example, current general circulation models predict *decreased* precipitation over Central America and the southern Andes. Moreover, the total amount of rainfall is of lower relevance for epiphytic plants than seasonality or rainfall frequencies, and there is a consensus that the number and/or intensity of extreme events such as droughts will increase in the future (Solomon et al. 2007). Particularly relevant for this review are direct long-term observations (Pounds et al. 1999) and modelling exercises (Still et al. 1999) that consistently suggest a change in the altitude of cloud formation in tropical mountains. This finding is alarming because cloud forests harbour a stunning diversity of epiphytic plants. Upward shifts in the height of the orographic cloud bank formation and increased evapotranspiration will impose their strongest effect during the dry season, when cloud forests depend on the deposition of cloud droplets.

The simulations performed by Still et al. (1999) suggested a shift of several hundred metres assuming doubled atmospheric CO_2 levels. The consequences of such a shift have been explored in an experimental study in a montane setting in Costa Rica. Nadkarni and Solano (2002) transplanted epiphytes and their arboreal soil from upper cloud forest trees to trees at slightly lower elevations which were naturally exposed to smaller amounts of cloud water: leaf mortality increased, while leaf production and longevity decreased.

On the other hand, changes in precipitation in the lowlands with a more pronounced seasonality may have a rather limited impact. This is indicated by the results of a number of long-term studies in the lowlands of Panama. One study found a 40% *increase* in the number of epiphyte individuals in the period from 1995 to 2002 in spite of the extreme 1997–1998 El Niño drought (Laube and Zotz 2007). Annual censuses of populations of two epiphyte species corroborate that the impact of this extreme event was rather low (Zotz et al. 2005; Zotz and Schmidt 2006), which contrasts strongly with the severe impact of El Niño droughts on trees in this and other tropical forests (Condit et al. 1995; Williamson et al. 2000). Population growth rates did not correlate with precipitation, but were more affected by the dynamics of the substratum. Dismissing the possibility that the species in these studies are exceptional, these results suggest that a moderate change in precipitation regimes in this and other seasonal tropical lowland forests will hardly affect the majority of vascular epiphyte species.

2.3.3 Extreme Events

A possible increase in the number of El Niño events has already been discussed in the preceding section. A further consequence of climate change, a predicted increase in the number of intense hurricanes and other tropical storms (Solomon et al. 2007), could affect epiphytes more than other components of tropical vegetation. For example, Hurricane Andrew destroyed ca. 90% of all canopy epiphytes in Southern Florida, making epiphytes suffer more mortality than any other plant group (Loope et al. 1994). Although exposed growing sites should make epiphytes particularly susceptible to increased wind damage, there are other reports which are not consistent with this expectation (Oberbauer et al. 1996)

2.4 *Direct Effects of Increased CO_2 on Plant Metabolism*

By the second half of this century, atmospheric CO_2 concentrations are expected to exceed 550ppm (Solomon et al. 2007). Noteworthy, photosynthesis is not carbon-saturated at the current ca. 380ppm, yet plants respond to elevated CO_2 levels not merely with increased assimilation rates. Acclimations include, to a varying degree, a down-regulation of photosynthesis, reduced N-content, reduced stomatal conductance, and increased water-use efficiency (Urban 2003): these responses will in turn influence leaf energy balance and plant nutrient relations. Although these generic physiological responses are well understood mechanistically (Gutschick 2007), acclimations are species- or even genotype-specific, making any change in performance measures of photosynthesis such as water or N-use efficiencies difficult to predict for a particular taxon or functional group. Since water is by far the most limiting factor for growth and survival among vascular epiphytes (Zotz and Hietz 2001), while competition is usually assumed to be of minor importance (Hietz and Briones 1998), increases in water use efficiency should be much more relevant than any increase in carbon gain or nutrient use efficiency. Actual data to test this notion are unavailable.

Several studies have investigated physiological changes of epiphytic plants under elevated CO_2, addressing changes in photosynthesis and growth (Li et al. 2002; Ong et al. 1998), changes in the activity of the carboxylating enzymes RuBisCO or PEP-C (Li et al. 2002), or changes in nocturnal acidification (Li et al. 2002; Nowak and Martin 1995). Growth was similarly stimulated in C3 and CAM species in a comparative study with several bromeliads and orchids (Monteiro et al., in press). Although surprising, because due to their biochemistry CAM plants are expected to benefit less from elevated CO_2, this result agrees with findings from terrestrial CAM plants (Drennan and Nobel 2000).

All mentioned studies provide some insight into the physiological response of vascular epiphytes to elevated CO_2, but neither were experimental conditions realistic nor were the selected plants very representative: all experiments were conducted with well-watered, fertilised individuals, frequently cloned nursery material studied in the greenhouse or in the laboratory. In contrast, the important ecological questions concerning the response of vascular epiphytes are related to the interaction of elevated CO_2 and changes in water supply and, to a lesser extent, poor nutrient supply (Zotz and Hietz 2001). The negative effects of reduced water supply due to climatic changes should be at least partly alleviated by elevated atmospheric CO_2 both in C3 and CAM epiphytes.

2.5 Biotic Exchange

Globally, invasive species pose a major threat to biodiversity (Sala et al. 2000), but this does not seem to be the case among epiphytes. There are very few reports of epiphytic plants escaping cultivation (Tejedor and McAlpin 2000), let alone indications of invasive epiphytes threatening local flora (Benzing 2004). This is surprising in view of the large-scale global trade of literally thousands of species of orchids, bromeliads or cacti as ornamentals, and suggests that the invasiveness of vascular epiphytes is very low.

Biotic exchange may affect epiphytes in other ways. First, host trees differ in their suitability for the establishment of epiphytes (Laube and Zotz 2006), because an introduced tree species may drastically change local epiphyte communities. A case in point is *Psidium cattleianum*, a tree that has invaded many of Hawaii's cloud forests, grows in dense thickets, often excluding other woody plants, and which supports many fewer epiphytes than native *Metrosideros polymorpha* due to its smooth and peeling bark (Mudd 2004). Likewise, the invasive exotic tree fern *Cyathea cooperi* supports many fewer epiphytes than native Hawaiian tree ferns (Medeiros et al. 1993). Reduced epiphyte cover, in turn, now substantially reduces the forest stand water retention capacity for precipitation (Mudd 2004). Secondly, epiphytes may be harmed by introduced pathogens or herbivores. There is at least one well-documented case with disastrous effects for epiphytic bromeliad populations in the southern USA: the introduction of the weevil *Metamasius callizona*, a native of Central America (Frank et al. 2006).

2.6 Nitrogen Deposition, Acid Rain and Other Atmospheric Pollution

There are no systematic studies on the effects of nitrogen deposition, acid rain or other forms of air pollutants on the geographic distributions of vascular epiphytes. However, atmospheric bromeliads are sometimes used as bioindicators in urban or industrial areas (e.g. Wannaz and Pignata 2006). Although species such as *Tillandsia recurvata* have proven to be quite tolerant of pollution (Graciano et al. 2003), taxa do of course differ in their physiological response (Wannaz and Pignata 2006). Overall, however, there is no published information that air pollution is currently a relevant factor for vascular epiphyte distributions beyond the immediate vicinity of larger urban aggregations or industrial plants.

3 Non-Vascular Epiphytes (Bryophytes and Lichens)

3.1 General Ecology and Distribution

Non-vascular epiphytes occupy even more diverse habitats than vascular epiphytes, including for instance desert cactuses, taiga trees and alpine shrubs. The proportion of epiphytes among bryophytes and lichens differs between habitats and climate zones. For bryophytes, it decreases from about 60% of ca. 9,000 species in the tropics to about 10% of ca. 7,000 species in temperate zone (Robbert Gradstein, personal communication), and for lichens it decreases from over 80% of ca. 20,000 species in the tropics to less than 20% of ca. 10,000 species outside the tropics (André Aptroot, personal communication). Still, species richness can be high even in boreal forest: e.g. over 100 epiphytic lichen species were found in 15 1-ha *Picea abies*-dominated forest plots in southern Finland (Kuusinen and Siitonen 1998). Temperate rain forests, notably those of New Zealand and the Pacific northwest of America, also provide very rich and productive non-vascular epiphyte habitats, with biomass amounting to about 2,600 kg ha^{-1} in some old-growth forests (McCune 1993).

In the tropics, the diversity of non-vascular epiphytes increases with altitude. Maximum diversity is generally reached in upper montane forest, at a somewhat higher altitude than the maximum in vascular epiphyte diversity (Grau et al. 2007; Wolf 1993). The altitudinal gradient in abundance is even stronger and is also stronger than that for vascular epiphytes. While in the lowlands bryophytes and lichens are generally small and inconspicuous, with biomass values well under 10 kg ha^{-1}, non-vascular epiphytic mass can amount to over 1,000 kg ha^{-1} just below the forest line, and there are even estimates of over 10,000 kg ha^{-1} (Frahm 1990; Pócs 1980). Not surprisingly, these forests are often referred to as "mossy forests".

Bryophytes and lichens belong to very different taxonomic groups, but are very similar in their ecology and ecophysiology. For one, they are generally small. Second, they are poikilohydric. When dehydrated, they are physiologically inactive, and many species can survive extended dry periods. Bryophytes and lichens occupy similar habitats, but to different extents. Lichens are generally adapted to drier circumstances than bryophytes, and in tropical forests they usually show a preference for the outer tree crowns, while bryophytes dominate the understorey and inner crowns (Cornelissen and Ter Steege 1989; Forman 1975; McCune 1993; Wolf 1994). Similarly, in both tropical and temperate zones, bryophytes tend to dominate in the wettest forests, whereas lichens are more common in drier forests (Cornelissen and Gradstein 1990; Frahm 2003).

3.2 *Land Use Change*

If areas cleared of forest are left to regenerate, the emerging secondary forest will soon be invaded by epiphytes. Secondary upper montane oak forests in Costa Rica harbored similar diversity of non-vascular epiphytes as primary forest in the same region, even though the secondary forest was only 10–15 years old. The species composition, however, was very different and even in 40-year-old secondary forest almost half of the primary forest species were still missing (Holz and Gradstein 2005). Similar slow recovery rates were recorded in other tropical forests (Acebey et al. 2003; Chapman and King 1983; Pinheiro Da Costa 1999), as well as in temperate (Norris 1987) and boreal forests (Kuusinen and Siitonen 1998). Most of these studies suggest that a regeneration period of at least 100 years is needed before most non-vascular epiphyte species have returned to the regrowing forest.

The changes in microclimate in remnant forest patches or along forest edges affect non-vascular epiphytes similarly to or more than vascular epiphytes (Belinchón et al. 2007; Kantvilas and Jarman 1993). Excessive light and drought are the strongest threats, so that understorey species, adapted to low light and relatively moist conditions, generally suffer most (Gradstein 1992). Because of their higher average drought tolerance, lichens are generally less affected than bryophytes (Lõhmus et al. 2006; Nöske et al. 2008). A species shift to lower positions on host trees is also often observed in disturbed forest and is again related to the higher light and lower humidity levels in these more open forests (Acebey et al. 2003; Andersson and Gradstein 2005). The effects of altered microclimates due to deforestation are in many ways similar to the effects of projected climatic changes, which also involve drier conditions (see Sect. 3.3)

The other main effect of longer forest edges, increased forest dynamics, may affect non-vascular epiphytes less than vascular epiphytes. Compared to vascular epiphytes, the small size of mosses and lichens makes them less susceptible to being knocked off by falling branches, but increased tree mortality affects all dependent flora in a similar fashion.

Remnant trees and forest patches play a vital role in the re-colonisation of secondary forests by epiphytes (Lõhmus et al. 2006), which, by definition, cannot rely on soil seed or spore banks. Although the spores of mosses and lichens are small and some can travel long distances, dispersal can limit species occurrence and genetic exchange between populations (Gradstein 1992; Sillett et al. 2000; Snäll et al. 2005). In general, understorey species, which are more specialised in local dispersal by larger vegetative gemmae (Richards 1984), may depend especially on local propagule sources and may therefore be dually affected by habitat fragmentation.

As for vascular epiphytes, selective logging is clearly less disastrous than complete forest clearance, but shade-adapted species from the understorey are still quite vulnerable, as are slow dispersers that thrive only in the canopy of large old trees (Friedel et al. 2006; Holz and Gradstein 2005). Non-vascular epiphytes are generally more host-specific than vascular epiphytes (Cornelissen and Ter Steege 1989; Slack 1976), perhaps due to their closer contact with the substrate. The selective logging of certain tree species could therefore pose a greater threat to non-vascular than to vascular epiphytes.

Where forests are substituted by agriculture, epiphytes may still do relatively well if the crops are sensibly managed woody perennials, such as cacao, coffee, citrus, and some timber species (Pócs 1982; Van Dunne and Wolf 2001). However, plantation trees/shrubs as potential hosts are often relatively young and simply structured and may even be 'cleaned' periodically of epiphytes to avoid real or imagined damage to the yield (Andersson and Gradstein 2005; Sporn et al. 2007). Better refuges are plantations of shade-dependent crops. In traditional systems, cacao, coffee and bananas are often grown under shade trees, which provide a habitat for many organisms including epiphytes. Epiphyte species numbers in such shaded plantations are often comparable to those in primary forests, although the species composition does change and typical primary forest species are often lacking (Hietz

2005; Steffan-Dewenter et al. 2007). The conversion from traditional shade-grown crops to sun-grown varieties is therefore an additional threat to epiphytes (Siebert 2002; Steffan-Dewenter et al. 2007).

3.3 Climatic Change

3.3.1 Warming

Non-vascular epiphytes are already affected by global warming. This is suggested by new and increased occurrences of (sub-)tropical epiphytic lichen species in The Netherlands between 1980 and 2001 (Aptroot and van Herk 2007; van Herk et al. 2002), and by northward and eastward range extensions of Atlantic and Mediterranean bryophyte species in Central Europe over the same period (Frahm 2001). Changes in the lichen flora from other parts of Europe may also be attributable to climatic warming (references in Aptroot and van Herk 2007).

Experimental studies including lichen and bryophyte responses to climatic change have focussed mostly on terrestrial ecosystems at high latitudes. The response of these terrestrial species is largely interdependent on the response of the vascular plants: vascular plant biomass generally increases with warming and shade-intolerant lichens and bryophytes decrease (e.g. Walker et al. 2006). Clearly, such a mechanism is unlikely to play a role for epiphytes.

We know of no observations of the effects of on-going climatic warming on non-vascular epiphyte distributions from the tropics, but two experimental studies show pronounced short-term effects. In a recent study in Bolivia, branches with dense bryophyte cover were transplanted to lower altitudes (from 3,000 to 2,700 and 2,500 m), where the climate was both warmer (+2.5°C at 2,500 m compared to 3,000 m) and drier. After 2 years, the relative abundance of species in the community had changed considerably (Jacome et al., in preparation, in Gradstein 2008). In a similar study in Costa Rica, where epiphyte mats were transplanted, the fate of the abundant mosses and lichens was not quantified (Nadkarni and Solano 2002), but after 1 year a clear deterioration of the entire epiphyte mat was apparent (Nalini Nadkarni, personal communication). However, this experiment covered only 150 altitudinal metres, so that the main change was the lower cloud water input and not temperature.

For poikilohydric plants like bryophytes and lichens even more than for homoiohydric plants, the effects of warming will interact strongly with effects of changed precipitation. Increased temperatures affect poikilohydric organisms in two ways: (1) through direct effects on metabolic rates, and (2) by increasing evaporation, thus reducing activity time or increasing desiccation damage in sensitive species. Increasing temperatures could thus shift species distributions to (previously) cooler as well as moister habitats. But what happens in regions that are already at an extreme end of either temperature or moisture? In tropical lowland forests, no better heat-adapted bryophytes and lichens are available to replace those species that may disappear. In cloud forests, no wetter habitats are available for species to escape to (Fig. 1). This is

Fig. 1 How will climate change affect regions at the extreme ends of temperature or moisture conditions? Expressed here is an example: in tropical lowland forests, no species from warmer zones are available to replace those species that may disappear, while in subalpine forests, no wetter/cooler forest habitats are available for species to move to

a specific example of a much more general issue (Williams et al. 2007): what happens to ecosystems in novel climates and to species in disappearing climates?

The reasons for the limited development of non-vascular epiphytes in tropical lowlands are not known with certainty, but a popular hypothesis involves a combination of high night temperatures causing strong carbon losses to respiration and high day temperatures causing dehydration, thereby restricting the time available for carbon gain by photosynthesis (Zotz 1999). Thus, two crucial factors are night-time temperature and day-time activity time.

CO2-exchange data from tropical lowland bryophytes are not available, while two foliose lowland lichen species have been studied in some detail (Zotz et al. 2003; Zotz and Winter 1994). Diel (24-h) field measurements on *Parmotrema endosulphureum* and *Leptogium azureum* showed consistently that almost the entire daily carbon gain was respired at night. The temperature response of CO_2 exchange in *P. endosulphureum* (Fig. 2a) was used to develop a simple model (Fig. 2b), which analyses how future temperature change would shift this lichen's carbon balance. Assuming typical current lowland conditions, maximum net photosynthesis rates under otherwise optimal conditions are only twice as high as respiration rates (Fig. 2a). The ratio of daily carbon gain to respiration is much lower, because the former is often strongly reduced due to desiccation, while these epiphytes are usually moist and active during the entire night. Currently, *P. endosulphureum* must take up CO_2 at maximum rates for at least 40% of the light period just to balance 12-h nocturnal respiration. A predicted temperature increase of 3°C without acclimatisation would make it necessary to photosynthesise at maximum rates for >90% of the day to achieve a positive carbon balance. This is clearly impossible: even in somewhat cooler montane habitats, lichens show net photosynthesis for only 30–80% of the light period, and at mostly suboptimal rates (Lange et al. 2004). However, assuming

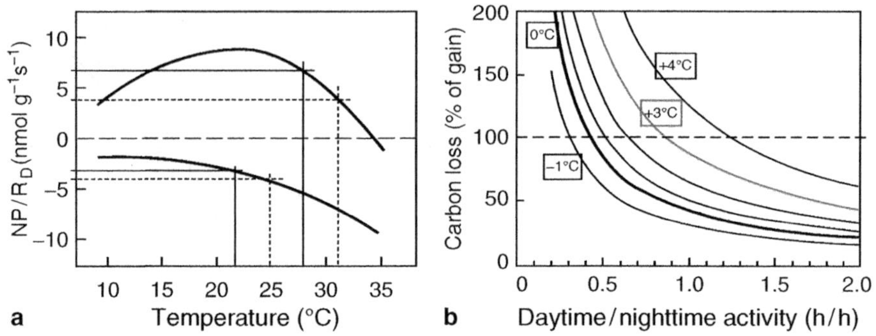

Fig. 2 a Temperature response of net photosynthesis under saturating light and at optimal moisture content (NP, *top graph*) and dark respiration (R_D, *bottom graph*) in *Parmotrema endosulphureum* from lowland Panama, adapted from Zotz et al. (2003). *Thin solid lines* indicate CO_2 exchange rates at current typical tropical lowland temperatures: 28°C day, 22°C night. *Dashed lines* indicate rates at 3°C warmer conditions: R_D then approximately equals NP. (**b**) Percentage of daily carbon gain lost during the night at different temperature conditions (1°C steps, *bold line* = current conditions: 28°C day, 22°C night; +1°C: 29°C day, 23°C night; etc.) as a function of physiological activity time. Note: this graph is based only on the data of plot a) and ignores the effects of suboptimal water contents and light intensities on NP

a shift in the temperature response of 2°C due to acclimatisation, a 3°C increase would effectively represent a 1°C increase, thus requiring photosynthetic activity during "only" 50% of the day for a balanced budget (Fig. 2b). The potential for acclimatisation of tropical lichens and bryophytes is obviously a crucial question, but completely unexplored.

The conclusions are dim: lowland lichens and bryophytes are, arguably, already living close to the edge of their physiological abilities. Thus, even a slight increase in temperature could make tropical lowlands completely uninhabitable to them.

3.3.2 Changes in Rainfall and Mist

The cool and very wet cloud forest habitat appears to be optimal for many non-vascular epiphytes, which is reflected by an enormous epiphytic biomass, up to 20 g per dm² branch area (Frahm 1990; Wolf 1993). Here, the most threatening aspect of climate change is increased dryness, both directly through higher temperatures and, especially, through related phenomena like lower cloud frequencies and a raising of the cloud base altitude (Pounds et al. 1999; Still et al. 1999).

The direct detrimental effect of increased dryness on cloud forest bryophytes and lichens was observed in the above-mentioned transplantation experiment in Costa Rica (Nalini Nadkarni, personal communication), while Jacome et al. (in preparation, in Gradstein 2008) observed only shifts in species dominances. Even in ecosystems less dependent on continuous moisture input, epiphyte species compositions changed drastically when conditions became drier after forest disturbance (Gradstein 1992). We therefore expect that strong responses of cloud forest epiphytes will occur everywhere. Some cloud forest species can be quite tolerant to desiccation (León-Vargas et al. 2006), but others probably are not – this is intuitively true, but unfortunately not a lot of factual information is available to test the assumption. Cloud forests often form the present-day treeline in tropical mountains. A warming-induced upward extension of these forests will probably be extremely slow (Bader et al. 2007), which means that an upward shift of epiphytes is impossible; no wetter epiphyte habitat will be available, and many cloud forest species may face extinction.

In temperate rain forests, another very productive habitat for non-vascular epiphytes, drying would also leave the now abundant epiphytes with no wetter place to go. Whether these forests really will become drier is region-specific: for the Pacific northwest of America drier summers are predicted by the end of the twentyfirst century, in contrast to an increase in rainfall for the New Zealand Alps (Solomon et al. 2007). A lack of habitats to escape to may also come to threaten, for instance, the already rare oceanic epiphyte flora along the west coast of Scotland (Ellis et al. 2007). In general, island ecosystems, be it in water or for instance lowland or desert matrices, will be very vulnerable. Additionally, islands often contain endemic species, which are thus at high risk of extinction (Aptroot, in preparation).

As for vascular epiphytes, increased drought will probably have the greatest effects in now continuously, or at least regularly, moist environments. In habitats with pronounced dry seasons, most bryophytes and lichens will be adapted to drought, and being inactive during drought periods, it will make little difference if these periods are somewhat extended.

3.3.3 Extreme Events

Although the small non-vascular epiphytes are less likely to be knocked off their hosts by strong winds than the larger vascular epiphytes, certain life forms, in particular hanging forms, could be sensitive. One of the few accounts of extreme weather effects on non-vascular epiphytes is that of the strong 1983 El Niño floods on the Galapagos Islands, which caused the waterlogged epiphytic lichens to rot or fall from their substrate (Weber and Beck 1985 in Gradstein 1992). Other effects of extreme weather will often be related to increased forest dynamics (e.g. through windfall and forest fires).

3.4 Direct Effects of Increased CO_2 on Plant Metabolism

The negative effects of warming on the carbon balance of bryophytes and lichens may be at least partly counteracted by increases in atmospheric CO_2 levels. In poikilohydric plants, the inability to regulate water loss limits the possible responses to CO_2 as compared to those of vascular plants, although the trade-offs related to, for instance, nitrogen allocation occur here just as much. In terrestrial bryophytes, the stimulating effect of higher CO_2 levels may be limited, because their close proximity to the respiring soil already exposes them to increased CO_2 levels

(DeLucia et al. 2003). This could also be the case for those epiphytes that grow closely attached to canopy soil or directly on living branches. For such species, increased respiration of the substrate due to higher temperatures may provide more extra CO_2 than the increased atmospheric levels (Sveinbjörnsson and Oechel 1992; Tuba et al. 1999). However, generally, epiphytes are much better coupled to the atmosphere than terrestrial species, so that global atmospheric CO_2 levels can certainly be of influence.

An additional advantage of higher CO_2 levels under a warming scenario is that the temperature optimum for net photosynthesis is shifted to higher temperatures (Sveinbjörnsson and Oechel 1992). Light compensation points are also lowered, so that species can grow at darker microsites, where they dry out more slowly and stay active for longer. Activity time is also prolonged because net photosynthesis can take place down to lower and up to higher water contents (Tuba et al. 1999). This can be particularly beneficial for tropical lowland species, for which we have hypothesised that carbon gain may become negative at higher temperatures due to lower net photosynthesis and through shorter activity times.

As in vascular plants, acclimatisation to high CO_2 levels can cause a return to the low-CO_2 photosynthesis rates. However, species differ in their acclimatisation potential. For instance, the moss *Polytrichum formosum* clearly down-regulated its chlorophyll and RuBisCO contents after several months at 700 ppm CO_2, while the moss *Tortula ruralis* and the lichen *Cladonia convulata* maintained their positive response (Tuba et al. 1999). Increased photosynthesis is also not necessarily invested in growth, for instance lichens may rather invest in phenolic compounds (Tuba et al. 1999). It seems unlikely that the response of epiphytic species to CO_2 would fundamentally differ from that of other bryophytes and lichens; species are probably similarly variable and predictions hard to make.

3.5 Biotic Exchange

Like vascular epiphytes, non-vascular epiphytes appear to have a very low invasive potential. There are few records of neophytic bryophytes or lichens becoming widely established (Söderström 1992) and no published records for epiphytes. We have also found no records of effects of other invasive species (e.g. herbivores, parasites, host trees) on non-vascular epiphytes, unless cultivated exotic trees are included. However, as for vascular epiphytes, the displacement of native trees by introduced tree species could have significant effects through altered substrate quality and microclimate.

3.6 Nitrogen Deposition, Acid Rain and Other Atmospheric Pollution

In contrast to the lack of information on pollution effects on vascular epiphytes, it is well established that distribution patterns of non-vascular epiphytes can be strongly influenced by airborne pollutants such as SO_2, NO_x and heavy metals (Farmer et al. 1992). The occurrence of non-vascular epiphytes in areas where the level of pollution as well as that of research resources are traditionally highest will partly account for this difference. On the other hand, the more "open" physiology of poikilohydric plants also makes them more susceptible to anthropogenic pollutants. Bryophytes and especially lichens are therefore frequently used as indicators for air pollution (Hawksworth and Rose 1970; Szczepaniak and Biziuk 2003), and bryophytes are even starting to be cultivated as pollution absorbers on city roofs and along motorways (Jan Peter Frahm, personal communication). Like vascular epiphytes, non-vascular species differ in their physiological response (Farmer et al. 1992). Some species may even indirectly benefit from air pollution. For instance, the epiphytic lichen *Tuckermanopsis sepincola* develops larger and more fertile thalli when exposed to air pollution because of the decline of its competitor, the lichen *Hypogymnia physodes* (Mikhailova 2007). Also, indirect effects of pollution on bark chemistry favour some species at the expense of others: nitrophytic species increase in response to NO_x pollution, while acidophytic species decline (Van Herk et al. 2003).

In Europe and North America, improved legislation and technical advancements have reduced air pollution compared to the 1970s in many areas, and this is reflected in a recovery of epiphytic mosses and lichens (Gilbert 1992; Van Dobben and De Barker 1996). However, in developing economies like in India and China, with growing industries and motorised traffic, air pollution is still increasing (Fowler et al. 1999; Richter et al. 2005) and could have significant negative effects on epiphyte populations, and especially on bryophytes and lichens (Farmer et al. 1992).

4 Synthesis

All drivers of biodiversity change identified by Sala et al. (2000) are expected to affect epiphytic flora. However, climatic and land use changes are by far the most important processes. A changing climate leaves species with three options (Holt 1990): (1) they remain stationary and evolve in situ, (2) they track appropriate niches spatially and migrate, or else (3) they go extinct. Although local adaptation to climatic change has been demonstrated repeatedly for plants and animals; even during extreme climatic shifts like the Pleistocene glaciations, species apparently did not change beyond their previous ranges of phenotypic variation (Parmesan 2006). Instead, species migrated and did so in a highly individualistic manner (Overpeck et al. 1992). If migration is the prime response to climatic change at ecological timescales, the likely synergism between climate change and habitat fragmentation is the most threatening aspect of the on-going climate change for global biodiversity (Lovejoy and Hannah 2005).

Does this general conclusion apply to epiphytic flora like it does to other species? A number of comparative analyses suggest that vascular epiphytic species tend to be more widespread than their terrestrial counterparts (Ibisch et al. 1996; Kessler 2001), which most likely suggests a higher dispersal ability of this life form. This in turn may make vascular epiphytes less susceptible to the interaction of habitat fragmentation and climate change. Bryophytes and lichens are generally also widely distributed (Schofield 1992), which as a group may make them less susceptible to fragmentation, although at least some epiphytic species are strongly dispersal-limited and these should be affected (Sillett et al. 2000; Snäll et al. 2005).

So are epiphytes really more threatened by climate change, as suggested by earlier reviews (Benzing 1998; Lugo and Scatena 1992)? In habitats which are characterised by continuously high moisture input such as tropical cloud forest, vascular as well as non-vascular epiphytes may indeed be particularly vulnerable. In those habitats, however, which are characterised by recurring seasonal drought such as seasonal lowland forests, epiphytes may actually be similarly or even less affected by severe drought, e.g. during El Niño events, than co-occurring trees. Interestingly, habitat fragmentation also has most disastrous effects on moisture-loving species, while more drought-resistant species sometimes even benefit. Thus, there is a common pattern: both climate change and land use change promote drought-resistant taxa among epiphytes, while the more mesic taxa suffer.

The consequences of temperature changes in the tropics for vascular epiphytes are very uncertain. For non-vascular epiphytes, we expect severe problems for tropical lowland species, because their carbon balance is precarious already and could easily become negative at higher temperatures. At higher latitudes, although still speculative, vascular epiphytes may benefit from rising temperatures by poleward range extension, while for the already widely distributed non-vascular epiphytes range shifts may bring about new species assemblages, as already observed in Europe (Aptroot and van Herk 2007).

Acknowledgements We thank Robbert Gradstein, André Aptroot, Robert Lücking, Jan Peter Frahm, Ryan Mudd, and Nalini Nadkarni for sharing ideas and unpublished material.

References

Acebey A, Gradstein SR, Krömer T (2003) Species richness and habitat diversification of bryophytes in submontane rain forest and fallows of Bolivia. Journal of Tropical Ecology 19:9–18

Andersson MS, Gradstein SR (2005) Impact of management intensity on non-vascular epiphyte diversity in cacao plantations in western Ecuador. Biodiversity and Conservation 14:1101–1120

Aptroot A (in preparation) Lichen as an indicator of climate and global change. In: Letcher TM (ed) Climate and global change: Observed impacts on planet earth

Aptroot A, van Herk CM (2007) Further evidence of the effects of global warming on lichens, particularly those with *Trentepohlia* phycobionts. Environmental Pollution 146:293–298

Archibold OW (1995) Ecology of world vegetation. Chapman & Hall, London

Bader MY, van Geloof I, Rietkerk M (2007) High solar radiation hinders tree establishment above the alpine treeline in northern Ecuador. Plant Ecology 191:33–45

Barlow J et al. (2007) Quantifying the biodiversity value of tropical primary, secondary, and plantation forests. Proceedings of the National Academy of Sciences of the United States of America 104:18555–18560

Barthlott W, Schmit-Neuerburg V, Nieder J, Engwald S (2001) Diversity and abundance of vascular epiphytes: A comparison of secondary vegetation and primary montane rain forest in the Venezuelan Andes. Plant Ecology 152:145–156

Belinchón R, Martínez I, Escudero A, Aragón G, Valladares F (2007) Edge effects on epiphytic communities in a Mediterranean *Quercus pyrenaica* forest. Journal of Vegetation Science 18:81–90
Benzing DH (1990) Vascular epiphytes. General biology and related biota. Cambridge University Press, Cambridge
Benzing DH (1998) Vulnerabilities of tropical forests to climate change: The significance of resident epiphytes. Climatic Change 39:519–540
Benzing DH (2004) Vascular epiphytes. In: Lowman MD, Rinker BH (eds) Forest canopies, 2nd edn. Elsevier, San Diego, pp 175–211
Campbell C, Atkinson L, Zaragoza-Castells J, Lundmark M, Atkin O, Hurry V (2007) Acclimation of photosynthesis and respiration is asynchronous in response to changes in temperature regardless of plant functional group. New Phytologist 176:375–389
Cardelús C, Colwell RK, Watkins Jr JE (2006) Vascular epiphyte distribution patterns: Explaining the mid-elevation richness peak. Journal of Ecology 94:144–156
Cascante-Marin A, Wolf JHD, Oostermeijer JGB, den Nijs JCM, Sanahuja O, Duran-Apuy A (2006) Epiphytic bromeliad communities in secondary and mature forest in a tropical premontane area. Basic and Applied Ecology 7:520–532
Chapman WS, King GC (1983) Floristic composition and structure of rainforest area 25 years after logging. Australian Journal of Ecology 8:415–423
Clark KL, Nadkarni NM, Gholz HL (2005) Retention of inorganic nitrogen by epiphytic bryophytes in a tropical montane forest. Biotropica 37:328–336
Condit R, Hubbell SP, Foster RB (1995) Mortality rates of 205 neotropical tree and shrub species and the impact of severe drought. Ecological Monographs 65:419–439
Cornelissen JHC, Gradstein SR (1990) On the occurrence of bryophytes and macrolichens in different lowland rain forest types of Mabura Hill, Guyana. Tropical Bryology 3:29–35
Cornelissen JHC, Ter Steege T (1989) Distribution and ecology of epiphytic bryophytes and lichens in dry evergreen forest of Guyana. Journal of Tropical Ecology 5:131–150
Cunningham SC, Read J (2003) Do temperate rainforest trees have a greater ability to acclimate to changing temperatures than tropical rainforest trees? New Phytologist 157:55–64
DeLucia EH et al. (2003) The contribution of bryophytes to the carbon exchange for a temperature rainforest. Global Change Biology 9:1158–1170
Diaz HF, Graham NE (1996) Recent changes in tropical freezing heights and the role of sea surface temperature. Nature 383:152–155
Downs RJ (1964) Photocontrol of germination of seed of the Bromeliaceae. Phyton 21:1–6
Drennan PM, Nobel PS (2000) Responses of CAM species to increasing atmospheric CO_2 concentrations. Plant, Cell and Environment 23:767–781
Ellis CJ, Coppins BJ, Dawson TP (2007) Predicted response of the lichen epiphyte *Lecanora populicola* to climate change scenarios in a clean-air region of Northern Britain. Biological Conservation 135:396–404
Farmer AM, Bates JW, Bell JNB (1992) Ecophysiological effects of acid rain on bryophytes and lichens. In: Bates JW, Farmer AM (eds) Bryophytes and lichens in a changing environment. Clarendon Press, Oxford, pp 284–313
Fitter AH, Fitter RSR (2002) Rapid changes in flowering time in British plants. Science 296:1689–1691
Forman RTT (1975) Canopy lichens with blue-green algae: A nitrogen source in a Columbian rain forest. Ecology 56:1176–1184
Foster PN (2001) The potential negative impacts of global climate change on tropical montane cloud forests. Earth Science Reviews 55:73–106
Fowler D et al. (1999) The global exposure of forests to air pollutants. Water, Air, and Soil Pollution 116:5–32
Frahm J-P (1990) Bryophyte phytomass in tropical ecosystems. Botanical Journal of the Linnean Society 104:23–33
Frahm J-P (2001) Bryophytes as indicators of recent climate fluctuations in Central Europe. Lindbergia 26:97–104
Frahm J-P (2002) Epiphytenwahnsinn. Bryologische Rundbriefe 53:7

Frahm J-P (2003) Climatic habitat differences of epiphytic lichens and bryophytes. Cryptogamie Bryologie 24:3–14

Frank JH, Cooper TM, Larson BC (2006) *Metamasius callizona* (Coleoptera: Dryophthoridae): Longevity and fecundity in the laboratory. Florida Entomologist 89:208–211

Friedel A, Oheimb GV, Dengler J, Härdtle W (2006) Species diversity and species composition of epiphytic bryophytes and lichens–A comparison of managed and unmanaged beech forests in NE Germany. Feddes Repertorium 117:172–185

Gentry AH, Dodson CH (1987) Diversity and biogeography of neotropical vascular epiphytes. Annals of the Missouri Botanical Garden 74:205–233

Gilbert OL (1992) Lichen reinvasion with declining air polution. In: Bates JF, Farmer AM (eds) Bryophytes and lichens in a changing environment. Clarendon Press, Oxford, pp 159–177

Graciano C, Fernández LV, Caldiz DO (2003) *Tillandsia recurvata* L. as a bioindicator of sulfur atmospheric pollution. Ecología Austral 13:3–14

Gradstein SR (1992) The vanishing tropical rain forest as an environment for bryophytes and lichens. In: Bates JW, Farmer AM (eds) Bryophytes and lichens in a changing environment. Clarendon Press, Oxford, pp 234–258

Gradstein SR (2008) Epiphytes of tropical montane forests – impact of deforestation and climate change. In: Gradstein SR, Gansert D, Homeier J (eds) The tropical mountain forest – patterns and processes in a biodiversity hotspot. University of Göttingen Press, Göttingen, pp 51–65

Grau O, Grytnes J-A, Birks HJB (2007) A comparison of altitudinal species richness patterns of bryophytes with other plant groups in Nepal, Central Himalaya. Journal of Biogeography 34:1907–1915

Gutschick VP (2007) Plant acclimation to elevated CO_2 – from simple regularities to biogeographic chaos. Ecological Modelling 200:433–451

Hawksworth DL, Rose F (1970) Qualitative scale for estimating sulphur dioxide air pollution in England and Wales using epiphytic lichens. Nature 227:145–148

Hietz P (2005) Conservation of vascular epiphyte diversity in Mexican coffee plantations. Conservation Biology 19:391–399

Hietz P, Briones O (1998) Correlation between water relations and within-canopy distribution of epiphytic ferns in a Mexican cloud forest. Oecologia 114:305–316

Hietz P, Buchberger G, Winkler M (2006) Effect of forest disturbance on abundance and distribution of epiphytic bromeliads and orchids. Ecotropica 12:103–112

Hofstede RGM, Wolf JHD, Benzing DH (1993) Epiphytic biomass and nutrient status of a Colombian upper montane rain forest. Selbyana 14:37–45

Holt RD (1990) The microevolutionary consequences of climate change. Trends in Ecology & Evolution 5:311–315

Holz I, Gradstein RS (2005) Cryptogamic epiphytes in primary and recovering upper montane oak forests of Costa Rica–Species richness, community composition and ecology. Plant Ecology 178:89–109

Ibisch PL, Boegner A, Nieder J, Barthlott W (1996) How diverse are neotropical epiphytes? An analysis based on the "Catalogue of the flowering plants and gymnosperms of Peru". Ecotropica 2:13–28

Jenkins DW (1999) Cold hardiness and cold sensitivity of bromeliads. Journal of the Bromeliad Society 49:32–41

Kantvilas G, Jarman SJ (1993) The cryptogamic flora of an isolated rainforest fragment in Tasmania. Botanical Journal of the Linnean Society 111:211–228

Kessler M (2001) Pteridophyte species richness in Andean forests in Bolivia. Biodiversity and Conservation 10:1473–1495

Knops JMH, Nash Iii TH, Schlesinger WH (1996) The influence of epiphytic lichens on the nutrient cycling of an Oak Woodland. Ecological Monographs 66:159–179

Krömer T, Gradstein SR (2003) Species richness of vascular epiphytes in two primary forests and fallows in the Bolivian Andes. Selbyana 24:190–195

Kuusinen M, Siitonen J (1998) Epiphytic lichen diversity in old-growth and managed *Picea abies* stands in southern Finland. Journal of Vegetation Science 9:283–292

Lange OL, Büdel B, Meyer A, Zellner H, Zotz G (2004) Lichen carbon gain under tropical conditions: Water relations and CO 2 exchange of Lobariaceae species of a lower montane rainforest in Panama. Lichenologist 36:329–342

Lasso E, Ackerman JD (2003) Flowering phenology of *Werauhia sintenisii*, a bromeliad from the dwarf montane forest in Puerto Rico: An indicator of climate change? Selbyana 24:95–104

Laube S, Zotz G (2006) Neither host-specific nor random: Vascular epiphytes on three tree species in a Panamanian lowland forest. Annals of Botany 97:1103–1114

Laube S, Zotz G (2007) A metapopulation approach to the analysis of long-term changes in the epiphyte vegetation on the host tree *Annona glabra*. Journal of Vegetation Science 18:613–624

Laurance WF, Delamônica P, Laurance SG, Vasconcelos HL, Lovejoy TE (2000) Rainforest fragmentation kills big trees. Nature 404:836

Laurance WF, Powell G, Hansen L (2002) A precarious future for Amazonia. Trends in Ecology & Evolution 17:251–252

Lawton RO, Nair US, Pielke RA, Welch RM (2001) Climatic impact of tropical lowland deforestation on nearby montane cloud forests. Science 294:584–587

León-Vargas Y, Engwald S, Proctor MCF (2006) Microclimate, light adaptation and desiccation tolerance of epiphytic bryophytes in two Venezuelan cloud forests. Journal of Biogeography 33:901–913

Lewis SL et al. (2004) Concerted changes in tropical forest structure and dynamics: Evidence from 50 South American long-term plots. Philosophical Transactions of the Royal Society of London Series B - - Biological Sciences 359:421–436

Li CR, Gan LJ, Xia K, Zhou X, Hew CS (2002) Responses of carboxylating enzymes, sucrose metabolizing enzymes and plant hormones in a tropical epiphytic CAM orchid to CO_2 enrichment. Plant, Cell and Environment 25:369–377

Lõhmus P, Rosenvald R, Lõhmus A (2006) Effectiveness of solitary retention trees for conserving epiphytes: Differential short-term responses of bryophytes and lichens. Canadian Journal of Forest Research 36:1319–1330

Loope L, Duever M, Herndon A, Snyder J, Jansen D (1994) Hurricane impact on uplands and freshwater swamp forest: Large trees and epiphytes sustained the greatest damage during Hurricane Andrew. BioScience 44:238–246

Lopez RG, Runkle ES (2005) Environmental physiology of growth and flowering of orchids. Hortscience 40:1969–1973

Lovejoy TE, Hannah L (eds) (2005) Climate change and biodiversity. Yale University Press, New Haven

Lugo AE, Scatena FN (1992) Epiphytes and climate change research in the Caribbean: A proposal. Selbyana 13:123–130

Malcolm JR, Liu C, Neilson RP, Hansen L, Hannah L (2006) Global warming and extinctions of endemic species from biodiversity hotspots. Conservation Biology 20:538–548

Malhi Y, Phillips O (eds) (2005) Tropical forests and global atmospheric change. Oxford University Press, Oxford

McCune B (1993) Gradients in epiphyte biomass in 3 *Pseudotsuga-Tsuga* forests of different ages in western Oregon and Washington. Bryologist 96:405–411

Medeiros AC, Loope LL, Anderson SJ (1993) Differential colonization of epiphytes on native (*Cibotium* spp.) and alien (*Cyathea cooperi*) tree ferns in a Hawaiian rain forest. Selbyana 14:71–74

Mikhailova IN (2007) Populations of epiphytic lichens under stress conditions: Survival strategies. Lichenologist 39:83–89

Monteiro JA, Zotz G, Körner Ch (in press) Tropical epiphytes in a CO_2-rich atmosphere. Acta Oecologica.

Mudd RG (2004) Significance of the epiphyte layer to stem water storage in native and invaded tropical montane cloud forests in Hawai'i. B.Sc. Thesis, Department of Global Environmental Science, University of Hawai'i at Manoa, Honolulu

Nadkarni NM, Solano R (2002) Potential effects of climate change on canopy communities in a tropical cloud forest: An experimental approach. Oecologia 131:580–586

Nascimento HEM, Laurance WF (2004) Biomass dynamics in Amazonian forest fragments. Ecological Applications 14:S127–S138

Norris DH (1987) Long-term results of cutting on the bryophytes of the *Sequoia sempervirens* forest in northern California. Symposia Biologica Hungarica 35:467–473

Nöske NM et al. (2008) Disturbance effects on diversity of epiphytes and moths in a montane forest in Ecuador. Basic and Applied Ecology 9:4–12

Nowak EJ, Martin CE (1995) Effect of elevated CO_2 on nocturnal malate accumulation in the CAM species *Tillandsia ionantha* and *Crassula arborescens*. Photosynthetica 31:441–444

Oberbauer SF, von Kleist K, Whelan KRT, Koptur S (1996) Effects of Hurricane Andrew on epiphyte communities within cypress domes of Everglades National park. Ecology 77:964–967

Ong BL, Koh CK-K, Wee YC (1998) Effects of CO_2 on growth and photosynthesis of *Pyrrosia piloselloides* (L.) price gametophytes. Photosynthetica 35:21–27

Overpeck JT, Webb RS, Webb T (1992) Mapping Eastern North American vegetation change of the past 18 Ka – no-analogs and the future. Geology 20:1071–1074

Parmesan C (2006) Ecological and evolutionary responses to recent climate change. Annual Review of Ecology Evolution and Systematics 37:637–669

Peres CA, Barlow J, Laurance WF (2006) Detecting anthropogenic disturbance in tropical forests. Trends in Ecology & Evolution 21:227–229

Phillips OL et al. (2002) Increasing dominance of large lianas in Amazonian forests. Nature 418:770–774

Pierce S, Winter K, Griffiths H (2002) The role of CAM in high rainfall cloud forests: An *in situ* comparison of photosynthetic pathways in Bromeliaceae. Plant, Cell and Environment 25:1181–1189

Pinheiro Da Costa D (1999) Epiphytic bryophyte diversity in primary and secondary lowland rainforests in southeastern Brazil. Bryologist 102:320–326

Pócs T (1980) The epiphytic biomass and its effect on the water balance of two rain forest types in the Uluguru Mountains (Tanzania, East Africa). Acta Botanica Academiae Scientiarum Hungaricae 26:143–167

Pócs T (1982) Tropical forest bryophytes. In: Smith AJE (ed) Bryophyte ecology. Chapman & Hall, London, New York, pp 59–104

Pounds JA, Fogden MPL, Campbell JH (1999) Biological response to climate change on a tropical mountain. Nature 398:611–615

Pypker TG, Unsworth MH, Bond BJ (2006) The role of epiphytes in rainfall interception by forests in the Pacific Northwest. II. Field measurements at the branch and canopy scale. Canadian Journal of Forest Research 36:832

Richards PW (1984) The ecology of tropical forest bryophytes. In: Schuster RM (ed) New manual of bryology, vol 2. The Hattori Botanical Laboratory, Nichinan, pp 1233–1270

Richardson BA, Rogers C, Richardson MJ (2000) Nutrients, diversity and community structure of two phytotelm systems in a lower montane forest, Puerto Rico. Ecological Entomology 25:348–356

Richter A, Burrows JP, Nüß H, Granier C, Niemeier U (2005) Increase in tropospheric nitrogen dioxide over China observed from space. Nature 437:129–132

Sala OE et al. (2000) Biodiversity – global biodiversity scenarios for the year 2100. Science 287:1770–1774

Schofield WB (1992) Bryophyte distribution patterns. In: Bates JF, Farmer AM (eds) Bryophytes and lichens in a changing environment. Clarendon Press, Oxford, pp 103–130

Siebert SF (2002) From shade- to sun-grown perennial crops in Sulawesi, Indonesia: Implications for biodiversity conservation and soil fertility. Biodiversity and Conservation 11:1889–1902

Sillett SC, McCune B, Peck JE, Rambo TR, Ruchty A (2000) Dispersal limitations of epiphytic lichens result in species dependent on old-growth forests. Ecological Applications 10:789–799

Slack NG (1976) Host specificity of bryophytic epiphytes in Eastern North America. Journal of the Hattori Botanical Laboratory 41:107–132

Snäll T, Ehrlén J, Rydin H (2005) Colonization-extinction dynamics of an epiphyte metapopulation in a dynamic landscape. Ecology 86:106–115

Söderström L (1992) Invasions and range expansions and contractions of bryophytes. In: Bates JF, Farmer AM (eds) Bryophytes and lichens in a changing environment. Clarendon Press, Oxford, pp 131–158
Solomon S et al. (eds) (2007) Climate Change 2007: The Physical Science Basis. Contribution of Working Group I to the Fourth Assessment Report of the Intergovernmental Panel on Climate Change. Cambridge University Press, Cambridge
Sporn SG, Bos MM, Gradstein SR (2007) Is productivity of cacao impeded by epiphytes? An experimental approach. Agriculture, Ecosystems and Environment 122:490–493
Steffan-Dewenter I et al. (2007) Tradeoffs between income, biodiversity, and ecosystem functioning during tropical rainforest conversion and agroforestry intensification. Proceedings of the National Academy of Sciences of the United States of America 104:4973–4978
Still CJ, Foster PN, Schneider SH (1999) Simulating the effects of climate change on tropical montane cloud forests. Nature 398:608–610
Sveinbjörnsson B, Oechel WC (1992) Controls on growth and productivity of bryophytes: Environmental limitations under current and anticipated conditions. In: Bates JW, Farmer AM (eds) Bryophytes and lichens in a changing environment. Clarendon Press, Oxford, pp 77–102
Szczepaniak K, Biziuk M (2003) Aspects of the biomonitoring studies using mosses and lichens as indicators of metal pollution. Environmental Research 93:221–230
Tejedor A, McAlpin BW (2000) *Ophioglossum pendulum* L. naturalized in Miami, Dade County, Florida. American Fern Journal 90:46–47
Timmermann A, Oberhuber J, Bacher A, Esch M, Latif M, Roeckner E (1999) Increased El Niño frequency in a climate model forced by future greenhouse warming. Nature 398:694–697
Tuba Z, Proctor MCF, Takács Z (1999) Desiccation-tolerant plants under elevated air CO_2: A review. Zeitschrift fur Naturforschung - – Section C Journal of Biosciences 54:788–796
Urban O (2003) Physiological impacts of elevated CO_2 concentration ranging from molecular to whole plant responses. Photosynthetica 41:9–20
Van Dobben HF, De Bakker AJ (1996) Re-mapping epiphytic lichen biodiversity in the Netherlands: Effects of decreasing SO2 and increasing NH3. Acta Botanica Neerlandica 45:55–71
Van Dunne HJF, Wolf JHD (2001) Development of epiphytic bryophyte and lichen vegetation on plantation coffee trees. In: Van Dunne HJF (ed) Epiphytes in secondary tropical rain forests. Ph.D. Thesis, University of Amsterdam, pp 95–111
van Herk CM, Aptroot A, van Dobben HF (2002) Long-term monitoring in the Netherlands suggests that lichens respond to global warming. Lichenologist 34:141–154
Van Herk CM, Mathijssen-Spiekman EAM, De Zwart D (2003) Long distance nitrogen air pollution effects on lichens in Europe. Lichenologist 35:347–359
Vaz APA, Figueiredo-Ribeiro RDL, Kerbauy GB (2004) Photoperiod and temperature effects on in vitro growth and flowering of *P. pusilla*, an epiphytic orchid. Plant Physiology and Biochemistry 42:411–415
Veneklaas EJ, Zagt RJ, Van Leerdam A, Van Ek R, Broekhoven AJ, Van Genderen M (1990) Hydrological properties of the epiphyte mass of a montane tropical rain forest, Colombia. Vegetatio 89:183–192
Walker MD et al. (2006) Plant community responses to experimental warming across the tundra biome. Proceedings of the National Academy of Sciences of the United States of America 103:1342–1346
Wannaz ED, Pignata ML (2006) Calibration of four species of *Tillandsia* as air pollution biomonitors. Journal of Atmospheric Chemistry 53:185–209
Williams JW, Jackson ST, Kutzbach JE (2007) Projected distributions of novel and disappearing climates by 2100 AD. Proceedings of the National Academy of Sciences of the United States of America 104:5738–5742
Williams-Linera G, Sosa V, Platas T (1995) The fate of epiphytic orchids after fragmentation of a Mexican cloud forest. Selbyana 16:36–40
Williamson GB et al. (2000) Amazonian tree mortality during the 1997 El Niño drought. Conservation Biology 14:1538–1542

Winter K, Smith JAC (1996) An introduction to crassulacean acid metabolism: Biochemical principles and biological diversity. In: Winter K, Smith JAC (eds) Crassulacean acid metabolism. Biochemistry, ecophysiology and evolution. Springer, Berlin, pp 1–13
Wolf JHD (1993) Diversity patterns and biomass of epiphytic bryophytes and lichens along an altitudinal gradient in the northern Andes. Annals of The Missouri Botanical Garden 80:928–960
Wolf JHD (1994) Factors controlling the distribution of vascular and non-vascular epiphytes in the northern Andes. Vegetatio 112:15–28
Wolf JHD (2005) The response of epiphytes to anthropogenic disturbance of pine-oak forests in the highlands of Chiapas, Mexico. Forest Ecology and Management 212:376–393
Zotz G (1999) Altitudinal changes in diversity and abundance of non-vascular epiphytes in the tropics – an ecophysiological explanation. Selbyana 20:256–260
Zotz G (2005) Vascular epiphytes in the temperate zones – a review. Plant Ecology 176:173–183
Zotz G, Hietz P (2001) The ecophysiology of vascular epiphytes: Current knowledge, open questions. Journal of Experimental Botany 52:2067–2078
Zotz G, Schmidt G (2006) Population decline in the epiphytic orchid, *Aspasia principissa*. Biological Conservation 129:82–90
Zotz G, Winter K (1994) Photosynthesis and carbon gain of the lichen, *Leptogium azureum*, in a lowland tropical forest. Flora 189:179–186
Zotz G, Schultz S, Rottenberger S (2003) Are tropical lowlands a marginal habitat for macrolichens? Evidence from a field study with *Parmotrema endosulphureum* in Panama. Flora 198:71–77
Zotz G, Laube S, Schmidt G (2005) Long-term population dynamics of the epiphytic bromeliad, *Werauhia sanguinolenta*. Ecography 28:806–814

Resolving the Dryland Decomposition Conundrum: Some New Perspectives on Potential Drivers

Heather L. Throop(✉) and Steven R. Archer

Contents

Abstract Decomposition of organic matter is a crucial component of biogeochemical cycles that strongly controls nutrient availability, productivity, and community composition. The factors controlling decomposition of litter in arid and semi-arid systems remain poorly understood, with an unresolved disconnect between measured and modeled decay rates. In contrast, decay rates in mesic systems are generally quite successfully predicted by models driven by climatic variables. Here, we explore the reasons for this disconnect by reviewing literature on the biotic and abiotic controls over dryland decomposition. Recent research on decomposition in drylands suggests that several key drivers of dryland decomposition have been historically overlooked and not included in models. In particular, UV photodegradation and soil transport processes, both a function of vegetation structure, may strongly influence dryland decomposition dynamics. We propose an expanded framework for studying dryland decay that explicitly addresses vegetation structure and its influence on decomposition. Spatial heterogeneity of vegetation in dryland systems necessitates considering how the spatial and temporal context of vegetation influences soil transport patterns and UV photodegradation, both of which may in turn affect abiotic and biotic decomposition processes.

H.L. Throop
Department of Biology, New Mexico State University, Las Cruces, NM 88005, USA
e-mail: throop@nmsu.edu

U. Lüttge et al. (eds.), *Progress in Botany 70*,
DOI:10.1007/978-3-540-68421-3,

1 Introduction

The concentration and distribution of nutrients in the soil fundamentally affect plant community composition and productivity. In most circumstances, the proximate source of nutrients for plant growth is the decomposition of organic matter. Via influences on soil fertility, decomposition rates may affect plant community composition and production, which in turn, feed back to affect the timing, quality and quantity of litter inputs (Aerts 1997b; Hobbie 1992). The litter and soil organic matter pools account for a large portion of C in terrestrial ecosystems (Schimel 1995; Schlesinger 1997), and the relatively rapid turnover of the leaf litter pool makes leaf litter decomposition one of the most dynamic components of biogeochemical cycles (Aerts 1997a). As such, understanding patterns and controls over decomposition rates are crucial to understanding C and nutrient balance.

The controls and dynamics of litter decomposition in arid and semi-arid ecosystems (hereafter "drylands"), are considerably less well understood than in mesic systems. This discrepancy may reflect the paucity of research in drylands relative to mesic systems and/or failure to include or appropriately represent key drivers. The fact that decomposition rates in mesic systems are reasonably well predicted by climatic variables, while measured decomposition rates in drylands are typically much greater than would be predicted based on climatic variables (Meentemeyer 1978; Whitford et al. 1981; Parton et al. 2007), suggests the latter. The mechanisms responsible for this disconnect are unclear. Here, we (1) review literature on controls over dryland decomposition, (2) discuss mechanisms related to UV photodegradation and soil transport processes that may account for discrepancies between predicted and measured decomposition rates in drylands, and (3) propose an expanded framework for studying decomposition in dryland systems.

2 Patterns of Dryland Decomposition

Decomposition rates are typically determined by quantifying differences in mass loss from mesh litterbags deployed in the field for varying lengths of time. As such, measured litter decay embodies a set of ecological processes spanning from comminution and fragmentation to mineralization. Decay rates can be expressed either in terms of percent mass loss or with the decay constant, K, which is the exponent of a single exponential decay model and thus represents mass loss over time (Olson 1963). Dryland decomposition is generally slower and more variable than that in mesic systems, reflecting variability in both litter composition and environmental conditions Whitford (2002).

Our mechanistic understanding of controls over decomposition has progressed (Meentemeyer 1978; Couteaux et al. 1995; Aerts 1997a), but predicting decomposition dynamics in dryland systems remains problematic (Kemp et al. 2003; Parton et al. 2007). Direct abiotic forces broadly predict decomposition rates, with regional/global patterns predicted by climate variables such as actual evapotranspiration (Aerts 1997a; Meentemeyer 1978). Decay rates over 5- and 10-year periods in the Long-term Intersite Decomposition Experiment Team (LIDET) study, in which common litter was tracked across sites in North and Central America, were best predicted with synthetic indices of climate variables (Gholz et al. 2000; Parton et al. 2007). Models based on climatic drivers consistently underestimate decay rates in drylands, suggesting decomposition controls differ from those in mesic systems (Parton et al. 2007). At local scales, decomposition is sensitive to initial litter chemistry (e.g., C:N ratios and lignin content; Hobbie 1992) and to indirect influences of vegetation (e.g., plant structural influences on microclimate; Mack and D'Antonio 2003). It is not yet clear how to resolve these broad- and fine-scale perspectives.

Recent research has advanced our understanding of the driving forces in organic matter decay in drylands. It appears that mechanisms driving decomposition in drylands may fundamentally differ from those operating in mesic systems. As such, different approaches may be needed to quantify cause–effect relationships and to develop robust generalizations.

3 Drivers of Decomposition

Local drivers of decomposition are either biotic (e.g., litter quality and decomposer organisms) or abiotic (e.g., temperature, moisture, UV radiation). These interact and may affect decomposition either directly or indirectly (e.g., soil moisture may affect decomposition via mediating decomposer community composition; Fig. 1). Although the prevailing dogma has been that the primary drivers in drylands are abiotic and that the importance of biotic drivers is reduced relative to mesic systems (MacKay et al. 1994; Osler et al. 2004), both biotic and abiotic drivers can strongly influence dryland decomposition rates.

3.1 Abiotic Drivers

3.1.1 Temperature

Temperature may affect decomposition via influences on enzyme kinetics or the activities and populations of decomposer organisms. Temperature is an important regulator of decomposition in some temperate and high latitude systems (Aerts

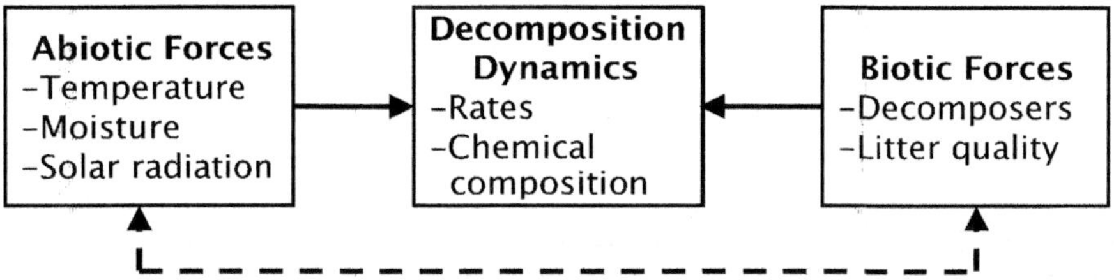

Fig. 1 Direct and indirect drivers of decomposition as traditionally addressed by decomposition studies. Biotic and abiotic facts are assumed to affect decomposition dynamics either directly (*solid lines*) or indirectly by affecting other processes (*dashed lines*)

2006; Hobbie 1996; McHale et al. 1998), and strong links between temperature and microbial respiration (Kirschbaum 2000) suggest temperature should strongly affect biologically-driven decomposition when moisture is not limiting. However, because optimal temperatures for microbial activity are often exceeded during the infrequent periods of growing season moisture availability in drylands, temperature-driven responses may be truncated (MacKay et al. 1986). Few studies have directly studied the relationship between temperature and decomposition in drylands. A shade treatment in the Chihuahuan Desert that decreased surface temperatures stimulated microarthropod populations, but this did not translate into subsequent changes in decay rates (MacKay et al. 1986). Temperature could also influence decomposition via its affect on near-surface soil and litter moisture content.

3.1.2 Precipitation/Moisture Availability

Precipitation and soil moisture availability also control decomposition in many systems (Austin and Vitousek 2000; McCulley et al. 2005), with moisture influencing the activities and population dynamics of decomposers. In addition, the moisture-limited nature of drylands may influence macro-decomposers (see Sect. 3.2.2.). Substantial physical decomposition can occur by leaching of water-soluble compounds (Sullivan et al. 1999) and, although seldom quantified, raindrop impact may cause substantial physical fragmentation of litter (Whitford 2002). Low annual precipitation in drylands suggests decomposition from these processes would be lower than in mesic systems, although moisture pulses associated with high-intensity convective storms typical of drylands may cause episodes of high litter fragmentation and leaching. Furthermore, standing dead may persist for long periods in drylands before it is transferred to surface litter pools, allowing considerable opportunities for leaching and fragmentation losses from raindrop impact.

Biological processes affecting dryland decomposition are generally limited by moisture availability. Two types of observational studies have been used to investigate moisture influences: decomposition in response to variation in rainfall patterns within a site and simultaneous litterbag deployments across a precipitation gradient. Studies assessing the role of within-site precipitation variability have observed decay rates in litterbags deployed at different times of the year or across years that differ in precipitation regime. Such studies have universally demonstrated a positive relationship between precipitation and decomposition rate (Ekaya and Kinyamario 2001; Pucheta et al. 2006; Strojan et al. 1987; Weatherly et al. 2003). Although such studies are confounded by intra-annual variability in temperature or decomposer communities, the consistent positive responses suggest local decay rates are limited by precipitation. Similarly, there was a positive correlation between decay rate and soil moisture for buried roots along a moisture gradient on a Chihuahuan Desert bajada (Mun and Whitford 1998). Manipulative studies using rainout shelters to reduce precipitation or sprinklers to increase precipitation have also demonstrated positive relationships between precipitation and decay rates (Brandt et al. 2007; Whitford et al. 1986; Yahdjian et al. 2006). The influence of precipitation may also depend on the stage of litter decomposition. In the Chihuahuan Desert, decomposition rates did not respond to altered precipitation until after 19 months in a 41-month study (Kemp et al. 2003).

Drylands also differ from mesic systems with respect to rainfall distribution, with rainfall pulses strongly controlling biological activities. Changes in the frequency, distribution, or size of rainfall events more strongly regulate ecological processes in drylands than does mean annual precipitation (Schwinning et al. 2004). The role of rainfall pattern was demonstrated with a Chihuahuan Desert study in which supplementation of 25 mm per month increased decomposition rates when added in 4 weekly aliquots, but not when added as a single monthly aliquot (Whitford et al. 1986). While links between precipitation distribution and decomposition have not been explicitly addressed, precipitation pulses are known to strongly control soil respiration and soil C fluxes (Huxman et al. 2004). We expect that decomposition should respond similarly.

3.2 *Biotic Drivers*

Biotic properties affecting decomposition include litter quality and the decomposer community present. While biotic drivers strongly affect decay rates in many mesic systems, they appear to be relatively less important in drylands (MacKay et al. 1994; Osler et al. 2004). In particular, litter fragmentation is primarily the result of decomposers in mesic systems, whereas abiotic processes appear to be more important in drylands (Whitford 2002). However, decomposer organisms can strongly affect litter breakdown via litter fragmentation and chemical transformations. The organisms present on decomposing material will vary with litter chemistry, microsite, temperature and moisture. Succession in decomposer community composition likely occurs as decomposition progresses from litter fragmentation through mineralization.

3.2.1 Litter Quality

Local rates of decomposition may be strongly controlled by the initial chemical composition of litter with N, lignin, and lignin:N being particularly important predictors of decomposition rates (Hobbie 1992). If the importance of biotic processes as decomposition drivers is lower in dryland than mesic systems (Whitford 2002), the importance of litter quality may be similarly reduced. However, there is little evidence in support of this inference. Consistent with studies in mesic systems, dryland studies quantifying litter quality and decomposition rates have also generally found positive influences of N and P and negative influences of lignin on decomposition rates (Table 1).

Table 1 Relationships between litter chemistry and decomposition rates in dryland studies

Litter chemistry variables	Influence on *K*	Study length	Location	Study species	References
Lignin	None	1 year	Chihuahuan Desert	*Yucca elata*	Schaefer et al. 1985
Lignin:N	None			*Larrea tridentata*	
C:N	None			*Flouresnia cernua*	
				Prosopis glandulosa	
				Chilopsis linearis	
				Mixture of annuals	
Lignin	–	21 months	Argentinean semi-arid grassland	*Poa ligularis*	Moretto et al. 2001
N	+			*Stipa gyneriodes*	
P	+			*Stipa tenuissima*	
C:N	–				
Lignin:N	–				
Lignin:P	–				
Lignin	–	2 years	Argentinean semi-arid grassland	*Poa ligularis*	Moretto and Distel 2003
N	+			*Stipa gyneriodes*	
P	+				
C:N	–				
Lignin:N	–				
Lignin:P	–				
Lignin	–	1 year	Sonoran Desert, Mexico	*Encelia farinose*	Martínez-Yrízar et al. 2007
N	–			*Olneya tesota*	
Lignin:N	–				
N	+	1 year	Sonoran Desert, Arizona	*Prosopis velutina*	Throop and Archer 2007
C:N	–			*Eragrostis lehmanniana*	

For each study, the decay rates (*K*) for at least two species with differing litter chemistry were compared. All studies are for leaf litter exposed on the soil surface in mesh litterbags

3.2.2 Decomposer Organisms

The role of organisms as drivers of decomposition and the identity of the functional groups involved are poorly understood and highly variable among drylands. Different suites of decomposers have very different impacts on litter breakdown, with macrofauna (2–20 mm; e.g., isopods, beetles) and mesofauna (100 μm–2 mm; e.g., mites, collembolans) generally involved with comminution of litter of different particle sizes, and microfauna (<100 μm; e.g., bacteria, fungi, nematodes) involved in chemical transformation of progressively smaller litter particles and molecules. Targeted biocides (Table 2) and litterbags with mesh sizes that discriminate based

Table 2 Summary of dryland studies using biocide treatments to explore decomposer impacts on decay rates

Treatments	Influence on *K*	Location	Time (months)	References
Biocide	–	Colorado short-grass steppe	9	Vossbrinck et al. 1979
Insecticide	–	Chihuahuan Desert[a]	1	Santos et al. 1981
Fungicide + insecticide	– –[b]			
Fungicide + insecticide + nematicide	– –[b]			
Insecticide	–	Chihuahuan Desert	6	Elkins and Whitford 1982
Biocide	–	Chihuahuan Desert[a]	4	Moorhead and Reynolds 1989
Fungicide	0	Chihuahuan Desert	5	MacKay et al. 1994
Insecticide	0			
Biocide	–			
Fungicide + Bacteriacide	0	Patagonian steppe	18	Austin and Vivanco 2006

Treatment influence on decomposition is expressed as the change in the decay constant, *K*, for treatments relative to the control. A decrease in K relative to the control is indicated by –; no influence is indicated by 0

[a] Buried litter

[b] *K* varied with treatments, control > insecticide > fungicide + insecticide = fungicide + insecticide + nematicide

on the size of decomposers have been used to assess the relative contribution of functional groups on decomposition.

The influence of macrofauna on dryland decomposition is highly variable in space and time. Macrofauna can be more important decomposers in drylands than mesic systems, with subterranean termites consuming >50% of the annual NPP in some dryland ecosystems (Johnson and Whitford 1975; Whitford et al. 1982; Silva et al. 1985) and redistributing large quantities of litter from the soil surface into subsurface galleries. In semi-arid Botswana, mass removal of wood was positively associated with the presence of termites that translocated surface litter to subterranean galleries where it is colonized by fungi (Schuurman 2005). In contrast, mass removal rates were not affected by the presence of non-fungus growing termite species. Similarly, leaf-cutter ants (*Atta* spp. and *Acromyrmex* spp.) may move considerable amounts of litter to subsurface fungus gardens in some dryland systems (Tadey and Farji-Brener 2007; Wetterer et al. 2001). Along with termites, colonization by tenebrionid beetle larvae and Thysanurans occurred on buried filter paper and cotton cloth in the Namib Sand Sea during dry periods (Jacobson and Jacobson 1998). However, macrofauna common in mesic systems (e.g., Annelids) are typically not abundant in drylands.

Mesofauna may also be important decomposers in drylands, although relationships between mesofaunal abundance and decay rates may be quite complex.

Several studies in the Chihuahuan Desert have documented a decrease in decomposition rates with chemical exclusion of microarthropods, particularly mites, although these exclusions may be the result of changing abundance of other decomposer organisms, such as bacteria and nematodes (Elkins and Whitford 1982; Santos et al. 1981). In contrast, other studies have found no microarthropod effects on decomposition (Silva et al. 1985). These discrepancies may simply reflect interannual variation and fluctuations in microarthropod abundance. When present, microarthopods may exert particularly strong indirect control over decomposition by regulating decomposer communities. For example, Santos et al. (1981) attributed the 40% decrease in decay rates for insecticide-treated litterbags to the elimination of predatory mites that allowed populations of bacteria-feeding nematodes to increase.

Experimental tests of the roles of microfauna activity in decomposition using chemical inhibitors have documented neutral to positive impacts of microbes on dryland decomposition rates (Table 2) in contrast to their more consistently positive impact in mesic systems. Strong temporal and spatial heterogeneity in the role of microbes in dryland decomposition are likely, however, given variation in soil moisture. For example, in the hyperarid Namib Sand Sea, decomposition rates of buried material were driven by precipitation-induced fungal colonization (Jacobson and Jacobson 1998).

3.3 Novel Drivers in Drylands

Decomposition studies in drylands have typically focused on controlling variables known to be important in more mesic systems (e.g., temperature, precipitation, soil or litter nutrient status, and microbial community), but robust generalizations for dryland decomposition remain elusive. Several recent studies suggest that mechanisms involved in dryland litter decay may differ substantially from that of mesic systems. A crucial distinction between drylands and mesic systems is the spatial heterogeneity of plant canopies in drylands; this difference in canopy cover may be at least in part responsible for some of these observed differences in driving mechanisms. Studies of canopy structure influence on decomposition in mesic systems have documented positive, negative, and neutral influences (Edmonds 1979; Binkley 1984; Zhang and Zak 1995; Hope et al. 2003). This variation in response may be due to system-specific differences in the nature of canopy influences on microclimate and subsequent microbial activity. In dryland systems, woody plant canopies may alter soil water availability by affecting canopy interception, stemflow, throughfall, evapotranspiration, and hydraulic redistribution, or decrease the intensity of solar radiation (including UV) and alter soil surface temperatures relative to inter-canopy areas (Breshears et al. 1997; Schlesinger and Pilmanis 1998; Zou et al. 2005) to influence decomposition dynamics. Furthermore, microbial pools are likely to be concentrated in subcanopy area with high organic matter relative to inter-canopy spaces (McCulley et al. 2004; Smith et al. 1994).

3.3.1 UV Photodegradation

Photodegradation by ultraviolet (UV) radiation was proposed over 40 years ago as a mechanism for litter breakdown in environments receiving high inputs of solar radiation (Pauli 1964). UV-B radiation is known to enhance decay of dissolved organic carbon and nitrogen compounds in aquatic systems (Zepp et al. 2007). Research on photodegradation of wood and paper products indicates absorption of radiative energy by lignin, cellulose, and hemicellulose leads to the formation of free and peroxy radicals (reviewed in Moorhead and Callaghan 1994). In addition to UV-B, shorter energy wavelengths (UV-A and visible light) may also drive photodegradation (Schade et al. 1999). While there are clear mechanisms by which photodegradation may occur, its contribution to dryland litter decomposition is currently not well understood.

The possibility of photodegradation in litter decomposition has led many to suggest this as an important driving mechanism in dryland systems with high radiative loads. Failure to explicitly account for photodegradation has been posited as an explanation for the disconnect between measured and modeled decay rates (Moorhead and Callaghan 1994; Parton et al. 2007; Whitford 2002). Photodegradation was also posited as the cause for decay in the absence of biotic activity following biocide application (MacKay et al. 1994) and for an unexpected positive correlation between initial litter lignin concentration and mass loss in a field study with six litter types in the northern Chihuahuan Desert (Schaefer et al. 1985). While the combination of greater solar radiation in most drylands and less total radiative interception by canopies suggest that photodegradation could be relatively more important in drylands than mesic systems, few studies have explicitly quantified UV impacts in drylands.

Three recent studies have offered, to our knowledge, the first manipulative tests of the role of photodegradation in low- to mid-latitude drylands. Each of these studies compared decay rates of leaf litter under UV-absorbing filters with those under UV-transmitting filters (Table 3). In the semi-arid Patagonian Steppe, Austin and Vivanco (2006) assessed decay of mixed grass litter in plastic-sided "litterboxes" covered with differing films to create three UV environments and found strong positive correlations between radiative load and decay rates. Similarly, Day et al. (2007) found greater rates of *Larrea tridentata* leaf mass loss under near ambient than reduced UV-B. Leaves exposed to near ambient UV-B also had greater mass loss of lignin, carbon, fats and lipids from than those in reduced UV-B settings. Finally, in the semi-arid Colorado shortgrass steppe, Brandt et al. (2007) found slower decay rates in litterbags deployed under shelters in which the majority of UV-A and UV-B was blocked. However, mass loss was greater in the near-ambient relative to reduced UV treatment only under dry conditions for litter with an initially high C:N. In this study, UV radiation enhanced loss of holocellulose, but did not affect lignin loss. Taken together, these manipulative studies provide strong support for the notion that UV radiation can accelerate litter breakdown under field conditions.

Photodegradation has also been invoked as a possible mechanism to explain microsite variation in decay rates in drylands, although evidence is not as clear as for the manipulative studies discussed above. Observations of greater decay

Table 3 Low- and mid-latitude dryland field studies using UV filters to manipulate solar energy environment

Radiation manipulation	Location	Mass loss from radiation	Litter chemistry variables measured	Chemical response to radiation	Time (months)	Area:litter mass ratio (cm^2:g)	References
Reduced UV-B	Patagonian steppe	+33%[a]	None		18	200	Austin and Vivanco 2006
Reduced total radiation	Patagonian steppe	+60%[a]	None		18	200	Austin and Vivanco 2006
Reduced UV-A and UV-B	Colorado short-grass steppe	5%	Holocellulose	+	36	33	Brandt et al. 2007
			Lignin	NC			
Reduced UV-B	Sonoran Desert, Arizona	+14–22%	Total organic C	NC	4 and 5	56	Day et al. 2007
			Total organic N	NC			
			C:N	NC			
			Lignin loss	+			
			Fats and lipids loss	+			
			Holocellulose loss	–			

Decomposition responses were investigated in terms of mass loss and litter chemistry. NC indicates no change in that response variable

[a]Change in the decay constant, *K*

constants on bare soil microsites compared to under shrub canopy microsites in desert scrub in Mexico were attributed to higher solar radiation on bare soil microsites (Arriaga and Maya 2007). However, there were no explicit measurements of radiation or other variables that might differ between the two microsites. Similarly, Martínez-Yrízar et al. (2007) attributed higher decay rates in "plains" sites relative to "hillslope" and "arroyo" sites in the Sonoran Desert to differences in UV photodegradation and termite colonization. Although the plains site had less aboveground biomass and only widely-spaced canopy cover, evidence for the role of UV was again circumstantial.

The risk of attributing microsite differences in decay rates to solar radiation while other factors co-vary was illustrated by a recent study of decomposition in an Arizona semi-desert grassland (Throop and Archer 2007). In a 1-year litterbag study, decay constants were significantly higher for bags deployed in bare soil microsites compared to under intact mesquite (*Prosopis velutina*) shrub canopies. However, decomposition rates were equally depressed under intact canopies as in sites where canopies were mechanically removed immediately prior to the start of the experiment. Solar radiation alone therefore did not explain differential decay rates in this study, with decay rates for litterbags in high UV environments spanning the range of K values observed in the study (Fig. 2). Definitive evidence for the role

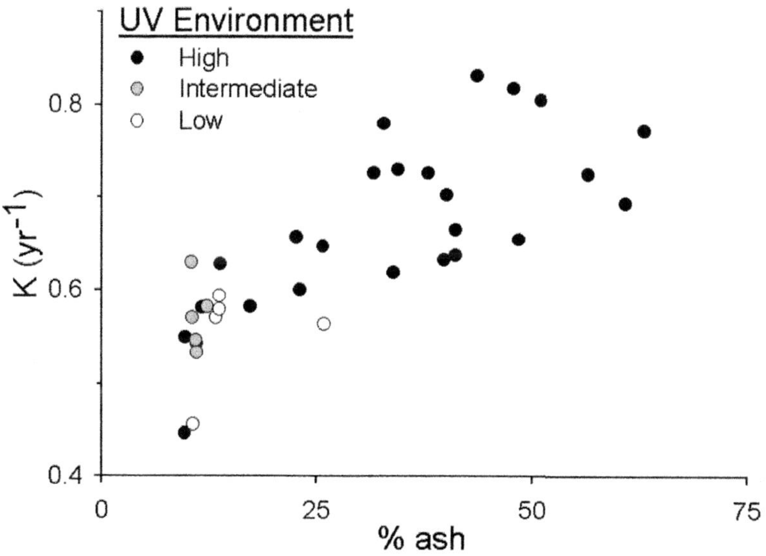

Fig. 2 Decomposition of mesquite leaflets (represented by K, the decay constant) in the Sonoran Desert, USA was strongly and positively correlated with soil deposition into litterbags (as indicated by % ash; Throop and Archer 2007). K varied two-fold, but not necessarily as a simple function of UV environment (high = open areas with no plant copy cover; low = under shrub canopies where light was attenuated ~55%). *High* UV environments are those with no woody canopy cover, *Intermediate* UV environments are under recently girdled mesquite shrubs, and *Low* are under mesquite shrubs with intact canopies

of photodegradation in dryland litter decay requires explicit manipulative experiments.

An alternative to field reductions of ambient radiative loads is experimentally exposing litter to differing radiative treatments with UV-emitting bulbs. In a 6-month experiment with no radiation or 12 h per day of radiation from UV-A and UV-B emitting bulbs, decay rates did not differ among radiation treatments for either juniper (*Juniperus monosperma*) or piñon pine (*Pinus edulis*) litter (Gallo et al. 2006). Changes in chemical composition appear to have been driven by moisture treatments rather than UV exposure. Several high latitude studies have also used UV-emitting bulbs, and the majority have found no significant UV effects on mass loss (Table 4). While manipulative experiments with artificial radiation sources provide highly controlled experimental conditions, a challenge with using experimental radiation sources is that these systems often do not realistically mimic natural solar radiation. Ratios of wavelengths that affect biological processes and photochemical reactions in natural systems may differ from those present in artificial sources (Caldwell and Flint 1997). Total radiation may differ as well, with Gallo et al. (2006) estimating that microcosm UV treatments provided only a third of the midday UV intensity that would have been experienced at their field site.

Although recent field manipulative studies provide evidence that UV photodegradation *can* play a role in drylands (Austin and Vivanco 2006; Brandt et al. 2007; Day et al. 2007), there is still considerable uncertainty regarding the specific biochemical pathways at play and the relative importance of photodegradation as a driver of dryland decomposition. Field experiments have attributed mass losses due to photodegradation from 5 to 60% (Table 3). Why might the results of these studies be so different? First, the study species varied, with two studies using grass litter and one study using material from an evergreen shrub. It is likely that tissues from different species vary considerably in susceptibility to photodegradation, with greatest rates of photodegradation likely in species that have high concentrations of photo-absorptive compounds such as lignin. Further, species may differ in the degree to which the cuticle provides protection from photodegradation, and the persistence of the cuticle after leaf senescence. The relative importance of photodegradation may also vary with time, depending on the susceptibility of chemical constituents present and the succession and activity of decomposer organisms. The field manipulative studies outlined above ranged from 4 to 36 months of UV exposure.

Another confounding factor in comparing study results is among-study variation in the wavelengths manipulated. By employing a treatment that blocked nearly all solar radiation, Austin and Vivanco (2006) found that although UV-B was responsible for the approximately half of the observed photodegradation, other wavelengths, including photosynthetically active radiation (400–700 nm), also significantly affected mass loss. Thus, the UV-specific filters used by Brandt et al. (2007) and Day et al. (2007) may underestimate total photodegradation effects as they exclude shorter wavelengths. However, blocking all solar radiation may have the complication of attenuating temperature (Austin and Vivanco 2006). Total solar radiation also varied among sites, with the higher latitudes of Colorado (40°49′N) and Patagonia (45°41′S)

Table 4 Growth chamber and high latitude studies investigating the influence of manipulated UV radiation on litter chemistry and mass loss

Radiation manipulation	Location	Litter chemistry variables measured	Chemical response to radiation	Mass loss from radiation	Time (months)	Area:litter mass ratio (cm^2:g)	References
No radiation vs UV-B irradiated	Growth chambers; Sweden	Lignin	–	None	2	Information not provided	Gehrke et al. 1995
		α-cellulose	–				
		Soluble carbohydrates	NC				
		Tannins	NC				
Enhanced UV-A; Enhanced UV-A and UV-B	Outdoor UV exposure bed; United Kingdom	C loss	–	None	16	125 (estimate)	Newsham et al. 1997
		N	NC				
		Lignin	NC				
		Holocellulose	NC				
		Ca	NC				
		Mg	NC				
		K	NC				
		P	NC				
Enhanced UV-B	The Netherlands	Soluble carbohydrates	NC	+8%[a]	2	36	Rozema et al. 1997
		Starch	NC				
		α-cellulose	NC				
		Tannins	NC				
		C	NC				
		Hemicellulose loss	+				
		Lignin Loss	+				
		N loss	+				
Enhanced UV-A; Enhanced UV-A and UV-B	The Netherlands	Leached nutrients	NC	None	6	44	Verhoef et al. 2000

(continued)

Table 4 (continued)

Radiation manipulation	Location	Litter chemistry variables measured	Chemical response to radiation	Mass loss from radiation	Time (months)	Area:litter mass ratio (cm^2:g)	References
Reduced UV-B	Tierra del Fuego, Argentina	Lignin	NC	–26%	4.6	50	Pancotto et al. 2003
		Cellulose	NC				
Reduced UV-B	Tierra del Fuego, Argentina	Soluble Carbohydrates	NC	None	29	19	Pancotto et al. 2005
		Cellulose	NC				
		Lignin	NC				
		Lignin:N	NC				
No radiation vs. UV-A and UV-B irradiated	Growth chambers with UV-emitting bulbs	Dissolved organic C	NC	None	6	Information not provided	Gallo et al. 2006
		Phenols	NC				
		C	NC				
		N[b]	NC				

NC indicates no change in response variable
[a]Estimated from figures
[b]Comparison between "wet-dark" and "wet-UV" treatments

likely experiencing reduced intensity of solar radiation relative to the Arizona (33.5°N) experiment. Finally, methods of containing litter varied widely among studies; these may be responsible for substantial among-study variation. Brandt et al. (2007) suggest that their lower percentage of mass loss ascribed to photodegradation relative to Austin and Vivanco (2006) was, at least in part, the result of interception of solar radiation by the mesh litterbags used to contain material. In contrast, the litterbox design of Austin and Vivanco (2006) and the envelopes made of UV filters by Day et al. (2007) minimized structural interception of radiation.

Methodological variation in the ratio of treatment area to litter mass would similarly affect photodegradation. As a rough index of litter exposure to solar radiation, we calculated the treatment area (cm^2)/litter mass (g) ratio for the three dryland experiments which manipulated UV radiation (Table 3). There was a large variation in ratios, ranging from 33 to 200, and a positive relationship between this index and the percentage of decomposition attributed to photodegradation.

We suggest photodegradation may be particularly important in drylands due to the persistence of standing dead in these systems. Synchronous drops of all tree and shrub leaves, or even predictable annual drop of leaf material, occur less frequently in drylands than in temperate mesic ecosystems. Similarly, grass material not consumed by herbivores may persist for several years as standing dead before being incorporated into the surface litter pool. This standing dead material would be subject to decomposition via photodegradation and leaching, while breakdown by decomposer organisms would be minimal.

UV radiation has also been proposed to affect decomposition rates by altering the activities, populations, or community composition of decomposer organisms by direct deleterious effects or indirectly, by changing litter quality (Duguay and Klironomos 2000; Pancotto et al. 2003, 2005; Rozema et al. 1997; Verhoef et al. 2000). Along these lines, accelerated decomposition of litter exposed to attenuated UV-B radiation in Tierra del Fuego was attributed to UV effects on fungal communities (Pancotto et al. 2003). Thin layers of litter or soil are likely to effectively shield decomposers from UV radiation, and thus UV effects on decomposers via modification of tissue quality may ultimately prove more important than deleterious effects on decomposers (Zepp et al. 2007). In another experiment, UV-B indirectly influenced barley decay as barley grown in an attenuated UV environment decomposed more quickly than did litter from plants grown under near-ambient UV-B. Attenuated UV exposure during barley growth led to reduced lignin:N, phosphorous, cellulose, and UV-B absorbing compounds (Pancotto et al. 2005). A meta-analysis of plant responses to artificial UV-B sources found a significant increase in UV-B absorbing compounds in response to elevated UV (Caldwell et al. 2003). The accumulation of such compounds would likely slow microbially-mediated decomposition.

3.3.2 Soil Deposition

The water-limited nature of dryland systems usually results in a patchy mosaic of vegetation and bare ground, with the relative abundance and stature of woody plants defining the general ecosystem type (e.g., grassland, shrubland, savanna, woodland). In addition to affecting solar radiation inputs to the litter layer, patterns of canopy cover in dryland systems lead to pronounced patterns of erosion and associated processes of soil transport and deposition (Fryrear 1985; Toy et al. 2002). Wind and water transport of soils are widely recognized as having a substantial influence on nutrient and vegetation distribution (Okin et al. 2006; Peters et al. 2006). By redistributing litter across the landscape, transport processes affect the location and microclimate in which litter decays. Importance of these redistribution processes on spatial patterns of decomposition is unknown, as decomposition studies typically use litterbags to constrain litter at a location.

In addition to re-distributing litter, transport processes may influence decomposition by affecting the rate and pattern of soil deposition onto litter (or burial of litter) and the creation of a "litter–soil matrix". Although a number of studies have explored decay rates of buried litter relative to surface litter in drylands (Moretto et al. 2001; Pucheta et al. 2006), the importance of soil mixing into litter has received little research attention. Indirect evidence that this mechanism may be important in dryland decomposition was recently presented in a study designed to tease apart the direct and indirect influences of plant canopies in a desert grassland invaded by shrubs (Throop and Archer 2007). In that study, litterbags were deployed in different microenvironments: beneath intact shrub canopies, inter-canopy areas, and in areas where shrubs were recently removed. If the direct influences of canopies on microclimate were a prevailing driver, faster decay rates would be expected in inter-canopy areas where solar radiation and temperature were greatest. In contrast, shrub-enhanced soil nutrient availability and decomposer communities would be expected to enhance decomposition under both intact shrub canopies and recently-removed canopies relative to inter-canopy areas if biological factors were a main driver of decay. Surprisingly, decomposition rates were equally depressed in areas of current or recently-removed shrub canopy cover relative to inter-canopy areas, indicting that neither decomposer communities/soil nutrient levels nor indirect canopy influences on microclimate had important influences on decomposition. Litterbag placements included low (under shrub canopies) and high (open areas between shrubs or settings where shrubs had been removed) light environments, with K (the decay constant) varying by a factor of two – but not in relation to the light environment. Instead, there was a strong positive correlation between the amount of soil deposited in litterbags and decomposition, with lowest deposition into litterbags in shrub-influenced sites (Fig. 2). The fact that soil accumulation in litterbags was lowest in shrub-influenced zones suggests that soil movement must be lowest in these areas. In this system, sub-canopy areas typically have higher litter and grass cover than inter-canopy areas (Tiedemann and Klemmedson 2004), and this likely reduces soil transport and increases deposition (Schlesinger and Pilmanis 1998; Tiedemann and Klemmedson 2004). It is interesting to note that the relationship between soil deposition and decomposition found by Throop and Archer (2007) was serendipitous. Without the manipulated canopy cover experiments, photodegradation would have been the most parsimonious explanation for accelerated decay rates in inter-canopy locations relative to subcanopy locations.

Although to our knowledge no other studies have explicitly addressed the role of litter–soil mixing on dryland decomposition, several studies provide circumstantial evidence suggesting that the soil–litter matrix may affect dryland decomposition. Day et al. (2007) suggest that contact with the soil may have sped decay rates of twigs deployed on the soil surface relative to twigs placed on other surfaces. However, this pattern was not observed for other litter type treatments in the same study. Photodegradation was invoked as a possible mechanism for greater decomposition in a "plains" microhabitat relative to "hillside" and "arroyo" sites in the Sonoran Desert (Martínez-Yrízar et al. 2007), but other potential among-site differences were not explored. We suggest that greater bare ground area in the plains may have been associated with enhanced soil transport and hence a higher probability of litter burial and subsequent increases in decay rates.

We hypothesize that soil deposition into the litter–soil matrix affects decomposition through several mechanisms. First, soil deposited onto litter may serve as a vector for microbial colonization of litter. Microbial colonization in litter may be limited in drylands, particularly if bacteria are dominant relative to fungi, and because of the heterogeneous distribution of microbes across the landscape (McCulley et al. 2004). Mechanisms that transport decomposers into litter may therefore be particularly important in drylands. If soils serve as an important microbial transport vector, then soils transported from areas of high microbial biomass, such as underneath shrub canopies (McCulley et al. 2004; Smith et al. 1994), may promote faster microbial litter colonization, and hence faster decomposition compared to soils transported from areas of low microbial biomass, such as inter-canopy areas. Second, soil deposition may effectively buffer litter, and resident decomposer organisms, from high temperatures and low moisture (Elkins and Whitford 1982; Whitford 2002), and prolong windows of environmental conditions suitable for microbial activity. This buffering may be particularly important in drylands because of the infrequency of suitable conditions for microbial activity and pulsed nature of precipitation events. Third, soil transport may cause physical abrasion to leaf litter, enhancing rates of fragmentation and increasing the surface area available for microbial attack. Explicit manipulative experiments are needed to determine which, if any, of these mechanisms play important roles in affecting dryland decomposition.

3.3.3 Combined Influences of Photodegradation and Soil Transport

While photodegradation clearly has the potential to affect decomposition rates, the impact of this driver has been highly variable when documented. We propose that interactions between photodegradation and soil transport may be responsible, at least in part, for the observed among-experiment variation. Soil deposition into the soil–litter matrix may inhibit photodegradation if soil effectively shields litter from UV radiation. In litterbag studies in the Sonoran Desert, we have observed that a thin film of soil frequently develops on litter within several months of exposure. The relationship between soil layer thickness and shielding is unclear; further research is needed to determine what, if any, shielding is provided by a thin soil film versus a thicker coverage of soil. If this interaction between photodegradation and transport processes occurs, soil deposition may enhance litter decomposition while simultaneously decreasing rates of photodegradation. The net outcome of these offsetting interactions is unknown, but would potentially influence both mass loss and chemical composition. In contrast, if the soil deposition into the litter–soil matrix enhances decay via abrasion and physical fragmentation, it is possible that soil deposition would intensify the effects of photodegradation, as it would expose unprotected new surface area of tissue that may have photo-reactive compounds. However, this enhancement would only occur on any tissue not shielded from UV by soil cover.

The temporal dynamics of photodegradation will be important in affecting the nature of interactions with soil deposition. If photodegradation is most important in catalyzing litter break down early in the decomposition process, there may be little influence with soil deposition as photodegradation will occur while leaf material remains as standing dead or on the soil surface prior to litter–soil matrix development. The rate at which the interaction between photodegradation and soil deposition occurs will also be a function of the rates of soil deposition which, in turn, is a function of vegetation structure.

3.3.4 Expanded Framework

We propose an expanded framework for understanding and predicting patterns and processes of decomposition in drylands (Fig. 3). This framework expands the traditional focus on abiotic and biotic controls over decomposition to explicitly include the litter–soil matrix, and the factors that influence its development. While biotic and abiotic drivers in the traditional framework have been generally successful at predicting decomposition dynamics in mesic systems, spatial heterogeneity of vegetation in dryland systems necessitates considering how the spatial and temporal context of vegetation influences wind and water soil transport patterns, which in turn affect development of the litter–soil matrix. The litter–soil matrix may then indirectly dictate decay by strongly mediating both abiotic and biotic processes.

Experimental tests of this expanded framework will require a reconsideration of traditional litterbag methods. Experimental approaches will need to be expanded to include vegetation structure and measurements of, and ultimately manipulations of, soil transport processes. Adaptations of the litterbag method may be appropriate for initial explorations of the expanded framework. For example, litterbags can be deployed under different patterns of vegetation structure that differ in soil transport processes, and soil accumulation into litterbags can be measured directly or estimated via ash content (Throop and Archer 2007). The contrasting results of Austin and Vivanco (2006) and Throop and Archer (2007) could be largely the result of different methods and associated study foci. Traditional litterbags may enhance soil accumulation, restrict UV transmission to litter, reduce contact between litter and the soil surface, and minimize photodegradation by virtue of the low ratio of treat-

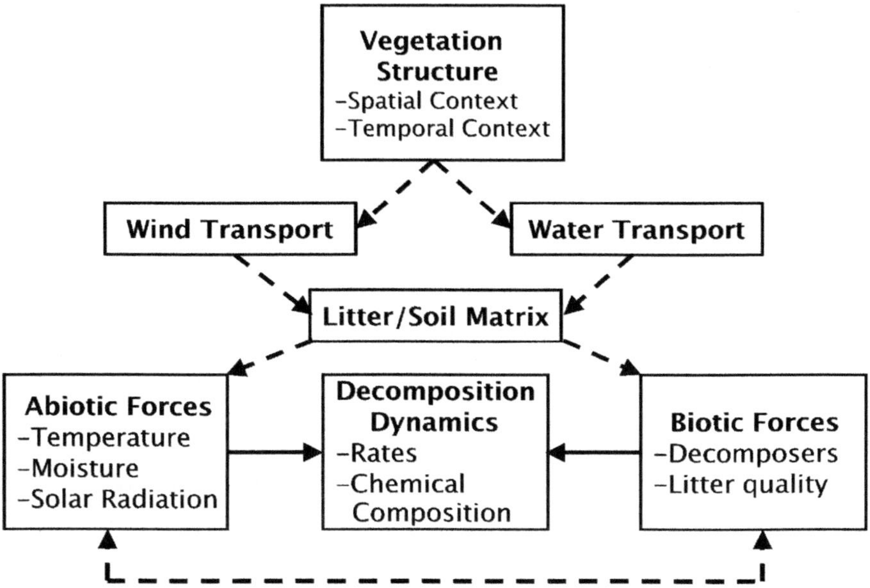

Fig. 3 Expanded framework for addressing decomposition in drylands. This framework includes erosion-based drivers, and proposes that development of the litter/soil matrix is a key but overlooked component of dryland decomposition. We hypothesize this matrix is controlled by wind/water transport of soil which, in turn, is controlled by vegetation structure. We further hypothesize that as the litter–soil matrix develops, biotic and abiotic drivers are strongly mediated

ment area:litter mass. Plastic-framed litterboxes covered with perforated UV filters (Austin and Vivanco 2006) minimize interception of radiation, but also minimize development of the litter–soil matrix by restricting air and water flow. Mesh cages (Torres et al. 2005) circumvent some of the aforementioned artifacts of litterbags and litterboxes, but careful consideration must be given to mesh size. Fine mesh will restrict movement of wind- and water-transported soil, whereas coarser mesh will be more susceptible to loosing litter fragments, hence confounding measurement of mass losses from decay with those due to fragmentation and export.

We suggest that understanding decomposition processes in drylands within our expanded framework will be facilitated by standardizing, or at least reporting, the ratio of treatment area to litter mass (see Table 3). This index can be used to facilitate cross-site comparisons, and using a standard ratio may be useful for cross-site comparisons. Depending on study objectives, however, system-specific differences in litterfall rates and standing pools of litter may make it desirable to use the litter layer typically present at the study site(s) as a guide for determining the appropriate area:mass ratio.

Ideally, assessments of decomposition dynamics will include analyses of how both litter and soil move across the landscape and interact at locations where litter accumulates to determine in situ decomposition. Although explicitly measuring both litter transport patterns and decay rates would be extremely challenging, linking soil and litter transport with decomposition dynamics will ultimately be crucial to understanding controls over dryland decomposition. Models based on a mechanistic understanding derived from key soil deposition–litter decomposition studies may be one method for making these links.

4 Conclusions

Litter decomposition dynamics exert strong controls over biogeochemical processes. These controls may be particularly important in drylands where litter, soil organic matter, and mineral nutrient pools are generally very small but may turn over rapidly. Short-term decomposition dynamics are particularly important in drylands because of the pulsed nature of precipitation drivers and the relatively small size of litter, soil organic matter, and mineral nutrient pools.

Recent studies have identified several processes historically overlooked as drivers in dryland decomposition dynamics. Incorporation of these processes may help explain the persistent disconnect between modeled and measured decomposition rates in drylands. Photodegradation, soil transport processes, and their interaction may be crucial determinants of decomposition. It is likely that photodegradation is more important in dryland than mesic systems due to the combination of greater solar radiative loads and less attenuation of incoming radiation from plant canopies. Transport of soil by wind and water is similarly enhanced in drylands due to low, discontinuous and patchy vegetative cover. We suggest that future research on dryland decomposition consider a framework that extends beyond traditional abiotic and biotic drivers to consider how vegetation structure mediates decomposition via its influence on soil transport.

Acknowledgements We have benefited from discussions about decomposition and drylands with P. Barnes, D. Breshears, and W. Whitford. This work was supported in part by USDA-NRI Managed Ecosystems 2005–35101–15408 to S.A., H.T., and M. McClaran and by the NSF-NMSU Advance Program to H.T.

References

Aerts R (1997a) Climate, leaf litter chemistry, and leaf litter decomposition in terrestrial ecosystems: a triangular relationship. Oikos 79:439–449

Aerts R (1997b) Nitrogen partitioning between resorption and decomposition pathways: a trade-off between nitrogen use efficiency and litter decomposibility? Oikos 80:603–606

Aerts R (2006) The freezer defrosting: global warming and litter decomposition rates in cold biomes. J Ecol 94:713–724

Arriaga L, Maya Y (2007) Spatial variability in decomposition rates in a desert scrub of northwestern Mexico. Plant Ecol 189:213–225

Austin AT, Vitousek PM (2000) Precipitation, decomposition and litter decomposability of *Metrosideros polymorpha* in native forests on Hawai'i. J Ecol 88:129–138

Austin AT, Vivanco L (2006) Plant litter decomposition in a semi-arid ecosystem controlled by photodegradation. Nature 442:555–558

Binkley D (1984) Does forest removal increase rates of decomposition and nitrogen release? Forest Ecol Manag 8:229–233

Brandt LA, King JY, Milchunas DG (2007) Effects of ultraviolet radiation on litter decomposition depend on precipitation and litter chemistry in a shortgrass steppe ecosystem. Glob Change Biol 13:2193–2205

Breshears DD, Rich PM, Barnes FJ, Campbell K (1997) Overstory-imposed heterogeneity in solar radiation and soil moisture in a semiarid woodland. Ecol Appl 7:1201–1215

Caldwell MM, Flint SD (1997) Uses of biological spectral weighting functions and the need of scaling for the ozone reduction problem. Plant Ecol 128:67–76

Caldwell MM, Ballare CL, Bornman JF, Flint SD, Björn LO, Teramura AH, Kulandaivelu G, Tevini M (2003) Terrestrial ecosystems, increased solar ultraviolet radiation and interactions with other climate change factors. Photochem Photobiol Sci 2:29–38

Couteaux MM, Bottner P, Berg B (1995) Litter decomposition, climate and litter quality. Trends Ecol Evol 10:63–66

Day TA, Zhang ET, Ruhland CT (2007) Exposure to solar UV-B radiation accelerates mass and lignin loss of *Larrea tridentata* litter in the Sonoran Desert. Plant Ecol 193:185–194

Duguay KJ, Klironomos JN (2000) Direct and indirect effects of enhanced UV-B radiation on the decomposing and competitive abilities of saprobic fungi. Appl Soil Ecol 14:157–164

Edmonds RL (1979) Decomposition and nutrient release in Douglas-fir needle litter in relation to stand development. Can J Forest Res 9:132–140

Ekaya W, Kinyamario J (2001) Production and decomposition of plant litter in an arid rangeland of Kenya. Afr J Range Forage Sci 18:125–129

Elkins NZ, Whitford WG (1982) The role of microarthropods and nematodes in decomposition in a semi-arid ecosystem. Oecologia 55:303–310

Field C, Behrenfeld M, Randerson J, Falkowski P (1998) Primary production of the biosphere: integrating terrestrial and oceanic components. Science 281:237–240

Fryrear DW (1985) Soil cover and wind erosion. Trans Am Soc Agric Eng 28:781–784

Gallo ME, Sinsabaugh RL, Cabaniss SE (2006) The role of ultraviolet radiation in litter decomposition in arid ecosystems. Appl Soil Ecol 34:83–91

Gehrke C, Johanson U, Callaghan TV, Chadwick D, Robinson CH (1995) The impact of enhanced ultraviolet-B radiation on litter quality and decomposition processes in *Vaccinium* leaves from the sub-Arctic. Oikos 72:213–222

Gholz HL, Wedin DA, Smitherman SM, Harmon ME, Parton WJ (2000) Long-term dynamics of pine and hardwood litter in contrasting environments: toward a global model of decomposition. Glob Change Biol 6:751–765

Hobbie SE (1992) Effects of plant species on nutrient cycling. Trends Ecol Evol 7:336–339

Hobbie SE (1996) Temperature and plant species control over litter decomposition in Alaskan tundra. Ecol Monogr 66:503–522

Hope GD, Prescott CE, Blevins LL (2003) Responses of available soil nitrogen and litter decomposition to openings of different sizes in dry interior Douglas-fir forests in British Columbia. Forest Ecol Manag 186:33–46

Huxman TE, Snyder KA, Tissue D, Leffler AJ, Ogle K, Pockman WT, Sandquist DR, Potts DL, Schwinning S (2004) Precipitation pulses and carbon fluxes in semiarid and arid ecosystems. Oecologia 141:254–268

Jacobson KM, Jacobson PJ (1998) Rainfall regulates decomposition of buried cellulose in the Namib Desert. J Arid Environ 38:571–583

Johnson KA, Whitford WG (1975) Foraging ecology and relative importance of subterranean termites in Chihuahuan desert ecosystems. Environ Entomol 4:66–70

Kemp PR, Reynolds JF, Virginia RA, Whitford WG (2003) Decomposition of leaf and root litter of Chihuahuan desert shrubs: effects of three years of summer drought. J Arid Environ 53:21–39

Kirschbaum MUF (2000) Will changes in soil organic carbon act as a positive or negative feedback on global warming? Biogeochemistry 48:21–51

Mack MC, D'Antonio CM (2003) The effects of exotic grasses on litter decomposition in a Hawaiian woodland: The importance of indirect effects. Ecosystems 6:723–738

MacKay WP, Silva S, Lightfoot DC, Pagani MI, Whitford WG (1986) Effect of increased soil moisture and reduced soil temperature on a desert soil arthropod community. Am Midl Nat 116:45–56

MacKay WP, Loring SJ, Zak JC, Silva SI, Fisher FM, Whitford WG (1994) Factors affecting loss in mass of creosotebush leaf litter on the soil surface in the northern Chihuahuan Desert. Southwestern Nat 39:78–82

Martínez-Yrízar A, Nuñez S, Burquez A (2007) Leaf litter decomposition in a southern Sonoran Desert ecosystem, northwestern Mexico: Effects of habitat and litter quality. Acta Oecol 32:291–300

McCulley RL, Archer SR, Boutton TW, Hons FM, Zuberer DA (2004) Soil respiration and nutrient cycling in wooded communities developing in grassland. Ecology 85:2804–2817

McCulley RL, Burke IC, Nelson JA, Lauenroth WK, Knapp AK, Kelly EF (2005) Regional patterns in carbon cycling across the Great Plains of North America. Ecosystems 8:106–121

McHale PJ, Mitchell MJ, Bowles FP (1998) Soil warming in a northern hardwood forest: trace gas fluxes and leaf litter decomposition. Can J Forest Res 28:1365–1372

Meentemeyer V (1978) Macroclimate and lignin control of litter decomposition rates. Ecology 59:465–472

Moorhead DL, Callaghan T (1994) Effects of increasing ultraviolet-B radiation on decomposition and soil organic matter dynamics – a synthesis and modeling study. Biol Fertil Soils 18:19–26

Moorhead DL, Reynolds JF (1989) The contribution of abiotic processes to buried litter decomposition in the northern Chihuahuan Desert. Oecologia 79:133–135

Moretto AS, Distel RA (2003) Decomposition of and nutrient dynamics in leaf litter and roots of *Poa ligularis* and *Steipa gyneriodes*. J Arid Environ 55:503–514

Moretto AS, Distel RA, Didone NG (2001) Decomposition and nutrient dynamic of leaf litter and roots from palatable and unpalatable grasses in a semi-arid grassland. Appl Soil Ecol 18:31–37

Mun HT, Whitford WG (1998) Changes in mass and chemistry of plant roots during long-term decomposition on a Chihuahuan Desert watershed. Biol Fertil Soils 26:16–22

Newsham KK, McLeod AR, Roberts JD, Greenslade PD, Emmett BA (1997) Direct effects of elevated UV-B radiation on the decomposition of *Quercus robur* leaf litter. Oikos 79:592–602

Okin GS, Gillette DA, Herrick JE (2006) Multi-scale controls on and consequences of aeolian processes in landscape change in arid and semi-arid environments. J Arid Environ 65:253–275

Olson JS (1963) Energy storage and the balance of producers and decomposers in ecological systems. Ecology 44:322–331

Osler G, Gauci C, Abbott LK (2004) Limited evidence for short-term succession of microarthropods during early phases of surface litter decomposition. Pedobiologia 48:37–49

Pancotto VA, Sala OE, Cabello M, Lopez NI, Robson TM, Ballare CL, Caldwell MM, Scopel AL (2003) Solar UV-B decreases decomposition in herbaceous plant litter in Tierra del Fuego, Argentina: potential role of an altered decomposer community. Glob Change Biol 9:1465–1474

Pancotto VA, Sala OE, Robson TM, Caldwell MM, Scopel AL (2005) Direct and indirect effects of solar ultraviolet-B radiation on long-term decomposition. Glob Change Biol 11:1982–1989

Parton W, Silver WL, Burke IC, Grassens L, Harmon ME, Currie WS, King JY, Adair EC, Brandt LA, Hart SC, Fasth, B (2007) Global-scale similarities in nitrogen release patterns during long-term decomposition. Science 315:361–364

Pauli F (1964) Soil fertility problem in arid and semi-arid lands. Nature 204:1286–1288

Peters DPC, Bestelmeyer BT, Herrick JE, Fredrickson EL, Monger HC, Havstad KM (2006) Disentangling complex landscapes: New insights into arid and semiarid system dynamics. Bioscience 56:491–501

Pucheta E, Llanos M, Meglioli C, Gaviorno M, Ruiz M, Parera C (2006) Litter decomposition in a sandy Monte desert of western Argentina: Influences of vegetation patches and summer rainfall. Aust Ecol 31:808–816

Rozema J, Tosserams M, Nelissen HJM, van Heerwaarden L, Broekman RA, Flierman N (1997) Stratospheric ozone reduction and ecosystem processes: enhanced UV-B radiation affects chemical quality and decomposition of leaves of the dune grassland species *Calamagrostis epigeios*. Plant Ecol 128:285–294

Santos PF, Phillips J, Whitford WG (1981) The role of mites and nematodes in early stages of buried litter decomposition in a desert. Ecology 62:664–669

Schade GW, Hoffmann RM, Crutzen PJ (1999) CO emissions from degrading plant matter (I). Measurements. Tellus B 51:889–908

Schaefer D, Steinberger Y, Whitford WG (1985) The failure of nitrogen and lignin control of decomposition in a North American desert. Oecologia 65:382–386

Schimel DS (1995) Terrestrial ecosystems and the carbon cycle. Glob Change Biol 1:77–91

Schlesinger WH (1997) Biogeochemistry: an analysis of global change, 2nd edn. Academic Press, San Diego

Schlesinger WH, Pilmanis AM (1998) Plant-soil interactions in deserts. Biogeochemistry 42:169–187

Schuurman G (2005) Decomposition rates and termite assemblage composition in semiarid Africa. Ecology 86:1236–1249

Schwinning S, Sala OE, Loik ME, Ehleringer JR (2004) Thresholds, memory, and seasonality: understanding pulse dynamics in arid/semi-arid ecosystems. Oecologia 141:191

Silva SI, MacKay WP, Whitford WG (1985) The relative contributions of termites and microarthropods to fluff grass litter disappearance in the Chihuahuan Desert. Oecologia 67:31–34

Smith JL, Halvorson JJ, Bolton JH (1994) Spatial relationships of soil microbial biomass and C and N mineralization in a semi-arid shrub-steppe ecosystem. Soil Biol Biochem 26:1151–1159

Strojan CL, Randall DC, Turner FB (1987) Relationship of leaf litter decomposition rates to rainfall in the Mojave Desert. Ecology 68:741–744

Sullivan NH, Bowden WB, McDowell WH (1999) Short-term disappearance of foliar litter in three species before and after a hurricane. Biotropica 31:382–393

Tadey M, Farji-Brener AG (2007) Indirect effects of exotic grazers: livestock decreases the nutrient content of refuse dumps of leaf-cutting ants through vegetation impoverishment. J Appl Ecol 44:1209–1218

Throop HL, Archer S (2007) Interrelationships among shrub encroachment, land management, and leaf litter decomposition in a semi-desert grassland. Ecol Appl 17:1809–1823

Tiedemann AR, Klemmedson JO (2004) Responses of desert grassland vegetation to mesquite removal and regrowth. J Range Manage 57:455–465

Torres PA, Abril AB, Bucher EH (2005) Microbial succession in litter decomposition in the semi-arid Chaco woodland. Soil Biol Biochem 37:49–54

Toy TJ, Foster GR, Reynard KG (2002) Soil erosion: processes, prediction, measurement and control. Wiley, New York

Verhoef HA, Verspagen JMH, Zoomer HR (2000) Direct and indirect effects of ultraviolet-B radiation on soil biota, decomposition and nutrient fluxes in dune grassland soil systems. Biol Fertil Soils 31:366–371

Vossbrinck CR, Coleman DC, Woolley TA (1979) Abiotic and biotic factors in litter decomposition in a sermiarid grassland. Ecology 60:265–271

Weatherly HE, Zitzer SF, Coleman JS, Arnone JA (2003) *In situ* litter decomposition and litter quality in a Mojave Desert ecosystem: effects of elevated atmospheric CO_2 and interannual climate variability. Glob Change Biol 9:1223–1233

Wetterer JK, Himler AG, Yospin MM (2001) Foraging ecology of the desert leaf-cutting ant, *Acromyrmex versicolor*, in Arizona (Hymenoptera: Formicidae). Sociobiology 37:633–649
Whitford W (2002) Ecology of desert systems. Academic Press, San Diego, CA
Whitford WG, Meentemeyer V, Seastedt TR, Cromack K, Crossley DA, Santos P, Todd RL, Waide JB (1981) Exceptions to the AET model – deserts and clear-cut forest. Ecology 62:275–277
Whitford WG, Steinberger Y, Ettershank G (1982) Contributions of subterranean termites to the "economy" of Chihuahuan Desert ecosystems. Oecologia 55:298–302
Whitford WG, Steinberger Y, MacKay W, Parker LW, Freckman D, Wallwork JA, Weems D (1986) Rainfall and decomposition in the Chihuahuan Desert. Oecologia 68:512–515
Yahdjian L, Sala O, Austin A (2006) Differential controls of water input on litter decomposition and nitrogen dynamics in the Patagonian Steppe. Ecosystems 9:128–141
Zepp RG, Erickson DJI, Paul ND, Sulzberger B (2007) Interactive effects of solar UV radiation and climage change on biogeochemical cycling. Photochem Photobiol Sci 6:286–300
Zhang QH, Zak JC (1995) Effects of gap size on litter decomposition and microbial activity in a subtropical forest. Ecology 76:2196–2204
Zou C, Barnes PW, Archer SR, McMurtry CR (2005) Soil moisture redistribution as a mechanism of facilitation in savanna tree-shrub clusters. Oecologia 145:32–40

Agricultural Crop Models: Concepts of Resource Acquisition and Assimilate Partitioning

Eckart Priesack(✉) and Sebastian Gayler

Contents

Abstract In this review, we analyze how the acquisition of resources, e.g. carbon (C) and nitrogen (N), and the distribution of assimilates between plant organs is described by common agricultural crop growth models. We consider agricultural crop growth models that are integrated into larger agro-ecosystem or agricultural soil-plant-atmosphere system models. These system models are developed to simulate not only plant growth processes but also energy and matter fluxes between atmosphere and soil including decomposition of plant residues and C- and N-turnover of soil organic matter. Within the crop models different approaches are used to up-scale eco-physiological processes from the plant-organ level to the plant and canopy level, they are discussed with respect to data requirement and adequate representation of resource acquisition. Considering mainly trees, basic concepts used to model assimilate partitioning in plants have been classified as empirical, teleonomic, based on source-sink relations or based on transport and transformation processes. Application of these concepts in agricultural crop models are presented and examined. Moreover, a survey of modeling approaches is given that consider the impact of different kinds of biotic and abiotic stresses on partitioning in crop growth models.

E. Priesack
Institute of Soil Ecology, Helmholtz Zentrum München, German Research Center for Environmental Health
e-mail: priesack@gsf.de

U. Lüttge et al. (eds.), *Progress in Botany 70*,
DOI:10.1007/978-3-540-68421-3,

1 Historical Background

1.1 Agro-Ecosystem Models

The development of numerical mathematical models to simulate plant growth and cycles of water and nutrients in soil-plant-atmosphere systems dates back to the late 1960s and early 1970s when by the introduction of computers with enough speed and memory it was possible to explicitly simulate plant processes such as photosynthesis (Goudriaan and van Laar 1994; Bouman et al. 1996) or soil transport processes including the processes of convection and diffusion (Frissel et al. 1970; Wierenga and de Wit 1970; van Genuchten et al. 1974). However, more complex models of the soil-plant system came up only after personal computers had become more and more common and widespread at the end of the 1970s and during the 1980s.

After also models to calculate the turnover of soil carbon (C) and nitrogen (N) had been developed in the 1970s (Dutt et al. 1972; Beek and Frissel 1973; Mehran and Tanji 1974; Tanji and Gupta 1978), at the beginning of the 1980s the first models result that combine plant growth with soil water flow, N transport and/or soil C- and N-turnover (Watts and Hanks 1978; Seligman and van Keulen 1981). Until mid of the 1980s several models were developed that include approaches still used nowadays as basis for the simulation of agricultural and forest systems such as CENTURY (Parton et al. 1994), NCSOIL (Molina et al. 1983), EPIC (Williams and Renard 1985), CERES (Jones and Kiniry 1986; Ritchie et al. 1987), LEACHM (Hutson and Wagenet 1992), SOILN (Johnsson et al. 1987) and ANIMO (Berghuijs-van Dijk et al. 1985). Whereas these older models were revised, from the end of the 1980s until the mid of the 1990s several new models emerged that address the increased request for different applications mainly in the field of agricultural production (Shaffer et al. 2001) with focus on the soil N-transport and N-turnover: HERMES (Kersebaum 1989, 1995), DAISY (Hansen et al. 1990), NLEAP (Shaffer et al. 1991), SUNDIAL (Smith et al. 1996), GLEAMS (Leonard et al. 1987), EXPERT-N (Engel and Priesack 1993), DNDC (Li et al. 1992), CANDY (Franko et al. 1995), WAVE (Vanclooster et al. 1995), CropSyst (Stockle et al. 1994).

Based on extensive experimental datasets several model comparisons were accomplished (de Willigen 1991; Diekkrüger et al. 1995; Tiktak and van Grinsven 1995; Kersebaum et al. 2007). By comparison of simulated with experimental results, in particular by comparing time series of crop biomass and contents of soil water and soil mineral N, general model deficits could be identified, that both concerned all models and only certain models or modeling approaches. However, because of the high model complexity it was in general difficult or even not possible to exactly specify the cause of a certain simulation inaccuracy or modeling error. Often also the resolution in time or space of the experimental data was not high enough to identify model wrongness (Diekkrüger et al. 1995). Therefore, among the models that delivered adequate simulation results were very simple, more empirical models as well as more complex, physically based models (de Willigen 1991; Diekkrüger et al. 1995). An integrative documentation of 23 agro-ecosystem models (Engel et al. 1993) gives a survey on the state of model development for application to agricultural crop systems at begin of the 1990s. A more recent survey can be found in Shaffer et al. (2001), where information on numerous models from Europe and North America is provided and additionally different model applications are presented.

1.2 Crop Models

Most of the agricultural crop models integrated into agro-ecosystem models have developed from two schools, founded by C.T. de Wit in the Netherlands and J.T. Ritchie in the US (Bouman et al. 1996; van Ittersum et al. 2003; Jones et al. 2003). The models SUCROS and ORYZA (Goudriaan and van Laar 1994), the model package WOFOST (World Food Studies) (Boogaard et al. 1998), and the recently developed model GECROS (Yin and van Laar 2005) belong to the Dutch school. The best-known models from the US are CERES, SOYGRO and CROPGRO which have been gathered with other models into the modeling package DSSAT (Decision Support System for Agrotechnology Transfer) (Jones et al. 2003). From the beginning the Dutch model developers had been concentrated on the accurate prediction of potential yield assuming optimal supply of nitrogen and water, while the American scientists had focused on practical applicability and analyzed more the impact of environmental conditions on crop growth and yield. As an example, Dutch scientists have early modeled assimilation and respiration including up-scaling from the leaf to the canopy level while the CERES model family is still based on the simple, but robust summary concept of radiation use efficiency (Yin and van Laar 2005). Only recently, a biochemical model for the photosynthesis process has been implemented in the CERES-Maize model (Lizaso et al. 2005). There are also examples of typical combinations of modeling approaches from both the Dutch and the US school such as the crop model SPASS (Wang and Engel 2000; Gayler et al. 2002; Priesack 2006) and the maize model Hybrid-Maize (Yang et al. 2004).

The classical models have meanwhile been rewritten following a strictly modular structure and are coupled to convenient modeling packages representing complete soil-plant-atmosphere systems like DSSAT (Jones et al. 2001; Jones et al. 2003), APSIM (Keating et al. 2003), STICS (Brisson et al. 2003) and Expert-N (Sperr et al. 1993; Priesack 2006). Therefore besides modules for the different crops, the packages contain also modules for management, soil water and thermal regime, carbon and nitrogen turnover and transport. New modeling approaches have tried to improve the classical models by replacing descriptions of individual processes and their interactions based on newly understood physiological mechanisms (Gayler and Priesack 2005; Yin and van Laar 2005). It is expected that these improved models will provide more reliable predictions of the impact of environmental change on ecosystems. Further examples for recent developments in cropping system models are given by van Ittersum et al. (2003), Jones et al. (2003), Keating et al. (2003), and Brisson et al. (2003). Moreover, White (2006) reviews recent progress in using genetic information in modeling wheat crops, also new attempts have been made to improve the modeling of environmental and genetic effects on protein concentration in wheat, e.g. Asseng and Milroy (2006), Martre et al. (2006), Weiss and Moreno-Sotomayer (2006), and Triboi et al. (2006).

2 Resource Acquisition

2.1 Carbon Assimilation

In the CERES crop model family the starting point for the calculation of the daily canopy dry matter production is the determination of the daily photosynthesis of the crop. It is assumed that 50% of the daily total incident solar radiation R_s^{day} [MJ ha^{-1} day^{-1}] is photosynthetically active radiation. The amount of intercepted photosynthetically active radiation R_{ipa}^{day} [MJ ha^{-1} day^{-1}] by the plant canopy is then computed by an exponential function to account for the shading using the leaf area index f_{LAI} [-] and the extinction coefficient α_{ext} [-], that depends on the plant species:

$$R_{ipa}^{day} = 0.5 R_s^{day} [1 - \exp(-\alpha_{ext} f_{LAI})] \tag{1}$$

From this daily intercepted radiation a potential daily dry matter production rate $\mu_{B,pot}^{day}$ [kg ha^{-1} day^{-1}] is determined by

$$\mu_{B,pot}^{day} = f_{RUE} R_{ipa}^{\alpha_r} \tag{2}$$

where f_{RUE} [kg MJ^{-1}] denotes the radiation use efficiency and α_r [-] the radiation exponent, each of both empirical parameters depend on the particular plant species that is considered. The actual daily dry matter production rate $\mu_{B,act}^{day}$ [kg ha^{-1} day^{-1}] of the canopy then results by considering the impacts of average daily air temperature, soil water and nitrogen availability on daily photosynthesis by reduction or stress factors related to temperature $f_{ph,T}$, respectively to soil water and nitrogen content $f_{ph,\theta,N}$:

$$\mu_{B,act}^{day} = f_{ph,T}\, f_{ph,\theta,N}\, \mu_{B,pot}^{day} \tag{3}$$

The concept used in the CERES models of simple scaling from leaf to canopy by treating the canopy surface as one leaf and solving equations for leaf energy balance and photosynthesis at the top of the canopy is usually called a big leaf model. By assuming an optimal distribution of leaf nitrogen the simple analytical expression of canopy photosynthesis allows for a fast numerical computation compared to numerical integration over leaf layers. However, the big leaf concept is considered to be too simple mainly because leaf photosynthesis depends non-linear on absorbed radiation and leaf nitrogen (Norman 1993; Friend 2001; Beyschlag and Ryel 2007) and the use of average absorbed radiation will over-estimate canopy photosynthesis (Spitters 1986). Furthermore, energy partitioning in the canopy between sunlit and shaded leaves is also non-linear and has to be considered to correctly estimate leaf temperature and heat fluxes for the canopy (Wang and Leuning 1998).

In contrast to the above summary concept of radiation use efficiency, in the crop models SUCROS (van Laar et al. 1997), WOFOST (Boogaard et al. 1998) and INTERCOM (Kropff and van Laar 1993) intercepted photoactive radiation and CO_2 assimilation are calculated by considering direct and diffusive radiation components, shaded and sunlit leaves and also different spatial layers within the canopy. Moreover, unlike the CERES models, which compute gross assimilation in a daily time step, the diurnal variation of incident radiation is explicitly considered.

The actual daily gross photosynthesis of the canopy is calculated following Spitters et al. (1989). It is assumed that the radiation use efficiency ϵ_{PAR} [kg (CO_2) MJ^{-1}] and the maximal gross leaf photosynthesis rate at light saturation P_{max} [kg(CO_2) ha^{-1} day^{-1}] are constant within the canopy and that the impact of senescence, temperature and availability of water and nitrogen can be described in a homogeneous way for the whole canopy by using

$$P_{\mathrm{gma}} = P_{\mathrm{max}} f_{\mathrm{ph,sen}} f_{\mathrm{ph,T}} f_{\mathrm{ph},\theta,\mathrm{N}} \tag{4}$$

where P_{gma} is the actual gross leaf photosynthesis rate [kg(CO_2) ha^{-1} day^{-1}] at light saturation, $f_{\mathrm{ph,sen}}$ [-] is the senescence reduction factor, $f_{\mathrm{ph,T}}$ [-] the temperature factor, $f_{\mathrm{ph},\theta}$ [-] the water stress factor and $f_{\mathrm{ph,N}}$ [-] the leaf-N reduction factor.

The photosynthesis rate $P_{\mathrm{g,shd,z}}$ [kg(CO_2) ha^{-1} day^{-1}] per shaded leaf area is then be estimated from the photosynthetic active part of the actual radiation $R^{\mathrm{act}}_{\mathrm{abs,shd,z}}$ [MJ m^{-2} day^{-1}] absorbed by shaded leaves at depth z:

$$P_{\mathrm{g,shd,z}} = P_{\mathrm{gma}}[1 - \exp(-0.5 R^{\mathrm{act}}_{\mathrm{abs,shd,z}} \epsilon_{\mathrm{PAR}} / P_{\mathrm{gm}})] \tag{5}$$

In a similar way the photosynthesis rate $P_{\mathrm{g,slt,z}}$ [kg(CO_2) ha^{-1} day^{-1}] of the sunlit leaves can be computed for a spherical leaf angle distribution where the sine of the leaf angle α is equally distributed between 0 and 1 independent of the sun height (Goudriaan and van Laar 1994):

$$P_{\mathrm{g,slt,z}} = P_{\mathrm{gma}} \int_0^1 [1 - \exp(-0.5 R^{\mathrm{act}}_{\mathrm{abs,slt,z}} \epsilon_{\mathrm{PAR}} / P_{\mathrm{gm}})] \, \mathrm{d} \sin\alpha \tag{6}$$

The average total photosynthesis rate in depth z of sunlit and shaded leaves $P_{\mathrm{g,lvs,z}}$ [kg(CO_2) ha^{-1} day^{-1}] results from the fraction $f_{\mathrm{slt,z}}$ of sunlit leaves at depth z:

$$P_{g,lvs,z} = f_{slt,z} P_{g,slt,z} + (1 - f_{slt,z}) P_{g,shd,z} \tag{7}$$

The numerical integration of $P_{g,lvs,z}$ over the leaf area index $f_{LAI,z}$ [-] above the canopy depth z from the top of the canopy to the soil surface yields the actual total gross photosynthesis of the canopy and the daily amount of CO_2 assimilated by the canopy finally results by integration over the daylength (Spitters et al. 1989).

This multi-layer concept of upscaling gross leaf photosynthesis first derived for the SUCROS model was later integrated into several crop models such as WOFOST (Boogaard et al. 1998), INTERCOM (Kropff and van Laar 1993), SPASS (Wang and Engel 2000; Gayler et al. 2002), APSIM-CPL (Wang et al. 2002), Hybrid-Maize (Yang et al. 2004) among others. A simplification of the multi-layer approach avoiding the over-simplification of the big-leaf concept is the two-leaf model (de Pury and Farquhar 1997; Wang and Leuning 1998) in which the canopy leaves are splitted into sunlit and shaded fractions and each leaf fraction is treated by an one-layered leaf model. The two-leaf model proved to be computationally ten times more efficient than the multi-layer model of Leuning et al. (1995), but resulted at nearly the same prediction of canopy photosynthesis (de Pury and Farquhar 1997). This two-leaf concept was adopted in the crop growth model GECROS (Yin and van Laar 2005).

As in the multi-layer approach the fraction $f_{slt,z}$ of sunlit leaves at canopy depth z is equal to the direct-beam radiation reaching depth z:

$$f_{slt,z} = \exp(-k_{dir,blk} f_{LAI,z}) \tag{8}$$

where $f_{LAI,z}$ [-] is the cumulative index of leaf area above canopy depth z and $k_{dir,blk}$ [-] is the direct beam radiation extinction coefficient for theoretical canopies with black leaves which depends on the position of the sun. Thus, the total fraction f_{slt} [-] of sunlit leaves of the canopy is calculated by:

$$\begin{aligned} f_{slt} &= \frac{1}{f_{LAI,z}} \int_0^{f_{LAI}} \exp(-k_{dir,blk} f_{LAI,z}) df_{LAI,z} \\ &= [1 - \exp(-k_{dir,blk} f_{LAI})] / (k_{dir,blk} f_{LAI}) \end{aligned} \tag{9}$$

and the total fraction of shaded leaves by $f_{shd} = 1 - f_{slt}$.

The actual radiation absorbed by the sunlit leaf fraction of the canopy is determined as the sum of direct-beam, diffuse and scattered-beam components and the radiation absorbed by the shaded leaf fraction of the canopy results as the difference between the total radiation absorbed by the canopy and the radiation absorbed by the sunlit leaves (de Pury and Farquhar 1997).

This calculation is not only applied to photosynthetically active radiation (PAR) but also to near-infrared radiation (NIR), since both radiation components are needed to estimate potential leaf transpiration and leaf energy balance (Leuning et al. 1995) and have different values for the reflection, scattering and extinction coefficients of direct-beam and diffusive radiation within the canopy (Goudriaan and van Laar 1994; Yin and van Laar 2005). It is assumed that half of the incident solar radiation is in the PAR and the other half in the NIR waveband (Leuning et al. 1995; Yin and van Laar 2005).

The leaf photosynthesis model of GECROS is based on the model of Farquhar et al. (1980) which describes photosynthesis by underlying biochemical mechanisms and allows to assess the interaction of environmental state variables, e.g. elevated atmospheric CO_2 concentrations combined with rising temperatures (Leuning et al. 1995; Nikolov et al. 1995; Yin and van Laar 2005). The potential gross leaf photosynthesis rate at light saturation P_{gmp} is given by:

$$P_{gmp} = P_{max}(1 - \Gamma_{*} / C_c)\min(V_c, V_j) \tag{10}$$

where Γ_* is the CO_2 compensation point in the absence of dark respiration, C_c is the CO_2 concentration at the carboxylation site, V_c and V_j are the carboxylation rates limited by Rubisco activity and by electron transport, respectively. V_c is calculated according to Farquhar et al. (1980) and V_j follows a generalized model of electron transport (Yin et al. 2004). The maximal carboxylation rates $V_{c\,max}$ and J_{max} depend on leaf temperature and leaf nitrogen, the dependence of leaf photosynthesis on C_c and its relation to intercellular CO_2 is differently modeled for C_3 and C_4 plants, see Yin and van Laar (2005), Appendix A. Moreover, since the Penman-Monteith equation (Monteith 1973) is applied to estimate potential leaf transpiration, the potential stomatal conductance of CO_2 in the absence of water stress is used to calculate stomatal resistance to water transfer. Additionally, the difference in saturated vapor pressure between leaf interior and external air divided by the difference between leaf and air temperature is taken as proxy for the slope of the water vapor curve (Yin and van Laar 2005).

Therefore, if the biochemical photosynthesis model of Farquhar et al. (1980) is applied, the spatial integration that up-scales leaf photosynthesis to the canopy level has to account not only for the depth dependency of radiation absorption but also for the change with canopy depth of model parameters for photosynthesis and transpiration. In the case of the GECROS model these are the maximal carboxylation rates $V_{c\,max}$ and J_{max} and the resistances to the transfer of heat r_{bh} and water r_{bw} from leaf to air. The carboxylation rates are related to the leaf nitrogen content and the transfer resistances to wind speed. It is assumed that the vertical decline of the photosynthetical active leaf N can be described by an exponential profile (Yin et al. 2000) such that the photosynthetical active N for the whole canopy N_c, for the sunlit leaf fraction $N_{c,slt}$ and for the shaded leaf fraction $N_{c,shd}$ of the canopy can be estimated by Yin et al. (2000) and Yin and van Laar (2005):

$$N_c = N_0[1 - \exp(-k_N f_{LAI})] / k_N - N_b f_{LAI} \tag{11}$$

$$\begin{aligned} N_{c,slt} &= \{1 - \exp[-(k_N + k_{dir,blk}) f_{LAI}]\} / (k_N + k_{dir,blk}) \\ &\quad - N_b[1 - \exp(-k_{dir,blk} f_{LAI}] / k_{dir,blk} \end{aligned} \tag{12}$$

$$N_{c,shd} = N_c - N_{c,slt} \tag{13}$$

where N_b is the base or minimum value at or below which the photosynthesis is zero, N_0 is the N content of the leaves at the top of the canopy, and k_N is the leaf N extinction coefficient (Yin et al. 2003). Similarly, by assuming an exponentially declining profile of the wind speed in the canopy (Leuning et al. 1995), the boundary layer conductances for the whole canopy and for the sunlit and shaded leaf fractions can be scaled up by integration and analytical closed-form expressions for boundary layer resistances to heat and water transfer are obtained (Yin and van Laar 2005).

The daily potential gross photosynthesis rate of the entire canopy results as in the multi-layer approach by numerical integration over the daylength using a Gaussian five-point method (Goudriaan 1986).

2.2 *Water and Nitrogen Uptake*

In most crop models the plant uptake of water and nitrogen from soil is mainly determined by the root distribution in the soil, atmospheric water demand, nitrogen demand of the plants and the availability of soil water and nitrogen. Usually the actual daily water uptake is estimated by taking the minimum of three daily rates (given as water amounts per day): the water uptake capacity of the roots, the soil water availability and the atmospheric water demand. See van den Berg and Driessen (2002) and van den Berg et al. (2002) for a review and comparison of several water uptake models. Similarly the actual daily N uptake is modeled by the minimum of the N uptake capacity of the roots, the soil N availability and the N demand of the crop.

In the CERES models the water and nitrogen that is potentially accessible by roots is described by availability factors per unit root length of the ith rooted soil layer, i.e. by $f_{W,i}^{avl}$ [-] for water, $f_{NO_3,i}^{avl}$ for nitrate-N and $f_{NH_4,i}^{avl}$ for ammonium-N:

$$f_{W,i}^{avl} = \min\{1; 0.00089 - \exp[62.0(\theta_i - \theta_{pwp,i})] / [6.68 - \ln(l_i)]\} \tag{14}$$

$$f_{NO_3,i}^{avl} = 1 - \exp(\gamma_{c,NO_3} C_{NO_3,i}), \quad f_{NH_4,i}^{avl} = 1 - \exp(\gamma_{c,NH_4} C_{NH_4,i}) \tag{15}$$

where the soil is assumed to be divided in discrete soil layers for which θ_i and $\theta_{pwp,i}$ are the volumetric soil water contents at actual time and permanent wilting point in the ith soil layer, l_i [mm mm^{-3}] is the root length density of the ith soil layer and $C_{NO_3,i}$ [kg(N) ha^{-1}], $C_{NH4,i}$ [kg(N) ha^{-1}] the nitrate-N and ammonium-N amounts in the ith soil layer. The equation for water availability was derived from a model describing radial water flow towards a single root assuming all soils have similar hydraulic conductivities near the permanent wilting point. It is also assumed that the water potential gradient between the root and the soil remains constant (Jones and Kiniry 1986; Ritchie et al. 1987; Wang and Smith 2004). The N availability factors are given by empirical relationships using crop specific nitrate-N and ammonium-N accessibility factors γ_{c,NO_3} [-] and γ_{c,NH_4} [-], respectively.

By the root resistance an upper limit to water uptake exists, which is considered by a maximal daily root water uptake per unit root length and soil depth of $A_{\mathrm{W}}^{\max} = 3.0\,\mathrm{mm}^3\ \mathrm{mm}^{-1}\ \mathrm{day}^{-1}$. There are also crop specific maximal uptake rates per unit root length $A_{\mathrm{NO3}}^{\max}$ [μg(N) cm^{-1} day^{-1}] of soil nitrate-N and $A_{\mathrm{NH4}}^{\max}$ [μg(N) cm^{-1} day^{-1}] of ammonium-N assumed. The daily potential root uptake rates $A_{W,i}^{pot}$ [mm day^{-1}] for water, $A_{\mathrm{NO3,i}}^{\mathrm{pot}}$ [kg(N) ha^{-1} day^{-1}] for nitrate-N and $A_{\mathrm{NH4,i}}^{\mathrm{pot}}$ [kg (N) ha^{-1} day^{-1}] for ammonium-N from the *i*th layer of thickness d_{i} [mm] are then defined by:

$$A_{\mathrm{W,i}}^{\mathrm{pot}} = A_{\mathrm{W}}^{\max} f_{\mathrm{W},i}^{\mathrm{avl}} d_i l_i \tag{16}$$

$$A_{\mathrm{NO_3},i}^{\mathrm{pot}} = A_{\mathrm{NO_3}}^{\max} f_{\mathrm{NO_3},i}^{\mathrm{avl}} f_{\theta,i} d_i l_i \tag{17}$$

$$A_{\mathrm{NH_4},i}^{\mathrm{pot}} = A_{\mathrm{NH_4}}^{\max} f_{\mathrm{NH_4},i}^{\mathrm{avl}} f_{\theta,i} d_i l_i \tag{18}$$

where $f_{\theta,i}$ [-] is the soil moisture deficit factor of the *i*th soil layer given by

$$f_{\theta,i} = (\theta_i - \theta_{\mathrm{pwp},i}) / (\theta_{\mathrm{fc},i} - \theta_{\mathrm{pwp},i}) \tag{19}$$

From (16) the total daily potential root water uptake $A_{\mathrm{pot,NO_3}}^{\mathrm{day}}$ is calculated as the sum of the daily potential root water uptake from each rooted layer. The daily actual root water uptake $A_{\mathrm{W},i}^{\mathrm{act}}$ from the *i*th soil layer is then given by:

$$A_{\mathrm{W},i}^{\mathrm{act}} = A_{\mathrm{W},i}^{\mathrm{pot}} \min(1; TR_{\mathrm{pot}}^{\mathrm{day}} / A_{\mathrm{pot,W}}^{\mathrm{day}}) \tag{20}$$

where $TR_{\mathrm{pot}}^{\mathrm{day}}$ [mm day^{-1}] denotes the daily potential transpiration rate describing the atmospheric water demand.

In a similar way the total daily potential nitrate-N uptake rate $A_{\mathrm{pot,NO_3}}^{\mathrm{day}}$ [kg(N) ha^{-1} day^{-1}] and the total daily potential ammonium-N uptake rate $A_{\mathrm{pot,NH_4}}^{\mathrm{day}}$ [kg(N) ha^{-1} day^{-1}] result as the sum of the corresponding uptake rates of each rooted soil layer, and the total daily potential mineral N uptake rate is obtained:

$$A_{\mathrm{pot,N_{min}}}^{\mathrm{day}} = A_{\mathrm{pot,NO_3}}^{\mathrm{day}} + A_{\mathrm{pot,NH_4}}^{\mathrm{day}} \tag{21}$$

Thus, an actual nitrogen uptake factor $f_{act,N}$ [-] can be defined that relates the total daily N demand $D_{\mathrm{c,N}}^{\mathrm{day}}$ [kg(N) ha^{-1} day^{-1}] of the crop to the total daily potential N supply:

$$f_{\mathrm{act,N}} = \min\{1, D_{\mathrm{c,N}}^{\mathrm{day}} / A_{\mathrm{pot,N_{min}}}^{\mathrm{day}}\} \tag{22}$$

which is used to reduce the potential N uptake if the N demand is lower than the potential uptake for computing the actual daily uptake of nitrate-N $A_{\mathrm{NO3,i}}^{\mathrm{act}}$ [kg(N) ha^{-1} day^{-1}] and ammonium-N $A_{\mathrm{NH4,i}}^{\mathrm{act}}$ [kg(N) ha^{-1} day^{-1}] from the *i*th soil layer:

$$A_{\mathrm{NO_3},i}^{\mathrm{act}} = \min\{A_{\mathrm{NO_3},i}^{\mathrm{pot}} f_{\mathrm{act,N}}, C_{\mathrm{NO_3},i} - C_{\mathrm{NO_3,min}}\} \tag{23}$$

$$A_{NH_4,i}^{act} = \min\{A_{NH_4,i}^{pot} f_{act,N}, C_{NH_4,i} - C_{NH_4,\min}\} \tag{24}$$

where $C_{NO_3,i}$, $C_{NH_4,i}$ denote the mineral N amounts [kg(N) ha^{-1}] in the ith soil layer and $C_{NO_3,min}$, $C_{NH_4,min}$ are the minimal mineral N amounts [kg(N) ha^{-1}] which cannot be extracted by the crop.

The pattern of water and nitrogen extraction among the rooted soil layers therefore is from each soil layer according to the ratio of available amounts within the root zone. Hence, uptake is from the layers with higher contents of water and nitrogen, respectively.

Moreover, the uptake depends on root growth and the downward root front velocity which are simulated differently in most crop models, albeit roots usually are assumed to be uniformly distributed in homogeneous soil layers and root growth is simulated almost without any effect of soil properties in addition to soil temperature and availability of soil water and nitrogen. Often a deep, well drained soil is assumed without any physical or chemical hindrance to root growth (Ritchie et al. 1987; Wang and Smith 2004). This reflects the fact that the complex interaction between root growth and soil properties is not fully understood and therefore empirical and simplified root growth models are used in most models (Eitzinger et al. 2004; Wang and Smith 2004). However, Jones et al. (1991) proposed a root growth model based on root length distribution following an exponential pattern, which decreases with soil depth and is subjected to constraint factors for each soil layer, a static and several dynamic factors. The static factor of each layer is a user defined input parameter that varies with depth and can account for effects of coarse fragments, plough layer, toxicity, extreme soil pH-values and nutrient deficiency other than nitrogen. The dynamic constraint factors include effects of soil strength, aeration, temperature, soil water and soil nitrogen contents.

Whereas many models including CERES (Jones and Kiniry 1986; Ritchie et al. 1987), SUCROS (Goudriaan and van Laar 1994), WOFOST (Boogaard et al. 1998), GECROS (Yin and van Laar 2005), STICS (Brisson et al. 2003), CropSyst (Stockle et al. 1994; Stockle et al. 2003), EPIC (Williams and Renard 1985), DNDC (Li et al. 1992; Zhang et al. 2002), HERMES (Kersebaum 1995) and CANDY (Franko et al. 1995) among others, simulate transport processes in the soil by use of capacity-type models, an alternative approach to simulate water flow and solute transport is based on numerical solutions of partial differential equations (pde), the Richards equation and the advection-dispersion transport equation (Gilding 1992; Engel et al. 1993; Priesack 2006; Yin and van Laar 2005). Among the models that apply the pde-type approaches are DAISY (Hansen et al. 1990), SIMULAT (Diekkrüger and Arning 1995), SOILN (Johnsson et al. 1987), SWAP (Kroes and van Dam 2003; Kroes and Roelsma 2007) and WAVE (Vanclooster et al. 1995). The model systems APSIM (Keating et al. 2003), Expert-N (Engel and Priesack 1993; Stenger et al. 1999; Priesack 2006) and THESEUS (Wegehenkel 2000) provide both, capacity-type and pde-type transport models, and allow users to choose the model complexity according to actual needs and input data availability.

The pde-type transport models include compatible sink terms to simulate water and nitrogen uptake by roots. Among the modeling approaches most often applied are the water uptake functions of the models SWACROP, SWATRE and SWAP

(Feddes et al. 1978; Belmans et al. 1983; Kroes and van Dam 2003; Feddes and Raats 2004) and the nitrogen uptake simulation method following the models NITWAT (McIsaac et al. 1985), WHNSIM (Huwe and van der Ploeg 1991) and WAVE (Vanclooster et al. 1995).

The water uptake approach defines the sink term S_w (t, z, h) [day^{-1}] of the Richards equation depending on time t, soil depth z and matric potential h by

$$S_w(t,z,h) = \alpha(h) S_{\max}(t,z) \tag{25}$$

where $0 \leq \alpha(h) \leq 1$ is a dimensionless water stress function which tends to zero at very wet or very dry soil, and where S_{max} (t, z) [day^{-1}] denotes the maximal water uptake rate from the rooted soil profile given by:

$$S_{\max}(t,z) = \beta(t,z) TR_{pot}(t) \tag{26}$$

using the potential transpiration rate TR_{pot} (t) [mm day^{-1}] at time t and the potential water uptake distribution β (t, z) [mm^{-1}], which is obtained from the root density distribution. The actual transpiration or soil water extraction rate TR_{act} (t) [mm day^{-1}] at time t is then calculated by integration to the actual root depth z_r:

$$\begin{aligned} TR_{act}(t) &= \int_0^{z_r} S_w(t,z,h)\,\mathrm{d}z = \int_0^{z_r} S_{\max}(t,z)\alpha(h)\,\mathrm{d}z \\ &= TR_{pot}(t)\int_0^{z_r} \alpha(h)\beta(t,z)\,\mathrm{d}z \end{aligned} \tag{27}$$

The sink term for the nitrogen transport model given by the N-uptake rate $\mathrm{d}N_{upt}^{act} / \mathrm{d}t$ [kg(N) ha^{-1} day^{-1}] results from the sum of the N-uptake by convection $\mathrm{d}N_{upt.conv}^{act} / \mathrm{d}t$ [kg (N) ha^{-1} day^{-1}], i.e. the mineral N-uptake from the soil solution by water uptake due to transpiration, and the N-uptake by diffusion $\mathrm{d}N_{upt.conv}^{act} / \mathrm{d}t$ [kg(N) ha^{-1} day^{-1}] due to diffusion of mineral N into the plant roots.

The actual convective N-uptake during the time interval Δt is calculated from the nitrate-N concentration c_{NO_3} and the ammonium-N concentration c_{NH_4} in the soil solution taken up by the roots with the transpiration flux represented by the sink term S_w [mm day^{-1}]:

$$\mathrm{d}N_{upt,conv}^{act,\Delta t} / \mathrm{d}t = \int_0^{z_r} \left[\int_{\Delta_t} S_w(t,z,h)(c_{NO_3} + c_{NH_4})\,\mathrm{d}t \right] \mathrm{d}z \tag{28}$$

The diffusive N-uptake results by considering the potential diffusive N-uptake $d\,N_{upt.dijj}^{pot,Dt} / \mathrm{d}t$ [kg(N) ha^{-1} day^{-1}] during Δt which is given by the difference between the N-demand of the crop, i.e. the total potential N-uptake $\mathrm{d}\,N_{upt}^{pot,Dt} / \mathrm{d}t$ [kg(N) ha^{-1} day^{-1}] during Δt and the actual convective N-uptake. By considering the maximal N-uptake by diffusion $\mathrm{d}\,N_{upt.dijj}^{max,Dt} / \mathrm{d}t$ [kg(N) ha^{-1} day^{-1}] that can occur during Δt, the actual diffusive N-uptake results from:

$$dN_{\text{upt,diff}}^{\text{act},\Delta t} / dt = \min(dN_{\text{upt,diff}}^{\text{pot},\Delta t} / dt; dN_{\text{upt,diff}}^{\max,\Delta t} / dt) \tag{29}$$

To this end the maximal diffusive N-uptake rate is determined by modeling the maximal possible radial diffusive flux of mineral N into a cylindrical root section under the actual soil conditions.

3 Assimilate Partitioning

3.1 Carbon and Nitrogen Allocation

Four main concepts to model the partitioning of assimilates have been used in plant and crop modeling as identified by Thornley and Johnson (1990) and subsequently confirmed by Lacointe (2000), LeRoux et al. (2001) and Genard et al. (2008). These concepts are either empirical, teleonomic, based on source-sink relations or based on transport and transformation processes.

A typical example of an empirical approach is the crop and variety specific assimilate partitioning scheme applied in the crop model SUCROS. Following Penning de Vries et al. (1989) the daily dry matter biomass growth rates of the different plant organs $\mu_{\text{B,plo}}^{\text{day}}$ [kg ha^{-1} day^{-1}], i.e. of leaves (*plo* = *lvs*), stems (*plo* = *stm*), storage organs (*plo* = *sto*) and roots (*plo* = *rts*) depend directly on the daily pool of carbohydrates $C_{\text{asm}}^{\text{day}}$ [kg(CH$_2$ O) ha^{-1} day^{-1}] that is available for growth. This pool will be partitioned to the plant organs according to their growth rates:

$$\mu_{\text{B,plo}}^{\text{day}} = C_{\text{asm}}^{\text{day}} f_{\text{ap,plo}} \zeta_{\text{B,crp}}^{-1} \tag{30}$$

where the efficiency coefficient of glucose to dry matter transformation $\zeta_{\text{B,crp}}$ [-] for the entire canopy is given by

$$\zeta_{\text{B,crp}} = f_{\text{ap,lvs}} \zeta_{\text{B,lvs}} + f_{\text{ap,stm}} \zeta_{\text{B,stm}} + f_{\text{ap,sto}} \zeta_{\text{B,sto}} + f_{\text{ap,rts}} \zeta_{\text{B,rts}} \tag{31}$$

resulting from the efficiency coefficients $\zeta_{\text{B,plo}}$ [-] and the partitioning coefficients $f_{\text{ap,plo}}$ [-] of the single plant organs. Whereas the efficiency coefficients are given by fixed empirical values, the partitioning coefficients can depend on the actual development stage s_{dev} via fixed tabular functions $f_{\text{plo}}(s_{\text{dev}})$ for the shoot (*plo* = *sht*), the leaves (*plo* = *lvs*) and the stems (*plo* = *stm*):

$$f_{\text{ap,sht}} = \zeta_{\theta} f_{\text{sht}}(s_{\text{dev}}), f_{\text{ap,rts}} = 1 - f_{\text{ap,sht}} \tag{32}$$

$$f_{\text{ap,lvs}} = f_{\text{ap,sht}} f_{\text{lvs}}(s_{\text{dev}}) \tag{33}$$

$$f_{\text{ap,stm}} = f_{\text{ap,sht}} f_{\text{stm}}(s_{\text{dev}}), f_{\text{ap,sto}} = f_{\text{ap,sht}} - f_{\text{ap,lvs}} - f_{\text{ap,stm}} \tag{34}$$

The additional factor ζ_θ [-] is a water stress factor, which represents the ratio of actual to potential transpiration and increases the proportion of the assimilates assigned to the root to promote root growth under conditions of water shortage.

Although the partitioning coefficients can vary during the season and can be modulated by external conditions such as water shortage, they can be applied only in the range of conditions for which they have been measured. However, if these conditions are actually met, this empirical allocation concept can be very efficient and usually gives reasonable predictions (LeRoux et al. 2001). The same or similar empirical allocation concepts are applied in the crop model packages WOFOST and APSIM, in the CERES crop model family and in the models SPASS and Hybrid-Maize among others. In the CERES crop models, this empirical approach is further extended by the application of priority schemes of resource partitioning. In this extension growth of some organs during certain development stages can only occur if enough assimilates are available after fulfilling the demand of the plant organs, which are favored in this stage of growth. Moreover, in the above crop models, also N allocation, if considered, follows the empirical approach by prescribing empirical values of the optimal N contents for shoot and root biomass, from which the N demand of the crop is derived.

To better describe the shoot-root partitioning which possibly responds most to environmental conditions, in the model GECROS a functional balance is assumed (Yin and van Laar 2005). This approach belongs to the teleonomic concepts where some intrinsic goal is prescribed, e.g. a functional balance between shoot and root or a fixed relationship between conducting tissue and foliage as in the pipe-model (Genard et al. 2008). According to the functional balance theory between root and shoot, the N acquisition by roots is proportional to the C gain by shoots. This concept is applied to relate the newly allocated C of shoots and roots

$$(\sigma_N / \eta)\frac{\mathrm{d}C_{\mathrm{rts}}}{\mathrm{d}t} = \sigma_C \frac{\mathrm{d}C_{\mathrm{sht}}}{\mathrm{d}t} \tag{35}$$

where C_{sht}, C_{rts} denote the respective C contents [kg(C) ha^{-1}] of shoot and root, where σ_C [kg(C) kg^{-1} (C) day^{-1}] is the relative shoot activity which is defined by $\sigma_C = (\mathrm{d}C / \mathrm{d}t) / C_{\mathrm{sht}}$, where σ_N [kg(N) kg^{-1} (C) day^{-1}] is the relative root activity given by $\sigma_N = (\mathrm{d}N / \mathrm{d}t) / C_{\mathrm{rts}}$ and where η is the N/C-ratio [-] in newly formed plant biomass. Then, by additionally assuming that specific root and shoot activities are independent of the root and shoot weights, Yin and Schapendonk (2004) derived equations for the fraction of newly assimilated C partitioned to the shoot $\lambda_{\mathrm{C,sht}}$ [-] and for the fraction of newly absorbed nitrogen N partitioned to the shoot $\lambda_{\mathrm{N,sht}}$ [-] (Yin and van Laar 2005):

$$\lambda_{\mathrm{C,sht}} = \frac{1}{1+\sigma_C \eta / \sigma_N} \tag{36}$$

$$\lambda_{\mathrm{N,sht}} = \frac{1}{1+[\sigma_C \eta N_{\mathrm{rts}} C_{\mathrm{sht}} / (\sigma_N N_{\mathrm{sht}} C_{\mathrm{rts}})]} \tag{37}$$

By further introducing an optimum principle, which provides an example for another type of teleonomic concept, Yin and van Laar (2005) incorporated an additional mechanism that allows plants to maximize their relative C gain by an appropriate control of the root-shoot partitioning. Based on Hilbert (1990) the following optimization criterion for maximum relative C gain is derived (Yin and van Laar 2005):

$$\sigma_N = \sigma_C^2 / (\mathrm{d}\sigma_C / \mathrm{d}\kappa) \tag{38}$$

and used to modify the C and N partitioning equations for $\lambda_{C,\mathrm{sht}}$ and $\lambda_{N,\mathrm{sht}}$ by substituting σ_N according to (38). The partitioning fractions then depend on the first order derivative of σ_C with respect to the N/C-ratio κ [-] of the total plant.

In the model GECROS the within-shoot C partitioning is determined by the strengths of growing organs as sinks for the available C and by assuming a priority scheme: "carbon goes first to the seed, next to the structural stem if carbon exceeds the demand of the seed and then to the leaf. Any further remaining carbon is transported to the shoot reserve pool" (Yin and van Laar 2005). Intra-shoot N partitioning is modeled by a target N concentration (input parameter). By assuming minimum leaf, stem and root N concentrations (input parameters) also the remobilization of N from leaves, stems and roots and the related stimulation of leaf and root senescence is simulated.

This approach for the within-shoot partitioning is an example for partitioning concepts that are based on the sink strength, i.e. the respective ability of different sinks to get available assimilates from the same source (Lacointe 2000). This sink strength results from the genetically determined potential growth and maintenance respiration rates and the potential net sink strength, which is defined by maximum rate at which the organ can accumulate dry matter (Genard et al. 2008).

The concepts to model assimilate partitioning between plant organs that are based on within-plant transport and transformation processes need an explicit consideration of plant architecture. Although many functional structural plant models have been developed during the last decade, these models are essentially single plant models and have not yet been applied at the whole canopy level. This is mainly due to the still weak understanding of the allocation among plant organs, but also due to the lack of appropriate up-scaling procedures that can consider the enormous complexity of plant-plant interactions at the canopy level (Marcelis and Heuvelink 2007; Vos et al. 2007).

3.2 Defense and Repair

Partitioning of assimilates and acquisition of resources may be further affected by different kinds of biotic and abiotic stresses like pathogen diseases or atmospheric pollutants (e.g. ozone). Consequently, in many crop growth models feedback mechanisms between the intensity of stress factors and plant growth or yield are considered. On the one side stress effects on plant growth are described at the process level. Examples are the reduction of the resource uptake capacities, the reduction of the light use efficiency, the acceleration of leaf senescence or the direct loss of active leaf area or root length. On the other side, stressors can cause additional assimilate sinks. Examples are the direct loss of assimilates by sap sucking insects, assimilate consumption in conjunction with repair of damaged tissue or enzymes, and the synthesis of defense-related secondary metabolites, which are not available for growth process.

The first models of the development of plant diseases simulate the impact of weather, host and pathogen genotypes, and crop management factors on disease severity but do not make the link to dynamic simulation models of crop growth and yield [e.g. Bruhn and Fry (1981)]. Modeling approaches that incorporate effects of insects or plant diseases in dynamic crop growth modules were developed since the 1980s (Boote et al. 1983; Rabbinge and Bastiaans 1989; Kropff et al. 1995). Such functions are available in several SUCROS and CERES derived models and in the models of the DSSAT package (van Ittersum et al. 2003; Jones et al. 2003). In these models, the intensities of different kind of stresses are simply introduced as time dependend forcing functions which can be specified by the model user. Feedback mechanisms are considered between the number of insects or the disease severity and the degree to which a process is affected by means of empirical relations. Due to the versatility of biotic stressors, many different approaches for the simulation of damage mechanisms were developed. Typical examples provides the model RICEPEST (Willocquet et al. 2002) which demonstrates how rice yield losses due to eight types of injuries can be simulated. The biomass sub-model of RICEPEST accounts for the daily accumulation and partitioning of assimilates towards plant organs. The partitioning coefficients determining the growth rates of leaves, stem, roots, and panicles depend on the development stage, corresponding to the SUCROS model. Different coupling points for damage mechanisms due to pests, diseases and weeds are included in the biomass growth sub-model. The dynamical development of vegetative and reproductive tiller numbers is calculated in a second sub-model.

1. Decrease of living leaf weight, B_{lvs} [kg ha^{-1}], may result from accelerated senescence caused by sheath blight or sap sucking brown plant hoppers, by die back of tillers due to different species of stem borers, or by defoliating insects.

$$\frac{\mathrm{d}B_{\mathrm{lvs}}}{\mathrm{d}t} = \mu_{\mathrm{B,lvs}} - \mu_{\mathrm{sen,lvs}} - \mu_{\mathrm{shb,lvs}} - \mu_{\mathrm{bph,lvs}} - \mu_{\mathrm{dh,lvs}} - \mu_{\mathrm{def,lvs}} \tag{39}$$

where $\mu_{\mathrm{B,lvs}}$ [kg ha^{-1} day^{-1}] denotes the growth rate of leaves, and $\mu_{\mathrm{sen,lvs}}$ [kg ha^{-1} day^{-1}] is the rate of leaf senescence depending only on the development stage. While the senescence rate due to sheath blight, $\mu_{\mathrm{shb,lvs}}$ [kg ha^{-1} day^{-1}], is assumed to be proportional to B_{lvs} multiplied with an user specified sheath blight severity factor, the senescence rate due to brown plant hopper, $\mu_{\mathrm{bph,lvs}}$ [kg ha^{-1} day^{-1}], is simulated proportional to the biomass of the insect, which also is given as a time dependent input quantity. Leaf damage due to stem borers, $\mu_{\text{dh,lvs}}$ [kg ha^{-1} day^{-1}], is accounted for by substracting dead tillers at a daily rate from the number of tillers and numerically linking this loss to the dry weight of leaves. The defoliation rate $\mu_{\text{def,lvs}}$ [kg ha^{-1} day^{-1}] corresponds to the amount of leaf weight which is removed daily by defoliating insects.

2. Loss of stem weight, B_{stm} [kg ha^{-1}], is accounted for in the model by stem borers. Corresponding to leaves, stem damage $\mu_{\text{dh,stm}}$ [kg ha^{-1} day^{-1}] is calculated from a daily rate of tiller die back:

$$\frac{dB_{\text{stm}}}{dt} = \mu_{\text{B,stm}} - \mu_{\text{dst}} - \mu_{\text{dh,stm}} \tag{40}$$

where $\mu_{\text{B,stm}}$ [kg ha^{-1} day^{-1}] is the growth rate of stem biomass and μ_{dst} [kg ha^{-1} day^{-1}] is the redistribution rate of biomass to panicles.

3. Reduced assimilate partitioning to panicles may be caused by stem borers and sheath rot lesions. Tillers infected by stem borers during the reproductive phase bear panicles in which grain filling does not occur. This is reflected in the model by yield losses proportional to the fraction of injured panicles. Sheath rot lesions also prevent grain filling of damaged panicles, which is reflected in the model by a second factor that decreases the rate of increase in panicle weight. By multiplying both effects it is assumed that one pathogen cannot affect a process that is already inhibited by another. Thus, panicle growth rate is calculated as:

$$\frac{dB_{\text{pan}}}{dt} = (\mu_{\text{B,pan}} + \mu_{\text{dst}})(1 - f_{\text{stb}})(1 - f_{\text{shr}}) \tag{41}$$

where $\mu_{\text{B,pan}}$ [kg ha^{-1} day^{-1}] is the panicle biomass growth rate, f_{stb} [-] is the maximum proportion of tillers infected by stem borers during the growing season and f_{shr} [-] is the fraction of panicles damaged by sheath rot.

4. Decrease in photosynthetic active green leaf area is considered in the model as consequence of sheath blight, brown spot lesions and the bacterial leaf blight:

$$f_{\text{LAI}} = f_{\text{sla}}(t) B_{\text{lvs}} (1 - f_{\text{sbs}})(1 - f_{\text{bsl}})^{\beta} (1 - f_{\text{blb}}) \tag{42}$$

where f_{LAI} [ha ha^{-1}] is the leaf area index, f_{sla} [ha kg^{-1}] is the specific leaf area, f_{sbs} [-] is the sheath blight severity fraction and f_{bsl} [-] and f_{blb} [-] are the fractions of leaf area covered by visual brown spot and bacterial leaf blight lesion. β [-] is the ratio between the leaf area covered by visual brown spot lesions and the leaf area covered by visual lesions plus the surrounding area that cannot photosynthesize due to the effect of fungal toxins.

5. Effects of competition by weeds for light, nutrients and water are simply introduced by a combined reduction factor affecting the crop growth rate $\mu_{\text{B}}^{\text{day}}$ [kg ha^{-1} day^{-1}], which is, like in CERES-models, calculated from the radiation interception of the canopy and the radiation use efficiency f_{RUE} [kg MJ^{-1}]:

$$\mu_{\mathrm{B}}^{\mathrm{day}} = f_{\mathrm{RUE}} R_{\mathrm{ipa}}^{\mathrm{day}} (1 - \exp^{-0.003 B_{\mathrm{weed}}}) \tag{43}$$

where α_{ext} [-] is the coefficient of light extinction and f_{LAI} is the green leaf area index and the variable B_{weed} is the total dry weight of weeds in the area. Examples for more explicit models of crop weed interactions, among many others, are given by Kropff and van Laar (1993) and Röhrig and Stützel (2001). In these models biomass of weeds may not be predefined by the model user, but is simulated also by an eco-physiological model. In this case the strength of competition effects depends on the distribution of the growth-determining or growth-limiting resources light, water and nitrogen to the competing species and the efficiency with which each species uses the available resources.

6. Direct loss of assimilates is considered in the model in case of infestation with the sap sucking brown plant hopper. The daily consumption rate of this stem borer is assumed to be proportional to its dry weight and is subtracted from the pool of assimilates that are available for further partitioning.

Similar to the effects of pathogens, increased levels of atmospheric ozone may cause damage to leaf tissue, which results in decreased canopy assimilation rates and increases maintenance respiration cost for ozone detoxification and repair of leaf tissue (Amthor 1988). A detailed mathematical model for the uptake and detoxification of ozone based upon the direct reaction of the pollutant with ascorbate localized within the apoplast is presented by Plöchl et al. (2000). This model takes into consideration the regeneration of dehydroascorbic acid in the cytosol, the rate of replenishment of cell wall ASC and the distribution of ASC between subcellular compartments. However, for the purpose of crop growth models simpler approaches are needed for calculating ozone effects on radiation use efficiency and assimilate partitioning, which addresses overall rates and costs of detoxification and repair. An approach for the estimation of metabolic costs of detoxification and repair, which reduce the amount of assimilates that are available for partitioning to plant organs is incorporated in two growth models for potato (Wolf and van Oijen 2003) and spring wheat (van Oijen et al. 2004). In this approach the rate of effective ozone uptake per day $F_{\mathrm{O_3}}$ [kg ha^{-1} day^{-1}] is calculated from the photoperiod average of atmospheric ozone concentration $c_{\mathrm{O_3}}$ [ml m^{-3}], the stomatal conductance g_{S} [kg ha^{-1} day^{-1} / (mlm^{-3})] and a factor f_{detox} [-] that describes the detoxified fraction of the ozone flux entering the stomata (van Oijen et al. 2004):

$$F_{\mathrm{O_3}} = g_{\mathrm{S}} c_{\mathrm{O_3}} (1 - f_{\mathrm{detox}}) \tag{44}$$

It is assumed that the overall costs for the detoxification process is proportional to the detoxification rate itself:

$$cost_{\mathrm{detox}} = g_{\mathrm{S}} c_{\mathrm{O_3}} f_{\mathrm{detox}} c_{\mathrm{detox}} \tag{45}$$

where $cos\ t_{\mathrm{detox}}$ [kg(CH$_2$O) ha^{-1} day^{-1}] is the amount of assimilates used for detoxification and c_{detox} [kg(CH$_2$O) kg^{-1} (detoxified O$_3$)] is the cost coefficient of detoxification.

As most effects of ozone on leaf tissue are consequences of a decrease of Rubisco content of leaves, direct effects on light capture and electron transport are not considered in the model. Assimilation rate per leaf area is not affected by ozone as long as the Rubisco content may be restored by repair processes. This is the case if

$$r_{\mathrm{repair}} > \varphi F_{\mathrm{O_3}} \tag{46}$$

where r_{repair} [kg(Rubisco) ha^{-1} day^{-1}] is the repair rate of Rubisco and φ [kg(Rubisco) kg^{-1} (O_3)] the ozone damage coefficient. Based on observations it is assumed that the repair rate is directly proportional to the amount of assimilates partitioned to leaves:

$$r_{\mathrm{repair}} = C_{\mathrm{ass}}^{\mathrm{day}} f_{\mathrm{lvs}} \frac{f_{\mathrm{repair}}}{c_{\mathrm{repair}}} \tag{47}$$

where $C_{\mathrm{ass}}^{\mathrm{day}}$ [kg(CH_2O) ha^{-1} day^{-1}] is the daily amount of carbohydrates that are available for growth and maintenance, f_{lvs} [-] is the fraction of carbohydrates partitioned to leaves, f_{repair} [-] is the fraction of assimilates that is used in repair and c_{repair} [kg(CH_2O) kg^{-1} (Rubisco repaired)] is the repair cost coefficient. Direct proportionality between repair rate and the amount of assimilates used for repair, cos ts_{repair} [kg(CH_2O) ha^{-1} day^{-1}], is assumed:

$$costs_{\mathrm{repair}} = r_{\mathrm{repair}} c_{\mathrm{repair}} \tag{48}$$

A somewhat different approach for the consideration of costs for defense is pursued in the plant growth model PLATHO (Gayler and Priesack 2005; Gayler et al. 2008). This model is designed for simulating growth of crops as well as woody plants and allows the consideration of stress effects on partitioning and resource acquisition in a very general way. Possible stress factors are leaf pathogens, root pathogens and atmospheric ozone. Contrary to other crop growth models, PLATHO considers explicitly partitioning of carbohydrates towards a pool of defense-related compounds which are available for plant reactions against the stress factors. Possible damages of resource uptake capacities and other physiological processes are decreased depending on the level of these compounds in plant tissues. The partitioning of assimilates to defense-related compounds is simulated depending on crop stage and environmental factors like nitrogen supply and water supply. This approach is based on the consideration that the partitioning of the mostly limited assimilates in plants is determined by the balance between the different, sometimes conflicting demands for maintenance and growth of organs on the one side and for defense against biotic and abiotic stresses on the other side (Matyssek et al. 2005). That such a trade-off exists and that there is a significant impact of environmental factors such as atmospheric carbon dioxide concentrations, water shortage, temperature, light conditions and nutrient supply on this trade-off was demonstrated in many studies during the last years [e.g. Bazzaz (1997), Coley et al. (1985), Fine et al. (2006), Glynn et al. (2003), Herms and Mattson (1992), Koricheva (2002), Mattson et al. (2005) and Matyssek et al. 2005)]. Most of these experiments focus on polyphenolic compounds which are known to be used for defense-related functions in numerous plant-pathogen interactions (Dixon 2001).

In PLATHO, a priority scheme of resource partitioning between "growth" and "defense" is realized, that is based on different plant-defense hypotheses, which were developed during the last 20 years to explain patterns and variations in concentration of carbon-based secondary compounds in plant tissues. The most prominent of these hypotheses are the "Carbon-Nutrient Balance" (CNB) hypothesis (Bryant et al. 1983; Tuomi et al. 1991), the "Growth-Differentiation Balance" (GDB) hypothesis (Herms and Mattson 1992) and the "Protein Competition" model (PCM) (Jones and Hartley 1999). A detailed presentation and comparison of these hypotheses can be found in Stamp (2003). The key assumptions of the partitioning routine of PLATHO are directly adopted from CNB, GDB and PCM: (1) Growth takes priority over defense (Tuomi et al. 1991). This is in accordance with the CNB and GDB hypotheses which predict that additional assimilates may be converted to secondary metabolites if carbohydrates accumulate in excess of growth demands or if availability of nitrogen is lower than the nitrogen demand required for growth processes. (2) The formation of carbon based defensive compounds (e.g. phenylpropanoids) requires sufficient nitrogen levels due to the requirements for biosynthesis of precursory compounds (Jones and Hartley 1999).

The equation that describes the trade-off in partitioning resources between growth and defense on the whole plant level is adopted from a theoretical consideration by Coley et al. (1985). The total demand of assimilates, D_{pot} [kg ha^{-1} day^{-1}], is divided into one part which is related to structural growth, D_W [kg ha^{-1} day^{-1}], and another part, which is related to defense, D_S [kg ha^{-1} day^{-1}]:

$$D_W = D_{\text{pot}}(1-\sigma), \quad D_S = D_{\text{pot}} \cdot \sigma \tag{49}$$

where the factor $\sigma = \sigma_C + \sigma_I$ [-], consists of a constitutive part, σ_C, and an inducible part, σ_I. σ_I is assumed to be non zero only in the case of actual stress, such as a pathogen attack or high atmospheric ozone concentrations. The constitutive part subsumes a minimal baseline of partitioning to defense related secondary compounds that are always needed for development, survival and health, and a dispensable part that is only allocated to defensive compounds if carbohydrates are accumulated in excess of growth demands. The realized conversion rate of assimilates to structural biomass, G_W [kg ha^{-1} day^{-1}] and defense-related compounds, G_S [kg ha^{-1} day^{-1}], result from the ratios between plant internal supply and demand of carbohydrates, φ_C [-] and nitrogen, φ_N [-]:

$$G_W = D_W \cdot \varphi_C \cdot \varphi_N^{\beta} \tag{50}$$

$$G_S / A_{\text{av}} = \begin{cases} \sigma \cdot \varphi_N^{\delta} / \varphi_C \; if \; A_{\text{av}} \geq G_W + D_S \\ [1-(1-\sigma)\varphi_N^{\beta}]\varphi_N^{\delta} \text{ else} \end{cases} \tag{51}$$

where for β [–] a value of 0.5 is assumed reflecting the fact that the growth rate is increased in a non linear way by additional nitrogen supply and δ [-] is set to 0.33 reflecting the fact that the metabolism of carbon – based secondary compounds is less affected by nitrogen deficiency than growth processes, due to the continuous regeneration of phenylalanine from a limited nitrogen pool during phenolic biosynthesis.

4 Summary and Conclusions

To simulate carbon assimilation the "big leaf" approach used in the CERES crop models turned out to be a robust concept to quantify the radiation-assimilation relationship at the canopy level. Even if the "big leaf" models overestimate photosynthesis by using an averaged absorbed irradiance, the empirical adjustment of the canopy photosynthetic capacity or the curvature of the light response of photosynthesis can lead to accurate computations of canopy photosynthesis (de Pury and Farquhar 1997). The more comprehensive approach of the SUCROS crop models separately considers beam and diffusive radiation, visible and near-infrared wavelengths and multiple layers with different leaf angle classes or with only two classes given by sunlit and shaded leaves (Goudriaan 1977). This approach, which avoids empirical adjustments, is now generally accepted as necessary prerequisite to correctly scale photosynthesis dynamics from leaves to canopies (Amthor et al. 1994; de Pury and Farquhar 1997; Chen et al. 1999). To incorporate the more complex but bio-chemically based leaf photosynthesis model of Farquhar et al. (1980) without loss of model robustness compared to "big leaf" approaches that are applicable at the regional and global scale, a two-leaf approach has been developed and successfully applied (de Pury and Farquhar 1997; Wang and Leuning 1998; Beyschlag and Ryel 2007). This two-leaf approach was integrated into the crop model GECROS (Yin and van Laar 2005) using robust closed form expressions to represent leaf nitrogen and wind velocity profiles within the canopy to correctly up-scale the relationships between absorbed irradiance, photosynthesis, transpiration and N supply from leaf to canopy. In this way, an efficient canopy photosynthesis model is made available which is based on a physically and bio-chemically appropriate description of leaf photosynthesis and its interaction with environmental factors. This considerably improves the reliability of agricultural crop growth simulations, in particular for applications to asses the response of crop production under changing climatic conditions.

In contrast to the modeling of C assimilation in leaves and its up-scaling to the canopy level, the acquisition of water and N from soil by roots is described by rather simple approaches. "In almost all the crop models, soil is considered to consist of horizontal layers, each layer being homogeneous and differences only existing between layers. The soil factors involved include temperature, soil water, and soil nitrogen content. The response of root growth to soil strength, soil structure, toxic chemical conditions, and soil microbiological conditions has been poorly treated or neglected" (Wang and Smith 2004). In particular, the spatial up-scaling in most crop models is achieved by simply assuming that already the average vertical root distribution of the canopy can represent the plant induced factor of the spatial pattern of water and N uptake. Similar to the "big leaf" summary concept this approach is not a scaled model in the sense that the variables have different definitions and interrelationships at each scale (Norman 1993). The approach may thus be named a "big root" approach. To derive more realistic models of water and N uptake by roots the plant-soil interactions have to be considered in more detail: The impact of different soil water flow regimes on water uptake such as preferential flow through macro-pores, bulk flow through the soil matric and eventually water vapor transport may be considered. For the N uptake, the convective and diffusive solute transport may play different important roles under different soil conditions. Certain roots may live under more favorable conditions and may be considered in analogy to sunlit leaves. Moreover, it is not clear which part and how much of the root system is active and potentially involved in uptake at a given time under given environmental conditions. Also the effects of diseases and pest on root growth and root activity need to be included (Wang and Smith 2004). The below ground modeling of plants' growth and resource acquisition often is not of the same quality as for the above ground and therefore, to keep crop models well-balanced, efforts to better describe canopy root systems and their abilities are urgently needed.

Different concepts to describe the partitioning of assimilates have been developed and mainly applied in tree growth models. In most agricultural crop models an empirical approach is pursued, which is based on fixed partitioning coefficients that may additionally depend on development stages. Because of their empirical nature these coefficients are only useful for similar sites and conditions at which they have been determined and their general applicability cannot be assumed. In the recently developed generic agricultural crop model GECROS more general concepts as functional balancing between root and shoot or an optimum principle have been applied to describe the allocation of assimilated C and absorbed N. During the model development it was tried to keep the experimental effort for parameterization as simple as possible, to avoid the dependence on local environment and to restrict crop growth model input data preferably to parameters that reflect the differences among genotypes (Yin and van Laar 2005).

The common way to describe pests in crop growth models is the approach where scouting data on pest damage are input into the crop model. There is still a lack of mechanistic simulation of pest dynamics, with concurrent coupling to crop models and it is a task for the future to create models of disease damage as a function of temperature, humidity, rainfall, and crop stage. These models could run coupled with the crop model to give predictions of disease effects on the crop without need for input of disease damage. More general models express the defense readiness of the plant by the level of defense-related metabolites in plant tissues. The main difficulty of these models that consider the dynamics of the mutual interaction between plant resistance and spread of diseases in crop canopies is due to the fact, that the total leaf concentration of defense-related compounds is not always crucial for the success of pathogen defense (Leser and Treutter 2005; Mittelstraß et al. 2006). Experiments with different cultivars of potato have shown that the velocity of the reaction rather than the level of defensive compounds is decisive for the success of the plant response (Ros et al. 2005).

To sum up, in contrast to the two decades before (Weiss 2003) during the last 5 years new efforts have been made to improve the model descriptions of assimilation and partitioning in agricultural crop canopies. General concepts often already applied in tree growth modeling have been introduced to canopy-level crop models to improve their general applicability. But, there is evidence that additional research is needed to also improve the modeling of below ground growth and resource acquisition of crops. The development and use of individual-based plant models, that explicitly consider 3D root architecture coupled to 3D soil models would help to shed some light on the dark and hidden nature of roots and soil.

References

Amthor JS (1988) Growth and maintenance respiration in leaves of bean (Phaseolus vulgaris L.) exposed to ozone in open-top chambers in the field. New Phytol 110:319–325

Amthor J, Goulden M, Munger J, Wofsy S (1994) Testing a mechanistic model of forest-canopy mass and energy exchange using eddy correlation: carbon dioxide and ozone uptake by a mixed oak-maple stand. Funct Plant Biol 21:623–651

Asseng S, Milroy SP (2006) Simulation of environmental and genetic effects on grain protein concentration in wheat. Eur J Agron 25:119–128

Bazzaz F (1997) Allocation of resources in plants: state of the science and critical questions. In: Bazzaz F, Grace J (eds) Plant resource allocation. Academic, San Diego, pp 1–37

Beek J, Frissel M (1973) Simulation of nitrogen behavior in soils. PUDOC, Wageningen

Belmans C, Wesseling J, Feddes R (1983) Simulation model of the water balance of a cropped soil: SWATRE. J Hydrol 63:271–286

Berghuijs-van Dijk J, Rijtema P, Roest C (1985) ANIMO Agricultural Nitrogen Model. NOTA 1671. Institute for Land and Water Management Research, Wageningen

Beyschlag W, Ryel R (2007) Canopy photosynthesis modeling. In: Pugnaire F, Valladares F (eds) Functional plant ecology, 2nd edn. CRC, Boca Raton, pp 627–653

Boogaard H, van Diepen C, Rötter R, Cabrera J, van Laar H (1998) WOFOST 7.1, User's guide for the WOFOST 7.1 crop growth simulation model and WOFOST control center 1.5. Tech. rep., DLO Winand Staaring Centre

Boote K, Jones J, Mishoe J, Berger R (1983) Coupling pests to crop growth simulators to predict yield reductions. Phytopathology 73:1581–1587

Bouman BAM, van Keulen H, van Laar HH, Rabbinge R (1996) The "School of de Wit" crop growth simulation models: A pedigree and historical overview. Agric Sys 52:171–198

Brisson N, Gary C, Justes E, Roche R, Mary B, Ripoche D, Zimmer D, Sierra J, Bertuzzi P, Burger P, Bussiere F, Cabidoche YM, Cellier P, Debaeke P, Gaudillere JP, Henault C, Maraux F, Seguin B, Sinoquet H (2003) An overview of the crop model. Eur J Agron 18:309–332

Bruhn J, Fry W (1981) Analysis of potato light blight epidemiology by simulation modelling. Phytopathology 71:612–616

Bryant J, Chapin F, Klein D (1983) Carbon/nutrient balance of boreal plants in relation to vertebrate herbivory. Oikos 40:357–368

Chen JM, Liu J, Cihlar J, Goulden ML (1999) Daily canopy photosynthesis model through temporal and spatial scaling for remote sensing applications. Ecol Model 124:99–119

Coley P, Bryant J, Chapin F (1985) Resource availability and plant antiherbivore defense. Science 230:895–899

de Pury D, Farquhar G (1997) Simple scaling of photosynthesis from leaves to canopies without errors of big-leaf models. Plant Cell Environ 20:537–557

de Willigen P (1991) Nitrogen turnover in the soil-crop system: comparison of fourteen simulation models. Fert Res 27:141–149

Diekkrüger B, Arning M (1995) Simulation of water fluxes using different methods for estimating soil parameters. Ecol Model 81:83–95

Diekkrüger B, Söndgerath D, Kersebaum KC, McVoy CW (1995) Validity of agroecosystem models a comparison of results of different models applied to the same data set. Ecol Model 81:3–29

Dixon RA (2001) Natural products and plant disease resistance. Nature 411:843–847

Dutt G, Shaffer M, Moore W (1972) Computer simulation model of dynamic biophysiochemical processes in soils. Univ. Arizona Agric. Exp. Stn. Tech. Bull. 196

Eitzinger J, Trnka M, Hosch J, Zalud Z, Dubrovsky M (2004) Comparison of CERES, WOFOST and SWAP models in simulating soil water content during growing season under different soil conditions. Ecol Model 171:223–246

Engel T, Priesack E (1993) Expert-N, a building block system of nitrogen models as a resource for advice, research, water management and policy. In: Eijsackers H, Hamers T (eds) Integrated soil and sediment research: a basis for proper protection. Kluwer Academic, Dordrecht, pp 503–507

Engel T, Klöcking B, Priesack E, Schaaf T (1993) Simulationsmodelle zur Stickstoffdynamik, vol. 25 of Agrarinformatik. Verlag Eugen Ulmer, Stuttgart

Farquhar GD, Caemmerer S, Berry JA (1980) A biochemical model of photosynthetic CO_2 assimilation in leaves of C3 species. Planta 149:78–90

Feddes RA, Raats P (2004) Parameterizing the soil-water-plant root system. In: Feddes RA, De Rooij G, van Dam JC (eds) Unsaturated-Zone Modelling, Wageningen UR Frontis Series 6. Kluwer Academic, Dordrecht, pp 95–141

Feddes R, Kowalik P, Zaradny H (1978) Simulation of field water use and crop yield. Simulation Monographs. Pudoc, Wageningen

Fine PV, Miller Z, Mesones I, Irazuzta S, Appel H, Stevens M, Sääksjärvi I, Schultz J, Coley P (2006) The growth-defense trade-off and habitat specialization by plants in Amazonian forests. Ecology 87(Suppl.):150–162

Franko U, Oelschlagel B, Schenk S (1995) Simulation of temperature-, water- and nitrogen dynamics using the model CANDY. Ecol Model 81:213–222

Friend AD (2001) Modelling canopy CO_2 fluxes: are "big-leaf" simplifications justified? Global Ecol Biogeogr 10:603–619

Frissel M, Poelstra P, Reiniger P (1970) Chromatographic transport through soils. III. A simulation model for the valuation of the apparent diffusion coefficient in undisturbed soils with tritiated water. Plant Soil 33:161–176

Gayler S, Priesack E (2005) PLATHO, a simulation model of resource allocation in the plant-soil system. Tech. rep., GSF – Institute of Soil Ecology. http://www.sfb607.de/english/projects/c2/platho.pdf, Cited 31 Dec 2007

Gayler S, Wang E, Priesack E, Schaaf T, Maidl FX (2002) Modelling biomass growth, N-uptake and phenological development of potato crop. Geoderma 105:367–383

Gayler S, Grams T, Heller W, Treutter D, Priesack E (2008) A dynamic model of environmental effects on allocation to carbon-based secondary compounds in juvenile trees. Ann Bot. 101, 1089–1098 doi:10.1093/aob/mcm169

Genard M, Dauzat J, Franck N, Lescourret F, Moitrier N, Vaast P, Vercambre G (2008) Carbon allocation in fruit trees: from theory to modelling. Trees 22, 269–282 doi: 10.1007/s00468-007-0176-5

Gilding B (1992) Mathematical modelling of saturated and unsaturated groundwater flow. In: Shutie X (ed) Flow and transport in porous media. Summer School, Beijing, China, 8.–26. August 1988. World Scientific, Singapore, pp 1–166

Glynn C, Herms DA, Egawa M, Hansen R, Mattson WJ (2003) Effects of nutrient availability on biomass allocation as well as constitutive and rapid induced herbivore resistance in poplar. Oikos 101:385–397

Goudriaan J (1977) Crop micrometeorology: a simulation study. Simulation Monographs. Pudoc, Wageningen

Goudriaan J (1986) A simple and fast numerical method for the computation of daily totals of crop photosynthesis. Agric Forest Meteorol 38:249–254

Goudriaan J, van Laar H (1994) Modelling potential crop growth processes. Textbook with exercises. Kluwer Academic, Dordrecht

Hansen S, Jensen H, Nielsen N, Svendsen H (1990) DAISY – Soil Plant Atmosphere System Model. The Royal Veterinary and Agricultural University, Copenhagen

Herms DA, Mattson WJ (1992) The dilemma of plants: to grow or defend. Q Rev Biol 67:283–335

Hilbert DW (1990) Optimization of plant root: shoot ratios and internal nitrogen concentration. Ann Bot 66:91–99

Hutson J, Wagenet R (1992) LEACHM: Leaching Estimation And Chemistry Model: A process-based model of water and solute movement, transformations, plant uptake and chemical reactions in the unsaturated zone. Version 3.0. Research Series No. 93–3. Cornell University, Ithaca, NY

Huwe B, van der Ploeg RR (1991) WHNSIM – a soil nitrogen simulation model for Southern Germany. Nutr Cycl Agroecosys 27:331–339

Johnsson H, Bergström L, Jansson P, Paustian K (1987) Simulated nitrogen dynamics and losses in a layered agricultural soil. Agric Ecosys Env 18:333–356

Jones C, Kiniry J (1986) CERES-Maize: a simulation model of maize growth and development. Texas A&M University Press, Temple, TX

Jones CG, Hartley SE (1999) A protein competition model of phenolic allocation. Oikos 86:27–44

Jones C, Bland W, Ritchie J, Williams JR (1991) Simulation of root growth. In: Hanks J, Ritchie J (eds) Modeling plant and soil systems, Agronomy 31. ASA, CSSA, SSSA, Madison, WI, pp 91–123

Jones JW, Keating BA, Porter CH (2001) Approaches to modular model development. Agric Sys 70:421–443

Jones JW, Hoogenboom G, Porter CH, Boote KJ, Batchelor WD, Hunt LA, Wilkens PW, Singh U, Gijsman AJ, Ritchie JT (2003) The DSSAT cropping system model. Eur J Agron 18:235–265

Keating BA, Carberry PS, Hammer GL, Probert ME, Robertson MJ, Holzworth D, Huth NI, Hargreaves JNG, Meinke H, Hochman Z, McLean G, Verburg K, Snow V, Dimes JP, Silburn M, Wang E, Brown S, Bristow KL, Asseng S, Chapman S, McCown RL, Freebairn DM, Smith CJ (2003) An overview of APSIM, a model designed for farming systems simulation. Eur J Agron 18:267–288

Kersebaum KC (1989) Die Simulation der Stickstoff-Dynamik von Ackerböden. Ph.D. thesis, Universität Hannover

Kersebaum KC (1995) Application of a simple management model to simulate water and nitrogen dynamics. Ecol Model 81:145–156

Kersebaum K, Hecker JM, Mirschel W, Wegehenkel M (2007) Modelling water and nutrient dynamics in soil-crop systems: a comparison of simulation models applied on common data sets. In: Kersebaum K, Hecker JM, Mirschel W, Wegehenkel M (eds) Modelling water and nutrient dynamics in soil-crop systems. Springer, Dordrecht, pp 1–17

Koricheva J (2002) Meta-analysis of sources of variation in fitness costs of plant antiherbivore defenses. Ecology 83:176–190

Kroes J, van Dam J (2003) Reference Manual SWAP 3.03. Alterra-report 773. Alterra, Green World Research, Wageningen, The Netherlands

Kroes J, Roelsma J (2007) Simulation of water and nitrogen flows on field scale; application of the SWAP-ANIMO model for the Müncheberg data set. In: Kersebaum K, Hecker JM,

Mirschel W, Wegehenkel M (eds) Modelling water and nutrient dynamics in soil-crop systems. Springer, Dordrecht, pp 111–128

Kropff MJ, van Laar HH (1993) Modelling crop-weed interactions. CAB International, Wallingford, UK

Kropff MJ, Teng PS, Rabbinge R (1995) The challenge of linking pest and crop models. Agr Syst 49:413–434

Lacointe A (2000) Carbon allocation among tree organs: a review of basic processes and representation in functional–structural tree models. Ann For Sci 57:521–533

Leonard R, Knisel W, Still D (1987) GLEAMS: grondwater loading effects of agricultural management systems. Trans ASAE 30:1403–1418

LeRoux X, Lacointe A, Escobar-Gutiérrez A, Dizès SL (2001) Carbon-based models of individual tree growth: a critical appraisal. Ann For Sci 58:469–506

Leser C, Treutter D (2005) Effects of nitrogen supply on growth, contents of phenolic compounds and pathogen (scab) resistance of apple trees. Physiol Plantarum 123:49–56

Leuning R, Kelliher FM, Pury DGG, Schultze ED (1995) Leaf nitrogen, photosynthesis, conductance and transpiration: scaling from leaves to canopies. Plant Cell Environ 18:1183–1200

Li C, Frolking S, Frolking T (1992) A model of nitrous oxide evolution from soil driven by rainfall events. 1. Model structure and sensitivity. J Geophys Res 97:9759

Lizaso JI, Batchelor WD, Boote KJ, Westgate ME (2005) Development of a leaf-level canopy assimilation model for CERES-maize. Agron J 97:722–733

Marcelis L, Heuvelink E (2007) Concepts of modelling carbon allocation among plant organs. In: Vos J, Marcelis L, de Visser P, Struijk P, Evers J (eds) Functional–structural plant modelling in crop production, Wageningen UR Frontis Series 22. Springer, Dordrecht, The Netherlands, pp 103–111

Martre P, Jamieson PD, Semenov MA, Zyskowski RF, Porter JR, Triboi E (2006) Modelling protein content and composition in relation to crop nitrogen dynamics for wheat. Eur J Agron 25:138–154

Mattson W, Julkunen-Tiitto R, Herms D (2005) CO_2 enrichment and carbon partitioning to phenolics: do plant responses accord better with the protein competition or the growth-differentiation balance model? Oikos 111:337–347

Matyssek R, Agerer R, Ernst D, Munch JC, Osswald W, Pretzsch H, Priesack E, Schnyder H, Treutter D (2005) The plant's capacity in regulating resource demand. Plant Biol 7:560–580

McIsaac G, Martin D, Watts D (1985) Users guide to NITWAT – a nitrogen and water management model. Agr. Eng. Dpt. University of Nebraska, Lincoln, NA

Mehran M, Tanji K (1974) Computer modeling of nitrogen transformations in soil. J Environ Qual 3:391–396

Mittelstraß K, Treutter D, Pleßl M, Heller W, Elstner E, Heiser I (2006) Modification of primary and secondary metabolism of potato plants by nitrogen application differentially affects resistance to phytophthora infestans and alternaria solani. Plant Biol 8:653–661

Molina J, Clapp C, Shaffer M, Chichester F, Larson W (1983) NCSOIL – a model of nitrogen and carbon transformations in soil: description, calibration and behavior. Soil Sci Soc Am J 47:85–91

Monteith J (1973) Principles of environmental physics. Edward Arnold, London

Nikolov NT, Massman WJ, Schoettle AW (1995) Coupling biochemical and biophysical processes at the leaf level: an equilibrium photosynthesis model for leaves of C3 plants. Ecol Model 80:205–235

Norman J (1993) Scaling processes between leaf and canopy levels. In: Ehleringer J, Field C (eds) Scaling Physilogical Processes: Leaf to Global. Academic, London, pp 43–75

Parton W, Ojima D, Cole C, Schimel D (1994) A general model for soil organic matter dynamics: sensitivity to litter chemistry, texture and management. In: Bryant R, Arnold R (eds) Quantitative modeling of soil forming processes. Soil Sci. Soc. Am., Madison, WI, pp 147–167

Penning de Vries F, Jansen D, ten Berge H, Bakema A (1989) Simulation of ecophysiological processes of growth in several annual crops. Simulation Monographs 29. Pudoc, Wageningen, NL

Plöchl M, Lyons T, Ollerenshaw J, Barnes J (2000) Simulating ozone detoxification in the leaf apoplast through the direct reaction with ascorbate. Planta 210:454–467

Priesack E (2006) Expert-N Dokumentation der Modell-Bibliothek. FAM Bericht 60. Hieronymus, München

Rabbinge R, Bastiaans L (1989) Combination models, crop growth and pests and diseases. In: Rabbinge R, Ward S, Van Laar H (eds) Simulation and systems management in crop protection, vols. 32 of Simulation Monographs. Pudoc, Wageningen, The Netherlands, pp 217–239

Ritchie J, Godwin D, Otter-Nacke S (1987) CERES-Wheat – A simulation model of wheat growth and development. Texas A&M University Press, College Station, TX

Röhrig M, Stützel H (2001) A model for light competition between vegetable crops and weeds. Eur J Agron 14:13–29

Ros B, Thümmler F, Wenzel G (2005) Comparative analysis of phytophthora infestans induced gene expression in potato cultivars with different levels of resistance. Plant Biol 7:686–693

Seligman N, van Keulen H (1981) PAPRAN: a simulation model of annual pasture production limited by rainfall and nitrogen. In: Frissel M, van Veen J (eds) Simulation of nitrogen behavior of soil-plant systems, Proc. Workshop. PUDOC, Wageningen, pp 192–221

Shaffer M, Halvorson A, Pierce F (1991) Nitrate leaching and economic analysis package (NLEAP): model description and application. In: Follet R, Keeney D, Cruse R (eds), Managing nitrogen for groundwater quality and farm profitability. Soil Sci. Soc. Am., Madison, WI, pp 285–322

Shaffer M, Ma L, Hansen S (eds) (2001) Modeling carbon and nitrogen dynamics for soil management. Lewis Publishers, Boca Raton

Smith JU, Bradbury N, Addiscott TM (1996) SUNDIAL: a PC-based system for simulating nitrogen dynamics in arable land. Agron J 88:38–43

Sperr C, Engel T, Priesack E (1993) Expert-N, Aufbau, Bedienung und Nutzungsmöglichkeiten des Prototyps. In: Engel T, Baldioli M (eds) Expert-N und Wachstumsmodelle. Referate des Anwenderseminars im März 1993 in Weihenstephan, Agrarinformatik 24. Verlag Eugen Ulmer, Stuttgart, pp 41–57

Spitters C (1986) Separating the diffuse and direct component of global radiation and its implications for modeling canopy photsynthesis. II. Calculations of canopy photosynthesis. Agric Forest Meteorol 38:231–242

Spitters C, van Keulen H, van Kraalingen D (1989) A simple and universal crop growth simulator: SUCROS87. In: Rabbinge R, Ward S, van Laar H (eds) Simulation and systems management in crop production. Simulation Monographs 32. Pudoc, Wageningen, pp 147–181

Stamp N (2003) Out of the quagmire of plant defense-hypotheses. Q Rev Biol 78:23–55

Stenger R, Priesack E, Barkle G, Sperr C (1999) Expert-N A tool for simulating nitrogen and carbon dynamics in the soil-plant-atmosphere system. In: Tomer M, Robinson M, Gielen G (eds) NZ Land Treatment Collective Proceedings Technical Session 20: Modelling of Land Treatment Systems. New Plymouth, New Zealand, pp 19–28

Stockle CO, Martin SA, Campbell GS (1994) CropSyst, a cropping systems simulation model: Water/nitrogen budgets and crop yield. Agric Sys 46:335–359

Stockle CO, Donatelli M, Nelson R (2003) CropSyst, a cropping systems simulation model. Eur J Agron 18:289–307

Tanji K, Gupta S (1978) Computer simulation modeling for nitrogen in irrigated croplands. In: Nielsen D, MacDonald AJ (eds) Nitrogen in the environment, vol. I. Academic, New York, pp 79–120

Thornley J, Johnson I (1990) Plant and crop modelling. A mathematical approach to plant and crop physiology. Clarendon, Oxford, UK

Tiktak A, van Grinsven HJM (1995) Review of sixteen forest-soil-atmosphere models. Ecol Model 83:35–53

Triboi E, Martre P, Girousse C, Ravel C, Triboi-Blondel AM (2006) Unravelling environmental and genetic relationships between grain yield and nitrogen concentration for wheat. Eur J Agron 25:108–118

Tuomi J, Fagerstrom T, Niemela P (1991) Carbon allocation, phenotypic plasticity, and induced defense. In: Tallamy D, Raupp M (eds) Phytochemical induction by herbivores. Wiley, New York, N.Y. USA, pp 85–104
van den Berg M, Driessen PM (2002) Water uptake in crop growth models for land use systems analysis: I. A review of approaches and their pedigrees. Agric Ecosys Env 92:21–36
van den Berg M, Driessen PM, Rabbinge R (2002) Water uptake in crop growth models for land use systems analysis: II. Comparison of three simple approaches. Ecol Model 148:233–250
van Genuchten MT, Davidson J, Wierenga P (1974) An evaluation of kinetic and equilibrium equations for the prediction of pesticide movement through porous media. Soil Sci Soc Am Proc 38:29–35
van Ittersum MK, Leffelaar PA, van Keulen H, Kropff MJ, Bastiaans L, Goudriaan J (2003) On approaches and applications of the Wageningen crop models. Eur J Agron 18:201–234
van Laar HH, Goudriaan J, van Keulen H (1997) SUCROS97: Simulation of crop growth for potential and water-limited production situations. Quantitative Approaches in System Analysis 14. C.T. de Wit Graduate School for Production Ecology and Resource Conservation, Wageningen
van Oijen M, Dreccer M, Firsching KH, Schnieders B (2004) Simple equations for dynamic models of the effect of CO_2 and O_3 on light-use efficiency and crop growth. Ecol Model 179:39–60
Vanclooster M, Viaene P, Diels J, Feyen J (1995) A deterministic evaluation analysis applied to an integrated soil-crop model. Ecol Model 81:183–195
Vos J, Marcelis L, de Visser P, Struijk P, Evers J (eds) (2007) Functional-Structural Plant Modelling in Crop Production. Wageningen UR Frontis Series 22. Springer, Dordrecht, The Netherlands
Wang E, Engel T (2000) SPASS: a generic process-oriented crop model with versatile windows interfaces. Env Model Softw 15:179–188
Wang E, Smith C (2004) Modelling the growth and water uptake function of plant root systems: a review. Aust J Agric Res 55:501–523
Wang YP, Leuning R (1998) A two-leaf model for canopy conductance, photosynthesis and partitioning of available energy I: Model description and comparison with a multi-layered model. Agric Forest Meteorol 91:89–111
Wang E, Robertson MJ, Hammer GL, Carberry PS, Holzworth D, Meinke H, Chapman SC, Hargreaves JNG, Huth NI, McLean G (2002) Development of a generic crop model template in the cropping system model APSIM. Eur J Agron 18:121–140
Watts D, Hanks J (1978) A soil-water-nitrogen model for irrigated corn on sandy soils. Soil Sci Soc Am J 42:492–499
Wegehenkel M (2000) Test of a modelling system for simulating water balances and plant growth using various different complex approaches. Ecol Model 129:39–64
Weiss A (2003) Introduction. Agron J 95:1–3
Weiss A, Moreno-Sotomayer A (2006) Simulating grain mass and nitrogen concentration in wheat. Eur J Agron 25:129–137
White JW (2006) From genome to wheat: emerging opportunities for modelling wheat growth and development. Eur J Agron 25:79–88
Wierenga P, de Wit C (1970) Simulation of heat transfer in soils. Soil Sci Soc Am Proc 34:845–848
Williams JR, Renard K (1985) Assessment of soil erosion and crop productivity with process models (EPIC). In: Follet R, Stewart B (eds) Soil erosion and crop productivity. Soil Sci. Soc. Am., Madision, WI, pp 68–102
Willocquet L, Savary S, Fernandez L, Elazegui F, Castilla N, Zhu D, Tang Q, Huang S, Lin X, Singh H, Srivastava R (2002) Structure and validation of RICEPEST, a production situation-driven, crop growth model simulating rice yield response to multiple pest injuries for tropical Asia. Ecol Model 153:247–268

Wolf J, van Oijen M (2003) Model simulation of effects of changes in climate and atmospheric CO_2 and O_3 on tuber yield potential of potato (cv. Bintje) in the European Union. Agr Ecosyst Environ 94:141–157

Yang HS, Dobermann A, Lindquist JL, Walters DT, Arkebauer TJ, Cassman KG (2004) Hybrid-maize–a maize simulation model that combines two crop modeling approaches. Field Crops Res 87:131–154

Yin X, Schapendonk AHCM (2004) Siulating the partitioning of biomass and nitrogen between roots and shoot in crop and grass plants. NJAS Wagen J Life Sci 51:407–426

Yin X, van Laar H (2005) Crop Systems Dynamics. Wageningen Academic, Wageningen

Yin X, Schapendonk AHCM, Kropff MJ, van Oijen M, Bindraban PS (2000) A generic equation for nitrogen-limited leaf area index and its application in crop growth models for predicting leaf senescence. Ann Bot 85:579–585

Yin X, Lantinga EA, Schapendonk AHCM, Zhong X (2003) Some Quantitative Relationships between Leaf Area Index and Canopy Nitrogen Content and Distribution. Ann Bot 91:893–903

Yin X, Van Oijen M, Schapendonk AHCM (2004) Extension of a biochemical model for the generalized stoichiometry of electron transport limited C3 photosynthesis. Plant Cell Environ 27:1211–1222

Zhang Y, Li C, Zhou X, Moore B III (2002) A simulation model linking crop growth and soil biogeochemistry for sustainable agriculture. Ecol Model 151:75–108

Clustered Distribution of Tree Roots and Soil Water Exploitation

M. Kazda(✉) and I. Schmid

Contents

Abstract A substantial number of experiments have shown root proliferation and concentration in nutrient-rich patches. This article focuses on root clustering observed on tree roots in forest stands and water uptake within these zones. Root clustering is seen as a rule in natural soils for optimized exploitation of aggregated resources. In summary, root distribution in nature is caused by two factors: (1) vertical gradients in soil organic mater and infiltration of precipitation, and (2) clusters of available nutrients accompanied by preferential uptake of the seepage water.

1 Introduction

Plant roots are in various interactions with their biotic and abiotic environment and their activities make the rhizosphere a unique living space. Architecture of the root system is designed to fulfill its major functions in exchange with the soil and the

---

M. Kazda
Institute of Systematic Botany and Ecology, Albert-Einstein-Allee 11, D-89081 Ulm, Ulm University, Germany
e-mail: marian.kazda@uni-ulm.de

U. Lüttge et al. (eds.), *Progress in Botany 70*,
DOI:10.1007/978-3-540-68421-3,

associated organisms as well as plant anchoring in the soil (Coutts et al. 1998). Architecture of the root system is closely linked to soil exploration and resource uptake. These processes together with microbial activities create several overlapping gradients in biochemical, chemical and physical soil properties. Furthermore, the characteristics of root systems vary among species (Kutschera 1960) and are further modified in response to the environment, in respect of physiological and mechanical constraints and, on the other hand, through root proliferation in nutrient-rich patches (for review, see Hodge 2006).

In natural soils, several processes during soil development and interactions with flora and fauna inevitably lead to spatial heterogeneities. They are caused by natural processes such as uprooted trees, water infiltration from stem flow, decaying logs, and the activity of the soil macro- and megafauna. Such soil heterogeneities can persist for decades, until they are homogenized by new overlaying processes. For plants living on these soils, it is the challenge to cope with these patchily distributed resources, and they proliferate their roots in zones of high availability (Schenk 2005; Hodge 2006).

In sight of the patchiness of resource distribution, the following subjects of root research are considered in this contribution: (1) heterogeneity of the exploitation of the belowground rooting space, and (2) influence of clustered root distribution on the soil water uptake.

2 Patterns of Soil Exploitation by Tree Roots

According to the resource availability, the highest root density was mostly found in the top soil layers (Jackson and Caldwell 1996; Lopez et al. 2001; Leuschner et al. 2004) and the vertical root distribution can be described as one-dimensional depth functions (Parker and van Lear 1996; Schmid and Kazda 2001; Schenk and Jackson 2002).

Besides the vertical distribution, early studies assumed that the rooting zone is completely and almost homogeneously exploited by roots (Krauss et al. 1939). Recent studies in natural ecosystems found, however, that fine roots concentrate in distinct soil patches (Caldwell et al. 1996, Hölscher et al. 2002; Pellerin and Pages 1996; Ryel et al. 1996). Such fine root clusters reflect the presence of nutrient patches, dead organic matter, zones of better water availability, mycorrhizal associations or specific microbial interactions (Hodge 2006). Root interactions with the associated biota and their influence on nutrient cycling (Bonkowski 2004) accompanied by changes in root architecture (Kreuzer et al. 2006) can be reflected in spatio-temporal root aggregation. Roots create radial heterogeneities on a small scale in the process of creating their own rhizosphere and many root functions stimulate biochemical, chemical and physical processes occurring in the rhizosphere and also along the roots, especially regarding the root apex. These processes were concisely reviewed by Hinsinger et al. (2005), who concentrated on rhizosphere geometry and heterogeneity in both space and time with the focus on the very fine roots.

2.1 Branching Patterns and Root Clustering

Branching patterns and root distribution largely influence the exploitation intensity and construction costs. Two concepts for root branching patterns are discussed in the literature (Richardson and zu Dohna 2003; Dunbabin et al. 2004): the herringbone (sparsely branched) root system and the dichotomous (highly branched) root system. Herringbone systems are preferential for the exploration of large or unrestricted volumes of soil. Dichotomous branching may result in a more locally intensive exploitation of the soil (Fitter et al. 1991; Spek and van Noordwijk 1994).

The basic question regarding investment in root structures is whether branching patterns preserve the cross-sectional root area (i.e., diameter exponent is kept constant at 2). Because the diameter exponent determines whether herringbone or dichotomous branching patterns are more costly, it is possible that plants differing in this exponent have very different ecological niches (Richardson and zu Dohna 2003). Furthermore, an inherent degree of plasticity in root topology may indicate that these different branching patterns are also related to soil conditions and resource availability (Fitter et al. 1991).

Analyses by Dunbabin et al. (2004) have suggested that highly branched root architectures (dichotomous-like) are efficient at capturing immobile ions (e.g., phosphate) from restricted soil volumes, providing the potential for intensive exploration of local soil volumes by maximized precision. These analyses also suggest that coarser, sparsely branched, root systems (herringbone-like) are more efficient at capturing mobile resources over large soil volumes. Dunbabin et al. (2004) modeled the root growth on a three-dimensional scale and tested the capturing of mobile ions from spatially and temporally heterogeneous soils. The simulated herringbone root system had a higher nitrate-uptake efficiency (nitrate-N m^{-3} soil) when supply varied spatially and temporally, also compared to spatial variability alone. In contrast, the efficiency of the dichotomous root system decreased under spatially and temporally dynamic nitrate supply. They concluded that localized root proliferation (i.e., dichotomous branching) can be efficient for foraging of non-uniform resources, but the efficiency of this investment strategy is not adequate for capturing nutrients from transient sources. The herringbone root system, on the other hand, was incapable of proliferating roots in nitrate patches. Thus, under variable nutrient supply, plants will maximize the total volume of soil explored, rather than intensify exploration in zones of resource availability. Because resource patches in natural soils are stable in size and time, the dichotomous root system once built will forage the resources better than a sparsely branched herringbone root system. This is also supported by Pregitzer et al. (1993), who found a longer life span of fine roots in nutrient patches.

2.2 Root Clusters in Monospecific Stands

A GIS-based investigation of root distribution on 76 m^2 trench profile walls (each 2 × 1 m in size) in *P. abies* and *F. sylvatica* stands provided a unique possibility to assess the root distribution on a large scale, based on x- and y-coordinates of each root greater than 2 mm diameter (Schmid and Kazda 2001). Evaluation of all roots between 2 and 5 mm diameter for both species showed maximum abundance in the top 20 cm of the soil, where pronounced root clusters occurred next to areas with only low root accumulation (Fig. 1).

Fleischer et al. (2006) applied methods of point process modelling (pair correlation function, L- and K-functions) on the same data and found stronger clustering in the case of spruce than in the case of beech. The estimated point process characteristics using the Matern-cluster model further differentiated between the species. The results showed a stronger clustering in a smaller range of attraction for spruce (calculated radius $r = 4.9$ cm), while the clustering was weaker for beech, but the range of attraction ($r = 7.4$ cm) was larger. Furthermore, the point density (i.e. number of roots) within these clusters was about three times higher within the spruce clusters compared to beech. The visualization of the Matern-cluster process

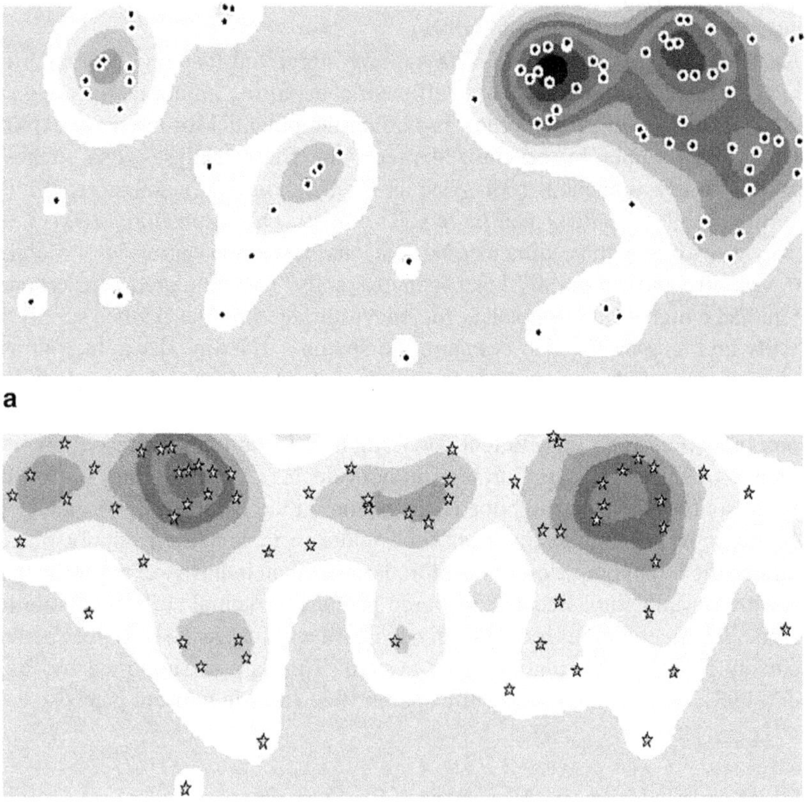

a

b

Fig. 1 Isolines of clustering intensity on soil trench walls (1 × 2 m) with points indicating the position of individual roots for *P. abies* (**a**) and *F. sylvatica* (**b**)

after inverse retransformation also suggested a depth-dependent size and shape of root clusters, in the resembling to the shapes generated by isolines of clustering intensity (cf. Schmid and Kazda 2005 and Fig. 1). Close to the soil surface, roots formed clusters along the horizontal axis. This shape also agreed with the horizontally distributed root points on the soil pits profiles. Roots accumulating along the horizontal axis as well as the shape of generated clusters may reflect attractive soil patches in the nutrient-rich topsoil layers. Deeper, the real size of clusters was larger and more circular.

Further tests confirmed that root clustering was not correlated with tree density around the investigated soil profiles and thus clusters of small roots are inherently present within the tree stands (Schmid and Kazda 2005). The results also indicated that the root system of spruce requires more roots to achieve a similar degree of space acquisition and thus beech exploits patchily distributed soil resources at lower root numbers. The dynamics of space exploitation can be extracted from root growth rates. Measurements on root sections larger than 5 mm in diameter revealed that radial growth of beech roots exceeded that of spruce significantly in both the pure and the mixed stands (Schmid and Kazda 2001).

Along with these results, there is increasing evidence that the belowground space in forests is exploited at a high degree of heterogeneity. Even in a quite regular man-made forest such as a clonal *Eucalyptus* plantation in Congo, Bouillet et al. (2002) found large spatial heterogeneity and root aggregation within the soil profile, independently from the distance to the planting row, especially for the very fine roots between 0.1 and 1 mm diameter. In a similar *Eucalyptus* plantation on nutrient-poor soil, horizontal distribution of superficial fine root mats was investigated by Laclau et al. (2004). They mapped fine root biomass on a 4-m^2 plot and found significant root aggregations. Tatarinov et al. (2007) estimated the root-inhabited soil volume as (1) a theoretical "pot" which enveloped the whole root system of individual trees, and (2) as "root-inhabited clumps" defined as soil volume around clusters of roots of second and higher orders. Their two-dimensional vertical cross-sections also depicted zones of high rooting intensity and soil parts much less inhabited by roots, a structural feature identical to that reported by Schmid and Kazda (2005). Tatarinov et al. (2007) calculated that about 73% of soil volume was filled with "pots" (whole-tree rootage) and that on average only 13% of soil available was inhabited by the intensively rooted clumps.

Models of root architecture and branching patterns increasingly attempt to account for uneven distribution of soil resources. These models aimed to consider soil physical and chemical properties or carbohydrate availability (Clausnitzer and Hopman 1994; Thaler and Pagès 1998). In the model by Pagès et al. (2004), the soil was described in a static way as horizontal layers of defined thickness. This modeling of root architecture in nature-like soil is a significant progress, even the philosophy behind the model was based on predictable vertically distributed resources. Despite these facts, model realization (Pagès et al. 2004) showed root aggregation, when implementing several parameters such as root emission, axial and radial growth, sequential branching, reiteration, transition, decay and abscission. Among these, the reiteration process which produces typical forks has a strong impact on soil colonization by multiplying the number of exploring axes. Together with sequential branching, reiteration can lead to observed multiple exploitation of the soil. Soil properties affect both root elongation and direction, as well as root branching density.

Coleman (2007) investigated the spatial and temporal patterns of root distribution of four tree species grown under drip tube irrigation and fertilization. The response to fertilization was greater than that to irrigation. The former significantly influenced tree root system, which developed in response to heterogeneity in resource supply created by drip tubes. Both fine and coarse (>5 mm) roots responded positively to this resource enrichment, indicating changes on the level of the whole root system. This dynamic experiment provided first experimental evidence on root clustering as response to heterogeneous supply in trees grown under field conditions and supports conclusions from field investigations by Bouillet et al. (2002) and Laclau et al. (2004) for *Eucalyptus* and Schmid and Kazda (2005) for *P. abies* and *F. sylvatica*.

2.3 Soil Exploitation and Root Clustering in Mixed Stands

The basic question in terms of root clustering is whether in mixed stands each individual forms its own separated clusters or whether roots of different individuals and/or species are present in zones of high rooting intensity. In the former case, the clustering will be more an expression of root architecture *per se*; the presence of several individuals within a cluster will clearly support the view of competition within zones of higher resource availability.

Root distribution in mixed stands of *P. abies* and *F. sylvatica* was evaluated in terms of root aggregation using the same procedure as described by Schmid and Kazda (2005). Figure 2 displays the isolines for rooting intensity for all roots above

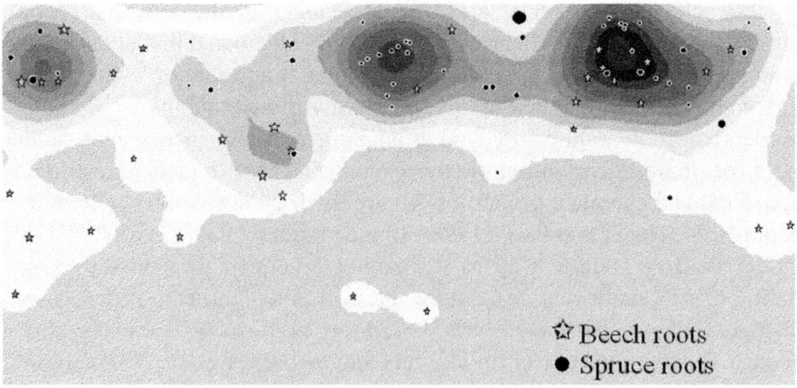

Fig. 2 Roots of beech (*stars*) and spruce (*dots*) form common clusters in areas of high rooting intensity

2 mm diameter on an exemplary pit wall. The darker color indicates higher root aggregation and the roots are superimposed over these areas. It is obvious that areas of high clustering intensity are made up from roots of both species. This finding confirms that the root clustering in specific soil parts is not caused by a tree individual growing just nearby the soil pit but is a clear indication of resources exploitation. Still, there remains a question about dynamics and competition in resources acquisition. More individual roots within a smaller range (see above) would emphasize the resource occupation by spruce. On the other hand, evaluation of the diameter growth of coarse roots (>5 mm) in mixed stands confirmed that growth rates of beech roots exceeded those of spruce by up to 25% (Schmid and Kazda 2001). Such findings suggest that European beech built a more sophisticated root system at higher root growth rates compared to Norway spruce, especially when competing for the same resources.

Even though roots of both species *P. abies* and *F. sylvatica* were present within the same cluster, a direct evidence for common resource use was still missing. Recent study by Göttlicher et al. (2008) estimated the lateral spread of roots in a *Pinus sylvestris* and *P. abies* mixed forest. Soil in 1-m^2 plots was labelled with ^{15}N and the uptake by surrounding trees was measured. The results showed extension of lateral roots up to 4–5 m and indicated overlapping of root systems, where up to 11 trees took up tracer from a 1-m^2 labelled area. A parallel study on girdled and non-girdled plots indicated the dependence of ectomycorrhizal fungi on photosynthates provided by trees. Overlapping root systems interconnected by fungal hyphea are thus presumably concentrated in nutrient-rich patches, prominent zones of high belowground competition.

Within stands of *Larix gmelinii* of different age (26, 105 and 220 years old) in the permafrost region of Siberia, Kajimoto et al. (2007) described horizontal distributions of both roots and crowns. They suggest that belowground inter-tree competition, evaluated by a "rooting area index," is more important than aboveground competition for light in the mature and overmature *L. gmelinii* stands. Root investigations by Tatarinov et al. (2007) also assessed the social status of the trees in the *Quercus robur* and *Fraxinus excelsior* forest. Trees which were classified as dominant, co-dominant and suppressed according to their aboveground parameters also retained their social status belowground. Both investigations clearly point out the strong link between competitive ability above and belowground, even though distribution patterns of resources are different.

2.4 Root Clustering as a Rule in Nature

Long-lasting patches of available resources will in consequence result in root aggregation, which can be seen as a rule in natural soils. Beyond controversy, root clustering in natural soils has been experienced by anybody who sampled soil cores for analysis of root data. Huge coefficients of variation, often above 100% (Schmid 2002), manifest a non-normal distribution of roots, with high abundances

Fig. 3 Schematic patterns of space exploitation belowground. *Grey scale* in the soil profile indicates the gradient in soil organic matter, and intensity of idealized patches demonstrates their different efficiency

in specific – but unrecognized – parts of the soil. On the other hand, root investigations in pots, nutrient solutions, etc. will hardly be able to produce representative branching patterns. In such experiments, nutrient addition and watering readily redistribute mobile nutrients such as nitrate through the entire rooting zone and thus make root aggregation less profitable (sensu Dunbabin et al. 2004). Furthermore, the sensing of mechanical obstacles by growing roots (Falik et al. 2005) will additionally modify root architecture.

As a summary of these studies carried out in completely different forest ecosystems, it is obvious that clustered root distribution is an inherent property of belowground space acquisition. This method of space sequestration is in contrast to the space exploitation above ground and is directly linked to gradients of resources (Schenk 2005). Another factor is the predictability of resources. Whereas photosynthetically active radiation and carbon dioxide concentration in the aboveground growing space follow clearly predictable gradients (Monsi and Saeki 1953; Hager 1975), roots must acquire multiple resources, the availability of which is often of low predictability. Still, positive feedback mechanisms, which promote root growth in soil parts already exploited by other roots (Schenk 2005) and sensing of roots growing nearby (Falik et al. 2003; Schenk 2006), can help to participate in the exploitation of a profitable nutrient patch. As a result, root distribution will show a vertical gradient (Schenk and Jackson 2002) which represents an integral of roots concentrated in patches of different exploitation intensity (Fig. 3). In addition, biological associations, mechanical constraints and oxygen limitation together with root competition further modify root distribution and architecture.

3 Heterogeneity in Soil Water Exploitation

3.1 Heterogeneity in Soil Water Content

Precipitation is intercepted by tree crowns and nutrients from wet deposition and plant surfaces are redistributed via troughfall and stemflow (Kazda 1990). Such redistribution together with soil morphology lead to spatial heterogeneities and fluctuating soil moisture content. High heterogeneity in soil water extraction was modeled from spatial 2D data on a 2.5 × 2.5 m scale for 0.55 m soil depth by Vrugt et al. (2001). Another comprehensive study by Schume et al. (2004) investigated soil water depletion and recharge in mature stands of *P. abies and F. sylvatica.* Results showed high spatial variability in water extraction, especially in the mixed stand. Both studies revealed large heterogeneities on the horizontal plane, but without assessing the root distribution. Laclau et al. (2004) studied the topography of root mats within the forest floor on nutrient poor *Eucalyptus* forest. The described patterns of root densities showed clear proliferation in zones of organic mater accumulation. Uptake from these fine root mats substantially utilized the released nutrients and prevented them from leaching.

Green et al. (2006) provided synthesis of measurements and models for root uptake and transpiration, mainly focusing on kiwifruit vines. Arrays of TDR sensors together with sap-flow measurement inside the kiwifruit vine's roots revealed the uptake dynamics during partial drying of the rooting zone. They formulated specific research needs for better understanding of the soil–root boundary and root resistance, and emphasize the need to link water uptake with root system architecture. This knowledge is crucial for the aim to reduce irrigation volumes. In this respect, preferential flow through the soil profile (Al Hagrey et al. 1999) as well as clustered root distribution together with preferential water uptake have to be considered.

3.2 Water Uptake from Root Clusters

Pierret et al. (2005) reviewed patterns of soil exploration regarding the complex interactions between soil structure, water, oxygen and nutrient availability. Initially, plants preferentially grow into previously formed macropores and finer roots extend into the surrounding soil matrix. These roots can penetrate smaller pores, especially at adequate moisture conditions, not restricting the flow of gases. The proximity of these initial forays to macropores ensures that oxygen supply is non-limiting, as the diffusion rate of oxygen in air-filled pores is about 10^4 times faster than in water. Concentration of fine roots along the well-aerated macropores is important regarding the high respiration rates of fine tree roots in zones of high nutrient availability (Pregitzer et al. 1998). A micromorphological study by van Noordwijk et al. (1992) also indicates that maize roots preferentially explore spaces between soil aggregates. However, the growth of small diameter roots into these aggregates was only rarely observed. The growth of the very fine roots adjacent to macropores is assured by well-balanced gas exchange by displacement of respirative CO_2 and simultaneous oxygen supply. Clustered root distribution in soil (see Sect. 2) may also influence the dynamics of water uptake and interactions with the water flow in macropores are the most likely.

Parallel to the investigations performed by Schmid and Kazda (2001) and Schmid (2002), irrigation experiments were initiated (Weinmeister, unpublished) in order to evaluate erosion and magnitude of the sub-surface flow. This experiment used the same soil pits where our assessment of root clustering was carried out (see Schmid 2002 for site description and root methodology). Here, we placed TDR soil moisture sensors within and outside intensively rooted patches at the same soil depth, mainly between 14 and 32 cm, depending on the position of root aggregation. Time course of soil moisture changes was recorded in 5-min intervals during and after the irrigation experiments in seven pure Norway spruce stands and in five mixed stands of spruce and beech.

Consistently over the irrigation experiment, soil moisture increased steadily. To identify short time dynamics, changes in soil moisture content were calculated for each 5-min interval. In both stands, soil moisture increase peaked earlier in soil areas without root clusters (Fig. 4). Thereafter, the increase of soil moisture was slower (declining values). Negative values for soil moisture change indicated drainage of respective soil parts. Both the delayed increase in intensively rooted soil patches and the negative values of soil moisture change (Fig. 4b) indicated preferential water uptake in these zones.

Doussan et al. (2006) investigated the spatial extension and propagation of water uptake front in narrow-leaf lupin root systems and measured soil–plant parameters.

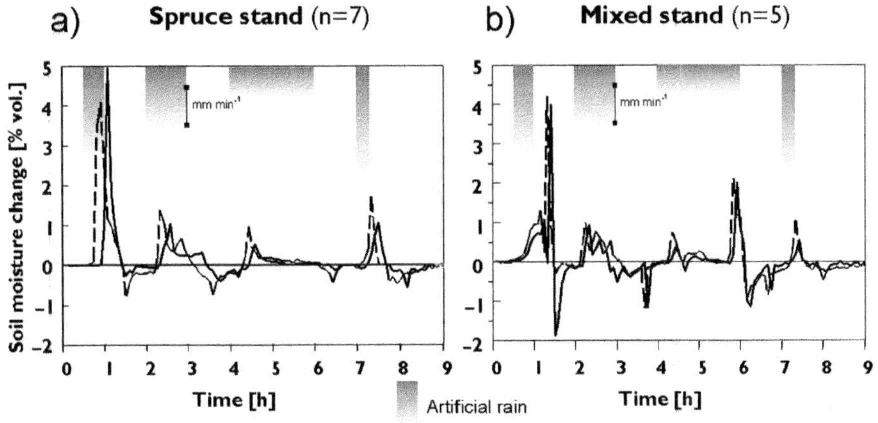

Fig. 4 Changes in soil moisture calculated for each 5-min sampling interval for pure Norway spruce stand (**a**) and for mixed stand with European beech (**b**) in root clusters (*solid line*) and in less rooted parts of the soil profile (*dashed line*)

Formation and propagation of a water extraction front in mature root systems was evident. Zones of water uptake were closely related to the soil hydraulic properties and root axial conductance. This is supported by Leuschner et al. (2004) who suggested root morphology, ultrastructure of the root cortex and root–soil interface conductivities as the main cause for the large spatial heterogeneity in root water uptake. Doussan et al. (2006) also showed that the taprooted architecture induced a more spatially concentrated uptake zone with higher flux rates. Their model, explicitly coupling root system and soil water transfers, can be useful to study water uptake in relation to root architecture. Especially, heterogeneous conditions as found under irrigation and clustered root distribution should be coupled to root hydraulic conductance in future research on root–soil water processes.

Fast water uptake from the water front within root clusters is supported by experiments performed by Bramley et al. (2007), who used root pressure probes to investigate root hydraulic conductivities. The first part of the root pressure relaxation curve was rapid (<1 s), the second phase was slower, in the order of several seconds, and the third phase was a long tail that was considered indicative of polarization effects in the xylem (Peterson and Steudle 1993). Such fast water uptake kinetics in plant roots will inevitably lead to instantaneous water uptake from the water front caused either by rain or by irrigation.

Tsutsumi et al. (2003) also modeled the interaction between the plant root system and soil moisture under consideration of inclined soil surface. The root system was calculated first and the intensity of water extraction was derived from these results. They visualized clear zones of high water uptake, with up to nine-fold relative spatial differences. In comparison to these models that were done on an area of ca. 0.5-m^2, al Hagrey (2007) measured by means of geophysical imaging the soil moisture heterogeneity in the rooting zone of oak trees over 16 days along a 60-m long gradient.

2D images of the soil profile displayed distinct patches of higher soil moisture content. Experimental felling and irrigation of selected trees was mirrored in appropriate changes in 2D water distribution. Both investigations strongly support our results reported above concerning preferential water extraction from specific parts of soil.

3.3 Water Extraction in Drying Soil

In our measurements of the soil water dynamics inside and outside the root clusters, two soil pits, one in each stand, remained instrumented with TDR sensors after the irrigation experiments were finished. Figure 5 shows the course of soil moisture content for the time span between 23 July and 2 August 1999. Steeper slope for the soil moisture decline in intensively rooted soil parts compared to "root free" soil revealed that soil parts with root clusters extract the water faster than less intensively rooted soil parts. Conspicuous is the wave-like shape of the soil moisture curve. Vertical lines in Fig. 5 mark midnight, and it is obvious that the sharpest decline of soil water occurs in the afternoon. This finding is in line with measurements by Čermák et al. (1984), who found a time lag between the onset of tree crown transpiration and the upward water movement in stem. This water movement in the plant is transduced via roots to the surrounding soil and is followed by declining soil water content. A recent study by Nadezhdina et al (2006) confirmed the widely reported

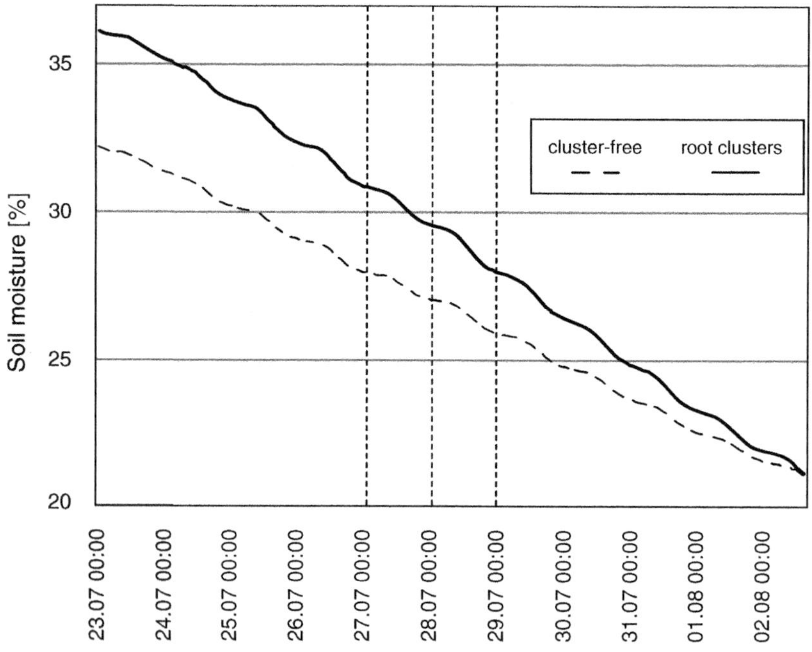

Fig. 5 Decline of soil moisture during a time span of 11 days without precipitation in root clusters (*solid line*) and in less rooted parts of the soil profile (*dashed line*)

finding that depletion of water stored in soil proceeds downward as the soil dries, but spatial heterogeneity related to water extraction by tree roots has to be considered.

Coners and Leuschner (2005) calculated real time water uptake rates per root surface area for individual fine root systems of 0.5–1.0 m in length. Water uptake and transport by roots followed the diurnal course of water vapour saturation deficit in the ambient air with highest uptake values during mid-day. This uptake is also reflected in the Fig. 5, showing the declining soil water content during afternoon. Coners and Leuschner (2005) reported a large spatial heterogeneity in uptake where neighboring roots of the same species differed up to 10-fold in their daily totals. It seems to be possible, that such differences are not only caused by root morphology and the root–soil interface itself (Leuschner et al. 2004) but are a result of the heterogeneity in soil water extraction and water movement into intensively rooted soil patches.

A comparison of soil moisture dynamics in root clusters with less rooted soil in the mixed stand (Fig. 5) by means of time series analysis revealed a significant ($p<0.01$) lag of eight measurement intervals, i.e. a time shift of ca. 40 min, before the water starts to decline in the less rooted soil parts. Similar differences between soil parts with high root abundance and apparently less rooted soil were also found for Norway spruce stands. There, the changes in soil moisture showed an even larger time lag ($p<0.001$) by about 80 min, i.e. 16 intervals per 5 min. These differences between the stands may result from smaller clusters of spruce compared to the beech stand (Fleischer et al. 2006) and therefore longer distances for soil water to reach zones of intensive water uptake.

The results shown in Figs. 4 and 5 point out heterogeneities in soil water use on the vertical plane, where fast uptake takes place in intensively rooted soil parts whereas the surrounding soil serves as a reservoir and/or is exploited at lower intensity. Water flow in soil occurs preferentially in macropores, and Stewart et al. (1999) found about 80% of the finest roots at a distance <1 mm from soil macropores (see also Pierret et al. 2005). Heterogeneity in the water uptake in the soil as reported above is directly linked to high rooting intensity and thus water uptake in root clusters. This view is supported by models calculated by Tsutsumi et al. (2003). On the other hand, models by Doussan et al. (2003) revealed a decreased volume of the rhizosphere by 30% when root architecture was more clumped due to larger overlap of the neighbouring depletion zones. The authors emphasized the importance of such overlaps when modeling nutrient acquisition. In the case of subsurface water flow in the macropores, the possibility of instant water uptake will foster such overlaps. In natural, structured soils, nutrients mineralized in the topsoil can be leached throughout the profile by heavy rain, and it is reasonable for plants to root prefe-rentially in zones of potential water and nutrient uptake.

4 Conclusions

Undoubtedly, there is a clear vertical gradient of soil organic matter which corresponds with the availability of most nutrients. At the first glance, it is preferential for the plant to concentrate its roots in the uppermost soil in order to acquire mineralized nutrients and precipitation water (Schenk 2005). However, the topsoil is prone to drought in many regions and it is also necessary to participate in the nutrient uptake when the water front moves rapidly through the macropores (Al Hagrey et al. 1999) and to prevent nutrient leaching from the system. This might be one of the reasons for high fine root densities around macropores (Stewart et al. 1999), modified further by soil-inherent heterogeneities in nutrient availability as well as by microbial and hyphal associations. As a result, patterns of root distribution are a combination of two processes: (1) vertical distribution according to gradients in soil organic matter and infiltration of precipitation (Schenk 2005), and (2) root clustering in zones of high nutrient availability and preferential uptake of the seepage water.

Acquisition, exploitation and maintenance of the growing space are of vital importance for species success in a specific environment. The processes linked to root distribution and dynamics are in this respect of a rather different nature compared to the aboveground growing space. Besides a clear vertical gradient in soil organic matter, availability of water and nutrients can vary considerably within time and space. Spatial heterogeneity ranges from a few mm (Parry et al. 2000) mainly as a result of microbial associations (Wanner 1999; Bonkowski 2004), to centimetres due to gradients in water availability, pH, nutrient mineralization and oxygen supply (Jackson and Caldwell 1993, Lynch 1995) to several meters (Kazda and Glatzel 1984). On such a scale, variations in nutrient availability as found under natural soil conditions will consequently lead to clustered root distribution patterns. Long-living trees have to maintain their root systems for a long time and have to assure a consistent exploitation of the resources as well as the long-range water and nutrient transport. Therefore, aggregations of nutrients and other resources in soil will result in long-lasting root structures, going along with competition within such root clusters.

Heterogeneity in soil exploitation by roots and interactions with soil biota are in the focus of recent scientific interest. New insights on the "hidden half" of the plant within the "soil black box" will unlock new fascinating knowledge on plants and their interactions with the biotic and abiotic environment.

Acknowledgements A substantial part of this research was supported by the Austrian Science Foundation within the Special Research Program, "Restoration of Forest Ecosystems", F008–08.

References

Al Hagrey SA (2007) Geophysical imaging of root-zone, trunk, and moisture heterogeneity. J Exp Bot 58: 839–854

Al Hagrey SA, Schubert-Klempnauer T, Wachsmuth D, Michaelsen J, Meissner R (1999) Preferential flow: first results of a full-scale flow model. Geophys J Int 138: 643–654

Bonkowski M (2004) Protozoa and plant growth: the microbial loop in soil revisited. New Phytol 162: 617–631

Bouillet JP, Laclau JP, Arnaud M, Thongo M'Bou A, Saint-André L, Jourdan C (2002) Changes with age in a spatial distribution of roots of Eucalyptus clone in Congo – Impact on water and nutrient uptake. For Ecol Manage 171: 43–57

Bramley H, Turner NC, Turner DW, Tyerman SD (2007) Comparison between gradient-dependent hydraulic conductivities of roots using the root pressure probe: the role of pressure propagations and implications for the relative roles of parallel radial pathways. Plant Cell Environ 30: 861–874

Caldwell MM, Manwaring JH, Durham SL (1996) Species interactions at the level of fine roots in the field: influence of soil nutrient heterogeneity and plant size. Oecologia 106: 440–447

Čermák J, Jenik J, Kucera J, Zidek V (1984) Xylem water flow in a crack willow tree (*Salix fragilis* L.) in relation to diurnal changes of environment. Oecologia 64: 145–151

Clausnitzer V, Hopmans JW (1994) Simultaneous modeling of transient three-dimensional root growth and soil water flow. Plant Soil 164: 299–314

Coleman MD (2007) Spatial and temporal patterns of root distribution in developing stands of four woody crop species grown with drip irrigation and fertilization. Plant Soil 299: 195–213

Coners H, Leuschner C (2005) *In situ* measurement of fine root water absorption in three temperate tree species - Temporal variability and control by soil and atmospheric factors. Basic Appl Ecol 6: 395–405

Coutts MP, Nielsen CCN, Nicoll BC (1998) The development of symmetry, rigidity and anchorage in the structural root systems of conifers. Plant Soil 217: 1–15

Doussan C, Pagès L, Pierret A (2003) Soil exploration and resource acquisition by plant roots: an architectural and modelling point of view. Agronomie 23: 419–431

Doussan C, Pierret A, Garrigues E, Pagès L (2006) Water uptake by plant roots: II – Modelling of water transfer in the soil–root system with explicit account of flow within the root system – comparison with experiments. Plant Soil 283: 99–117

Dunbabin V, Rengel Z, Diggle AJ (2004) Simulating form and function of root systems: efficiency of nitrate uptake is dependent on root system architecture and the spatial and temporal variability of nitrate supply. Funct Ecol 18: 204–211

Falik O, Reides P, Gersani M, Novoplansky A (2003) Self/non-self discrimination in roots. J Ecol 91: 525–531

Falik O, Reides P, Gersani M, Novoplansky A (2005) Root navigation by self inhibition. Plant Cell Environ 28: 562–569

Fitter AH, Strickland TR (1991) Architectural analysis of plant root systems. 2. Influence of nutrient supply on architecture in contrasting plant species. New Phytol 118: 383–389

Fitter AH, Stickland TR, Harvey ML, Wilson GW (1991) Architectural analysis of plant root system 1. Architectural correlates of exploitation efficiency. New Phytol 118: 375–382

Fleischer F, Eckel S, Schmid V, Schmidt I, Kazda M (2006) Point-process modelling of root distributions in pure stands of *Fagus sylvatica* and *Picea abies*. Can J For Res 36: 227–237

Göttlicher SG, Taylor AFS, Grip H, Betson NR, Valinger E, Högberg MN, Högberg P (2008) The lateral spread of tree root systems in boreal forests: estimates based on ^{15}N uptake and distribution of sporocarps of ectomycorrhizal fungi. For Ecol Manage 255: 75–81

Green SR, Kirkham MB, Clothier BE (2006) Root uptake and transpiration: from measurements and models to sustainable irrigation. Agric Water Manage 86: 165–176

Hager H (1975) Kohlendioxyd-Konzentration, -Flüsse und -Bilanzen in einem Fichtenwald. Münchener Universitäts-Schriften 26, 182p

Hinsinger P, Gobran GR, Gregory PJ, Wenzel WW (2005) Rhizosphere geometry and heterogeneity arising from root mediated physical and chemical processes. New Phytol 168: 293–303

Hodge A (2006) Plastic plants and patchy soils. J Exp Bot 57: 401–441

Hölscher D, Hertel D, Leuschner C, Hottkowitz M (2002) Tree species diversity and soil patchiness in a temperate broad-leaved forest with limited rooting space. Flora 197: 118–125

Jackson RB, Caldwell MM (1993) Geostatistical patterns of soil heterogeneity around individual perennial roots. J Ecol 81: 683–692

Jackson RB, Caldwell MM (1996) Integrating resource heterogeneity and plant plasticity: modelling nitrate and phosphate uptake in a patchy soil environment. J Ecol 84: 891–903

Kajimoto T, Osawa A, Matsuura Y, Abaimov AP, Zyryanova AO, Kondo K, Tokuchi N, Hirobe M (2007) Individual-based measurement and analysis of root system development: case studies for *Larix gmelinii* trees growing on the permafrost region in Siberia. J For Res 12: 103–112

Kazda M (1990) Sequential stemflow sampling for estimation of dry deposition and crown leaching in beech stands. In: Harison AF, Ineson P, Neal OW (eds) Field methods for studying nutrient cycling in terrestrial ecosystems. Elsevier, London, pp 46–55

Kazda M, Glatzel G (1984) Heavy metal enrichment and mobility in the infiltration zone of stemflow from beech (*Fagus sylvatica*) in the Vienna woods (in German). J Plant Nutr Soil Sci 147: 743–752

Krauss G, Müller K, Gärtner G, Härtel F, Schanz H, Blanckmeister H (1939) Standortsgemäße Durchführung der Abkehr von der Fichtenwirtschaft im nordwestsächsischen Niederland. Tharandter Forstl Jahrb 90: 481–715

Kreuzer K, Adamczyk J, Iijima M, Wagner M, Scheu S, Bonkowski M (2006) Grazing of a common species of soil protozoa (*Acanthamoeba castellanii*) affects rhizosphere bacterial community composition and root architecture of rice (*Oryza sativa* L.). Soil Biol Biochem 38: 1665–1672

Kutschera L (1960) Wurzelatlas Mitteleuropäischer Ackerunkräuter und Kulturpflanzen. DLG Verlag Frankfurt, Main

Laclau JP, Toutain F, Thongo M'Bou A, Arnaud M, Joffre R, Ranger J (2004) The function of the superficial root mat in the biogeochemical cycles of nutrients in congolese *Eucalyptus* plantations. Ann Bot 93: 249–261

Leuschner Ch, Hertel D, Schmid I, Koch O, Muhs A, Hölscher D (2004) Stand fine root biomass and fine roots morphology in old-growth beech forests as a function of precipitation and soil fertility. Plant Soil 258: 43–56

Lopez B, Sabate S, Gracia CA (2001) Vertical distribution of fine root density, length density, area index and mean diameter in *Quercus ilex* forest. Tree Physiol 21: 555–560

Lynch J (1995) Root architecture and plant productivity. Plant Physiol 109: 7–13

Monsi M, Saeki T (1953) Über den Lichtfaktor in den Pflanzengesellschaften und seine Bedeutung für die Stoffproduktion. Jpn J Bot 14: 22–52

Nadezhdina N, Čermák J, Gašpárek J, Nadezhdin V, Prax A (2006) Vertical and horizontal water redistribution inside Norway spruce (*Picea abies*) roots in the Moravian upland. Tree Physiol 26: 1277–1288

van Noordwijk M, Kooistra MJ, Boone FRFR, Veen BW, Schoonderbeek D (1992) Root–soil contact of maize, as measured by a thin-section technique. I. Validity of the method. Plant Soil 139: 109–118

Pagès L, Vercambre G, Drouet JL, Lecompte F, Collet C, Le Bot J (2004) Root Typ: a generic model to depict and analyse the root system architecture. Plant Soil 258: 103–119

Parker MM, van Lear DH (1996) Soil heterogeneity and root distribution of mature loblolly pine stands in piedmont soils. Soil Sci Soc Am J 60: 1920–1925

Parry S, Renault P, Chadoeuf J, Chenu C, Lensi R (2000) Particulate organic matter as a source of variation in denitrification in clods of soils. Eur J Soil Sci 51: 271–281

Pellerin S, Pages L (1996) Evaluation in field conditions of a three-dimensional architectural model of the maize root system: comparison of simulated and observed horizontal root maps. Plant Soil 178: 101–112

Peterson CA, Steudle E (1993) Lateral hydraulic conductivity of early metaxylem vessels in Zea mays L. Roots. Planta 189: 288–297

Pierret A, Moran CJ, Doussan C (2005). Conventional detection methodology is limiting our ability to understand the roles and functions of fine roots. New Phytol 166: 967–98

Pregitzer KS, Hendrick RL, Fogel R (1993) The demography of fine roots in response to patches of water and nitrogen. New Phytol 125: 575–580

Pregitzer KS, Laskowski MJ, Burton AJ, Lessard VC, Zak DR (1998) Variation in sugar maple root respiration with root diameter and soil depth. Tree Physiol 18: 665–670

Richardson AD, zu Dohna H (2003) Predicting root biomass from branching patterns of Douglas-fir root systems. Oikos 100: 96–104

Ryel RJ, Caldwell MM, Manwaring JH (1996) Temporal dynamics of soil spatial heterogeneity in sagebrush-wheatgrass steppe during a growing season. Plant Soil 184: 299–309

Schenk HJ (2005) Vertical vegetation structure below ground: scaling from root to globe. Prog Bot 66: 341–373
Schenk HJ (2006) Root competition: beyond resource depletion. J Ecol 94: 725–739
Schenk HJ, Jackson RB (2002) The global biogeography of roots. Ecol Monogr 72: 311–328
Schmid I (2002) The influence of soil type and interspecific competition on the fine root system of Norway spruce and European beech. Basic Appl Ecol 3: 339–346
Schmid I, Kazda M (2001) Vertical distribution and radial growth of coarse roots in pure and mixed stands of *Fagus sylvatica* and *Picea abies*. Can J For Res 31: 539–548
Schmid I, Kazda M (2005) Clustered root distribution in mature stands of *Fagus sylvatica* and *Picea abies*. Oecologia 144: 25–31
Schume H, Jost G, Hager H (2004) Soil water depletion and recharge patterns in mixed and pure forest stands of European beech and Norway spruce. J Hydrol 289: 258–274
Spek LY, Noordwijk M (1994) Proximal root diameters as predictors of total root system size for fractal branching models. II. Numerical model. Plant Soil 164: 119–128
Stewart JB, Moran CJ, Wood JT (1999) Macropore sheath: quantification of plant root and soil macropore association. Plant Soil 211: 59–67
Tatarinov F, Urban J, Čermák J (2007) Application of "clump technique" for root system studies of *Quercus robur* and *Fraxinus excelsior*. For Ecol Manage 255: 495–505
Thaler P, Pagès L (1998) Modelling the influence of assimilate availability on root growth and architecture. Plant Soil 201: 307–320
Tsutsumi D, Kosugi K, Mizuyama T (2003) Root-system development and water-extraction model considering hydrotropism. Soil Sci Soc Am J 67: 387–401
Vrugt JA, van Wijk MT, Hopmans JW, Simunek J (2001) One-, two-, and three-dimensional root water uptake functions for transient modeling. Water Resour Res 37: 2457–2470
Wanner M (1999) A review on the variability of testate amoebae: methodological approaches, environmental influences and taxonomical implications. Acta Protozool 38: 15–29

Quaternary Palaeoecology: Major Palaeoecological Problems of Europe

B. Frenzel

Contents

Abstract It is well known that the former view of three or four Quaternary glacial periods in northern or south-central Europe, respectively, cannot be kept any longer, because much more qualitatively different interglacial vegetation-sequences have since been discovered. Yet the question is how many of these different interglacial periods had existed in reality? The comparison with deep-sea isotope-curves, which is practised repeatedly now, cannot be accepted as a reliable way of argumentation, because firm stratigraphical connections between terrestrial and oceanic sediment sequences are generally lacking. Thus, one has to construct, on the continents, palaeoclimatological, geological and palaeoecological sequences independently from changes in the deep-sea curves of stable isotopes. Another problem is the timing and ways of immigration of the exacting taxa from their glacial refuge areas when at the beginning of interstadial or interglacial periods the climate began to improve. Here, too, the comparison with stable isotope curves of deep-sea foraminifera is of no help, because the dating quality of older deep-sea sediments is restricted. Finally, exact quantification of the palaeoecological consequences of early human activities has rarely been attempted in various regions of Europe. Thus, very often one cannot reliably differentiate between climate and man, both being factors which strongly influenced the palaeoecological situation in various regions of Europe at a very early period.

B. Frenzel
Institut für Botanik und Botanischer Garten 210, Universität Hohenheim, D-70593 Stuttgart, Germany
e-mail: bfrenzel@uni-hohenheim.de

U. Lüttge et al. (eds.), *Progress in Botany 70*,
DOI:10.1007/978-3-540-68421-3,

1 Exact Stratigraphical Position of Various Interglacial Periods

Analyzing the pollen-flora of the interglacial sequence at Surheide (southern part of Bremerhaven, northwestern Germany), Behre (2004) found that vegetation history was remarkably different here from that of other interglacial sequences. Although up to 1,000 pollen grains had been counted per horizon, only faint traces of *Taxus* and of *Ulmus* could be found, and *Carpinus* was completely lacking. Thus, although the general pollen sequence resembled that of the Cromerian-II Interglacial to some extent, there also existed remarkable differences. On the other hand, the Surheide pollen diagram shows similarities to those of the Ferdynandów profile in Poland and the Artern interglacial, situated near Voigtstedt in Central Germany. Nevertheless, the stratigraphical position of both of these interglacial periods poses some problems. Repeatedly, they are assumed to date from the Elsterian glaciation, and Behre states that the exact number of Middle Pleistocene interglacial periods is evidently not known. But an interglacial sequence within a glacial period? Krzyszkowski et al. (1996) very thoroughly investigated the Ferdynandovian interglacial at Belchatow, Central Poland, for pollen succession, stable isotopes and magnetic susceptibility of the sediments. They state that the stratigraphical position of this interglacial sequence is problematic. It is situated within the huge span of time between the Cromerian II and after the Holsteinian interglacials. The authors compare this interglacial sequence with that of the Shklov Interglacial in the former USSR. Yet, here, this "interglacial" poses serious problems regarding its stratigraphic position and its vegetation history (Frenzel 1982). Comparable warm-climate sequences, with unclear stratigraphical position, have been investigated by Nitychoruk et al. (2000) in the vicinity of Suwałki, and by Stuchlik and Wojcik (2001) at Łowisko, to the northeast of Rzeszów in southern Poland. Similar difficulties can be found comparing the papers of Litt et al. (2007) and Habbe et al. (2007) regarding the stratigraphical sequences in the formerly glaciated areas of Northern Europe and those in the South-German foreland of the Alps. Furthermore, here in the eastern Alps and in their northern foreland, Kovanda (2006), Kovanda et al. (1995a), and Havlíček et al. (1994) could show from well-exposed loess sequences that the glacial–interglacial sequences must have been much more complicated than was known before. Evidently, a similar situation exists for the Middle Pleistocene period of the South-German alpine foreland (Becker-Haumann 2004; Gerth and Becker-Haumann, 2007). Šibrava (2006) concluded from all these objections that the classical notions, introduced by Penck and Brückner (1909), i.e. Würmian, Rissian, Mindelian and Günzian glaciations, should be discarded. Yet, I feel that this will be impossible, due to the fact that these notions have meanwhile been distributed almost all over the Eurasian continent. To me, it seems better to clearly define what both authors had meant and to introduce new stratigraphical names when and where these notions are inadequate or meaningless. Similar problems are discussed by Billard (1995) showing that in the French and Italian Alps up to now, several fossil interglacial and interstadial soils have been lumped together within one special stratigraphical position, thus creating the impression of a "grand interglaciaire Mindel-Riss", although this big interglacial had never existed in reality. A comparable situation is discussed by Lippstreu and Stackenbrauch (2003), Meyer (2005), Weymann et al. (2005), and Nowel (2003a, b) concerning the sequence of events during the Middle Pleistocene Saalian Glaciation of Central Europe. Here, the problem is whether this glaciation was subdivided by stronger interstadial climatic improvements (Nowel 2003a, b) or only by faint warmings of the climate (Lippstreu and Stackenbrauch 2003; Meyer 2005), although the ice-front may have receded into the basin of the eastern Baltic Sea during this Drenthe–Warthe interval (Meyer 2005).

Kondratene (1996) published a very comprehensive analysis of the Pleistocene stratigraphy and palaeogeography of Lithuania, beginning with sediments of the final Pliocene period. All the crucial geological borings and palaeobotanical sequences are dealt with, together with several topographical maps. Repeatedly, one gets the impression that much material of the pollen flora has been redeposited, as has been shown for the till material by Mahaney and Kalm (1995). Unfortunately, the general lithostratigraphy of the sequences is only very briefly mentioned by Kondratene, although just these observations could help to better understand what might have happened. Instead, till layers or other minerogenic sediments are repeatedly ascribed to certain glacial periods, in a very hypothetical way, although there are serious doubts about this, as can be shown for the Kudre-915 sequence, situated in the vicinity of Anikščaj in northeastern Lithuania. The sequence is assumed to date from Lower Pleistocene times, although to me its interglacial pollen flora strongly resembles that of the typical Eemian, i.e. of the Last Interglacial. It is said that these interglacial layers are covered by three till layers, dating from three independent glaciations (Kondratene 1996). But how is the stratigraphical position of the three till layers proven? Much more convincing is the description of the Holsteinian interglacial at Stonava-Horní Suchá in the Ostrava region (Břízová 1994), although there the position of the "presumed" Elsterian and Saalian glaciations can only be given with some reservations. A very comprehensive analysis of the macro- and microfloras of the Mazovian (=Holsteinian) interglacial at Konieczki (Silesian–Cracovian Upland) is given by Nita (1999). Shalaboda (2001) pollen-analytically investigated seven last-interglacial localities, in various parts of Belarus. The regional pattern of vegetation history is given there for this interglacial, yet it seems that one of the sites studied (Vladyki, 54°27′00″N, 27°27′00″E) really dates from an older interglacial period. Houben (2003) published a very interesting analysis of the last interglacial mammalian fauna of the site of Lehringen, northern Germany. It seems that July temperatures had been higher there by at most 2–3°C than they are at present. This fits quite well into the well-known picture (Frenzel et al. 1992) and into the reconstructions given by Klotz et al. (2003). It may also parallel the findings of Rioual et al. (2001), although these authors did not quantify the climatic deviations of the last interglacial from present day conditions. Comparable observations are reported by Giunta et al. (2006) concerning temperatures of the east Mediterranean surface waters during the last interglacial period. Regarding the pedogenesis during interglacial times in Central Europe, see Stephan (2000). Robertsson (1997) reinvestigated the interglacial vegetation history at Leveäniemi, Swedish Lapland, which, regrettably, cannot be exactly dated, but is assumed to date from the Last Interglacial (but not from the Holsteinian) due to palaeoecological considerations. The observations mentioned by Gusskov and Levtchuk (1999), concerning the shape of the North-Eurasiatic continent during the last interglacial are welcome, because they help to better understand the migration facilities for plants and animals during that time in northern Europe.

From all the above, it can be seen that physical age data of the studied sediments are urgently needed. Several attempts at this already exist. However, the standard deviations of these data are very large, as can be seen in the studies of Balescu et al. (1997) for areas along the Seine valley near Rouen, and of Kovanda et al. (1995b) concerning two warm-climate periods near Poprad in Slovakia. This prevents any reliable analyses of what definitely happened at a certain time. So it seems that, without considerable improvements of the procedures of physically dating various types of sediments, progress in palaeoecological research will be very slow and rich in serious errors.

2 Immigration of Various Tree Taxa at the Beginning of Warm-Climate Periods

This problem has already very comprehensively been analysed by Lang (1994), Gliemeroth (1995), and Ralska-Jasiewiczowa et al. (2004) for the Late Glacial of the Last Glaciation. Most interestingly, Stefanova et al. (2006) found a needle of *Pinus peuce* which was already sedimented at an elevation of 2,124 m a.s.l., together with pollen of *Betula, Quercus* and *Alnus,* during Oldest Dryas times in the Pirin Mountains, Bulgaria, These findings are well dated. Thus, the relevant tree taxa obviously existed in the vicinity of this site. This can be proven by the ensuing steps in vegetation history, although distinct regional differences in the Late Glacial and Early Holocene vegetation history seem to have existed there. Further it could be shown that the Carpathian Mountains had been an important refuge area for *Picea* (Stefanova et al. 2006). A comparable situation could be shown by Bałaga (1998), concerning the Roztocze Mountains, southeastern Poland. Here, at an elevation of 300–414 m a.s.l., at about 11,800 B.P. (uncalibrated data) the share of *Pinus* and of *Betula* strongly had increased, while *Hippophaë* and *Larix* existed there at the same time. Yet for Younger Dryas times, only faint traces of *Picea, Ulmus, Corylus,* and *Quercus,* could be found there. Simultaneously, in the surroundings of the Pekárna Cave, Moravian Karst, only *Pinus* had lived in a species-rich steppe vegetation (Svoboda et al. 2000). No other trees seem to have occurred there at that time. On the other hand, Margilewski and Zernitskaya (2003) could show that at, approximately the same time, which is held to have been the Bølling/Allerød period (the ^{14}C-data from Kiew seem to be too old; no calibration), open forests composed of *Pinus sylvestris, Pinus cembra, Betula* sp., *Betula nana* and *Hippophaë rhamnoides* (in total 80% tree pollen) had already been formed in the Central Beskid Mountains, Outer Western Carpathians. A comparable pollen flora is described by Margilewski et al. (2003) for the Makowska Range in the Beskid Mountains, though here the picture seems to be complicated by a wealth of much older and redeposited pollen grains. Nevertheless, these relatively rich pollen floras point to possible Last Glacial forest refuge communities, which is in line with the very early Late Glacial immigration process of these taxa to Lower Austria and to Southern Germany (Frenzel 1983). At the sites mentioned (Central Beskid Mountains), their share seems to have diminished during Younger Dryas times. On the other hand, the mollusc and mammalian faunas of Martina Cave, Bohemian Karst, reflect quite well just an open Late Glacial parkland vegetation, with an increasing share of forest species beginning to occur during the Boreal times of the Holocene (Alexandrowicz 1997, 1999, 2004; Alexandrowicz and Kobojek 1997; Ložek and Horáček 2006; Margilewski 2006). For the late glacial vegetation history of Switzerland, see Burga (2005). Quite understandably, the contemporary vegetation (approx. 12,700 B.P.) in the vicinity of Namur, southeastern Belgium, strongly differs from the above mentioned vegetation types (Straus and Otte 1999). Here, the existing vegetation consisted of steppe vegetation on the hills or highlands, and of gallery forests, composed of birch, hazel, and alder along the rivers, together with a species-rich mammalian fauna. It is stressed that this type of vegetation has no modern analogues. This may be important for modelling past climates because, in doing so, it is generally assumed that past climates can be reconstructed from analogues in the modern biosphere. This problematical situation is profoundly described by Mathews (2000), who analyzed the late glacial and Holocene vegetation history in the surroundings of Havelberg, northwestern Germany. The correct interpretation of the well-dated steps in vegetation history there is impeded by a lot of redeposited pollen grains, including a wealth of thermophilic taxa. Nevertheless, it can be convincingly shown, that faint traces of approaching thermophilic trees are found there only from Preboreal times onwards. These processes are comprehensively described by Kaiser et al. (2002) for the surroundings of Lake Müritz, northeastern Germany, although the precise dating here is not as well done as by Mathews (2000) in the lower Elbe region. If the available ^{14}C-data are not sufficient, the Laacher See tephra may serve as a well-dated marker here (Juvigné et al. 1995; Michaelis and Skriewe 2004; Rowinsky and Strahl 2004). More information about the late glacial and the Holocene development of the bio- and hydrospheres in Mecklenburg-Vorpommern, northeastern Germany, can be found in Terberger et al. (2004). For the synchronously happening pedogenetic processes there see Kühn et al. (2002). Bratlund (1999) published a very welcome review of the problems posed by late glacial faunal remains in northern Europe: Difficulties exist in those cases, in which the factual material is insufficient, where the dating quality is poor, or where fossil thanatocoenoses are uncritically interpreted in terms of modern biocoenoses.

Ahlberg et al. (1996) analysed climate history and evolution of vegetation during the Late Glacial and the Early Holocene in western Ireland, based on $\delta^{18}O$-analyses, carried out on carbonates of lake sediments and pollen. It can be shown that at approximately 11,900 years B.P. (uncalibrated) *Pinus* had already reached 40% vegetation cover there, similar to *Juniperus*, accompanied by only faint traces of *Betula*. The final expansion of tree taxa occurred there after about 10,400 B.P., first very intensively *Juniperus*, followed by *Salix*, then by *Betula,* and finally by *Corylus*. Yet the share of *Pinus* was only very faint at that time. Does this mean that the much earlier traces of pine found there were only caused by long distance transport of pollen? These observations are paralleled by those of Mayle et al. (1997) on the palaeoecological development at Whitrig Bog, southeastern Scotland. It could be shown here that the climate improvement during the Alleröd period was followed by tree birches about 500–1,500 years later than by the carabid fauna. Comparable observations had already been published several years ago by other research groups working in Scotland. Thus, one has to accept that obviously various climate indicators had not simultaneously responded to climate change. Similar findings, together with considerations about former climate, are reported by Brooks et al. (1997) for the chironomid fauna of the Late Glacial of southeast Scotland. However, absolute age data are lacking. Kremenetski et al. (1997) have shown that pine had reached the modern shore-line of the western Kola Peninsula at about 7,000 B.P., followed by *Betula* at about 6,000 B.P. (uncalibrated values). Finally, Böcher and Bennike (1996) found that, on Jameson Land, East Greenland, 71°N, by about 7,920 ± 100 B.P. (uncalibrated date) 29 botanical and 9 insect taxa had already immigrated there, probably transported by winds from north-western Europe. In contrast to *Salix herbacea, Harimanella hypnoides, Vaccinium uliginosum, Empetrum nigrum* and *Dryas octopetala*, the very prominent species of today, *Salix arctica* and *Cassiope tetragona,* had not yet reached this region at that time. For late glacial changes of coastlines and the configuration of the North-Atlantic bottom layers, including glacial and periglacial sediments there, see Andersen et al.(1995), Berglund (1995), Jiang and Klingberg (1996), Knudsen et al. (1996), Lubinski et al. (1996), Svendsen et al. (1996), Hald and Aspeli (1997), Jensen et al. (1997), Lyså and Vorren (1997), Snyder et al. (1997), Madeyska (1999) and Gyllencreutz et al. (2006). Concerning the late glacial situation in Belgium, it has already been mentioned that modern analogues were lacking for the then existing vegetation. Aaris-Sørensen (1995) found just the same for the mammalian fauna of Allerød times (well dated) at Nørre Lyngby, Denmark. This clearly shows the difficulties for interpreting the former biosphere in a palaeoecologically correct way.

Precise age data are needed for all these investigations. We owe to Schaub (2007) and his colleagues (Schaub et al. 2004, 2005, 2007, 2008) a remarkable prolongation of the Central European dendroseries, up to 12,593 years cal. B.P. Based on this, even the marine reservoir effect of the Cariaco Basin, Venezuela (Frenzel 2007), could be corrected by an addition of 250 years (Schaub et al. 2008), thus helping to better understand vegetation development even in extra-European areas.

From all the above, it can be concluded that our knowledge of the late glacial immigration processes of exacting plant and animal taxa to Central and North Europe is comparatively good, insofar as the relevant sediment layers have been sufficiently dated. However, this situation worsens dramatically, if older warm climate periods are investigated.

Vegetation history of the older parts of the Quaternary is generally classified into phases of different ages according to the immigration characters and by the types of prevailing vegetation. This has always worked quite well, if a rich material is taken into consideration (Seret et al. 1992; Walkling and Coope 1996; Hoffmann et al. 1998; Peschke et al. 2000; Bibus et al. 2002; von Koenigswald and Menger 2002). However, a reliable synchronisation of sequences on land with those in deep-sea sediments (Guiot et al. 1993) may be questioned, if for the analysis of migration processes on land exact age data are missing. Considering marine sediments, this will work well at the coasts of the continents (Valen et al. 1995). Repeatedly, standard deviations of the obtained physical age data are very large (±7,400–4,700 years at ages of about 71,500 or 45,400 years; e.g., Cyrek et al. 2000). Thus, a critical analysis of immigration processes or of changes in the prevailing types of the biosphere is nearly impossible (Frechen et al., 2007). Similarly high values of standard deviations can be found in a study of the history of the southern Negev desert, Israel, by Greenbaum et al. (2006). This criticism does not belittle the scientific quality of the cited papers, but it means that the age data given are not precise enough to correctly understand when and why migration processes of various plant and animal taxa had happened and when exactly certain biocoenoses were formed. In other cases, the well carried out conventional procedure of Quaternary stratigraphy will clearly suffice for understanding which general changes happened in various aspects of Quaternary geology (Ložek and Cílek 2006; McCarroll et al. 1995; Robertsson et al. 1997; Ložek 2000, 2006a, b; Schirmer 2000). On the other hand, if several age data with larger standard deviations are given for geological processes, like ice advances (Link and Preusser 2005), this considerably helps to better understand what happened in the past, as long as these data are not used for proving when exactly and elicited by which climatical processes these advances happened.

Thus, the available physical dating methods should be seriously improved. Similarities in the form of some curves do not necessarily prove causes and consequences of certain changes in the past. On the other hand it would be very helpful, if the regional vegetation-patterns of European interstadial periods could be mapped, just as it was done earlier for the last two interglacials (Frenzel 1968).

Meanwhile, there are available very comprehensive analyses on Middle and Upper Pleistocene coleopteran faunas of the Vosges Mountains (Ponel 1995), on subfossil snail assemblages of Vistulian (Last Glacial) loesses in southern Poland (Łopuszyńska 2002), and on macrofossil plant remains of the Last Glaciation in Poland (Velichkevich and Mamakowa 1999).

3 Relative Importance of Climate and the Activity of Man for the Palaeoecological Changes of the Upper Holocene in Various Regions

It cannot be denied that several major palaeoecological changes happened during the Holocene in various regions of Europe. Yet the question is which factors triggered these changes. Of course, climate is no stable factor and it has been repeatedly found that several remarkable climate changes happened during the last 15,000 years. They can be analysed best where man has never been of major importance, like in the arctic and subarctic regions, e.g., Grosval'd et al. (1961), Mäkilä (1997), Vasilčuk et al. (2003), Kultti et al. (2006), Moros et al. (2006), or where the history of glaciers has intensively been studied (Wanner et al., 2005; Joerin et al. 2006; Kerschner et al. 2006). Another possibility for analysing influences of climate only was used by Krąpiec et al. (1998) in their studies of the $\delta^{13}C$ and δD values of the cellulose of oak tree-rings as climate-proxies in southern Poland. Climatic changes of this type have been mapped for the Northern Hemisphere (Klimenko and Klimanov 2003) or at least in northern Europe (Yu and Harrison 1995). Yet even in these regions palaeoecological changes may not necessarily depend solely on climate changes, as demonstrated by Holtmeier and Broll (2006) for a certain region of the Finnish subarctic, pointing to the detrimental influence of the moth *Epirritia autumnalis* and of strong grazing by herds of *Rangifer tarandus* on the vegetation. Of course, effects of climate or tectonic movements, without the influence of man, will be found in the history of marine tsunamis, as was shown by Ruiz et al. (2005) for south-western Spain. Further, these natural causes are responsible for eustatic or isostatic movements of shorelines (Ingólfsson et al. 1995; Korhola 1995; Lackschewitz et al. 1998; Boomer and Horton 2006; Gehrels et al. 2006), and may be for lake-level changes, which happened during the Older Holocene (Magny and Ruffaldi 1995; Baroni et al. 2006). Whether this can be accepted for the Late Holocene at the lower Arno River, Central Italy (Benvenuti et al. 2006) may be doubted, even if the floods of this river seem to have happened between 400 B.C. and 1,000 A.D. at a periodicity of about 200 years, which in general seems to be a periodicity of climatic change. The question is why similar changes could not be found there earlier or later.

Another problem is under which environmental conditions the earliest impact of man on the palaeoecological situation of his environment can be found. From my own observations on the Tibetan Plateau, I know that man can seriously affect a steppe or desert-steppe vegetation even if his population density is very low. Exactly this has been stressed recently by Collins et al. (2006) concerning the Late Glacial history of the River Thames, which was evidently strongly influenced by forest clearances even at that time. Moreover, Collins et al. (2006) point to the fact that one should not extrapolate observations made in one river system to others. Analysis of the late glacial situation is repeatedly and strongly impeded by wiggles in the ^{14}C-curves. Thus, a sound analysis of what might have been the causes and what the consequences faces several difficulties (Barton 1999; Burdukiewicz 1999; Kabaciński et al. 1999). Evidently such difficulties occurred in a south Swedish peat-bog, where, nevertheless, it was concluded that the decrease in thermophilic pollen at about AD 1,275–1,590 had been caused by declining solar activity (van der Linden and van Geel 2006).Yet even for neolithic times, it seems to be very difficult to clearly differentiate, even within the European forest regions, between climate and man as causes of ecological change, as discussed by Bauerochse and Metzler (2001) who analysed the history of neolithic plankways running through North German peat-bogs. Evidently, it is very important to clearly know from modern written reports how climate had changed during the last 250 years, including the end of the Little Ice Age (Schönwiese 2005; Rehfuess 2005) for getting a sound base for evaluating the intensity of Late Holocene climatic impact on the development of the biosphere. So, it becomes possible to avoid a highly theoretical reconstruction of former atmospheric CO_2 values by calculating this via the density of stomata and of epidermal cells (van Hoof et al. 2006) and then comparing the results of this construction with the development of former forests.

Although the problem of timing, intensity and regional patterns of human impact on vegetation and on the ecological situation of his environment is very old, it seems that much more exact palaeoecological work still has to be done to obtain a firm base for other investigations and conclusions. Evidently, very often the dating quality has to be considerably improved. This needs to be considered when the samples destined for dating are collected at the exposures. To me, it seems that many more samples per geological profile should be taken for physical dating than it is generally practised. Extrapolations for the probable ages of interesting levels are not always the best solution. Moreover, as has already been stated, new and much more precisely dating procedures have to be developed, because similarities in the shapes of curves do not necessarily prove a synchronicity, but may easily lead to wrong conclusions. The present pressure to "publish or perish" generally hampers the production of good, scientifically far-reaching publications. But only these will finally solve the above discussed problems.

References

Aaris-Sørensen K (1995) Palaeoecology of a Late Weichselian vertebrate fauna from Nørre Lyngby, Denmark. Boreas 24:355–365

Ahlberg K, Almgren E, Wright HE, Ito E, Hobbie S (1996) Oxygen-isotope record of Late-Glacial climatic change in Western Ireland. Boreas 25:257–267

Alexandrowicz WP (1997) Malakofauna osadów czwartorzędowych i zmiany irodowiska naturalnego Podhala w młodszym Vistulianie i Holocenie. Malacofauna and environmental changes. Fol Quatern 68:7–132

Alexandrowicz WP (1999) Evolution of the malacological assemblages in North Poland during the Late Glacial and Early Holocene. Fol Quatern 70:39–69

Alexandrowicz WP (2004) Molluscan assemblages of Late Glacial and Holocene calcareous tufas in Southern Poland. Fol Quatern 75:309

Alexandrowicz WP, Kobojek S (1997) The malacofauna of Upper Vistulian and Holocene sediments at Błeszno near Częstochowa (southern Poland). Malacofauna and environmental change. Fol Quatern 68:189–201

Andersen ES, Østmo SR, Forsberg CF, Lehman SJ (1995) Late- and post-glacial depositional environments in the Norwegian Trench, northern North Sea. Boreas 24:47–64

Bałaga K (1998) Post-glacial vegetational changes in the middle Roztocze (E Poland). Acta Palaeobotanica 38:175–192

Balescu S, Lamothe M, Lautridou JP (1997) Luminescence evidence for the Middle Pleistocene interglacial events at Tourville, northwestern France. Boreas 26:61–72

Baroni C, Zanchetta G, Fallick AE, Longinelli A (2006) Mollusca stable isotope record of a core from Lake Frassino, northern Italy: hydrological and climatic changes during the last 14 Ka. Holocene 16:827–837

Barton RNE (1999) Colonisation and resettlement of Europe in the Late Glacial: a view from the western periphery. Fol Quatern 70:71–86

Bauerochse A, Metzler A (2001) Landschaftswandel und Moorwegebau im Neolithikum in der südwestlichen Dümmer-Region. Telma 31:105–133

Becker-Haumann R (2004) Das Grönenbacher Feld bei Kempten: Neue Befunde zur Typusregion des Mindelglazials und zur Paareiszeit im Bayerischen Alpenvorland. Eiszeitalter Gegenwart 54:20–35

Behre KE (2004) Das mittelpleistozäne Interglazial von Surheide. Eiszeitalter Gegenwart 54:36–47

Benvenuti M, Mariotti-Lippi M, Pallecchi P, Sagri M (2006) Late-Holocene catastrophic floods in the terminal Arno River (Pisa, Central Italy) from the story of a Roman riverine harbour. Holocene 16:863–876

Berglund M (1995) Late Weichselian shore displacement in Halland, southwesern Sweden: relative sea-level changes and their glacio-isostatic implications. Boreas 24:324–344

Bibus E, Rähle W, Wedel J (2002) Profilaufbau, Molluskenführung und Parallelisierungsmöglichkeiten des Altwürmabschnittes im Lößprofil Mainz-Weisenau. Eiszeitalter Gegenwart 51:1–14

Billard A (1995) Le mythe du "grand interglaciaire Mindel-Riß"d'après l'étude des sols du Nord de Turin (Italie). Sborník geologických věd – Antropozoikum 22:5–62

Böcher J, Bennike O (1996) Early Holocene insect and plant remains from Jameson Land, East Greenland. Boreas 25:187–193

Boomer I, Horton BP (2006) Holocene relative sea-level movements along the north Norfolk coast, UK. Palaeogeog Palaeoclim Palaeoecol 230:32–51

Bratlund B (1999) Review of the faunal evidence from the Late Glacial in northern Europe. Fol Quatern 70:31–37

Břízová E (1994) Vegetation of the Holsteinian interglacial in Stonava-Horní Suchá (Ostrava region). Sborník geologických věd – Antropozoikum 21:29–56

Brooks StJ, Lowe JJ, Mayle FE (1997) The Late Devensian Lateglacial palaeoenvironmental record from Whritig Bog, SE Scotland. 2.Chironomidae (Insecta; Diptera). Boreas 26:297–308

Burdukiewicz JM (1999) Concerning chronology of the Hamburgian Culture. Fol Quatern 70:127–146

Burga CA (2005) Hauptetappen der Vegetationsentwicklung im Rhein-Gebiet seit dem Würm-Spätglazial (ab ca. 17 000 Jahre B.P.). In: Naturforschende Gesellschaft Zürich Der Rhein – Lebensader einer Region. Veröff Naturf Ges Zürich 208:124–134

Collins PhEF, Worsley P, Keith-Lucas DM, Fenwick IM (2006) Floodplain environmental change during the Younger *Dryas* and Holocene in Northwest Europe: Insights from the lower Kennet Valley, south central England. Palaeogeogr Palaeoclim Palaeoecol 233:113–133

Cyrek K, Nadachowski A, Madeyska T, Bocheński Z, Tomek T, Wojtal P, Miękina B, Lipecki G, Garapich A, Rzebik-Kowalska B, Stworzewicz E, Wolsan M, Godawa J, Kościów R, Fostowicz-Frelik Ł, Szyndlar Z (2000) Excavation in the Deszczowa Cave (Kroczyckie Rocks, Częstochowa Upland, Central Poland). Fol Quatern 71:5–84

Frechen M, Terhorst B, Rähle W (2007) The Upper Pleistocene loess/palaeosol sequence from Schatthausen in North Baden-Württemberg. Eiszeitalter Gegenwart 56:212–227

Frenzel B (1968) Grundzüge der pleistozänen Vegetationsgeschichte Nord-Eurasiens. Erdwissenschaftliche Forschungen 1. Steiner, Wiesbaden

Frenzel B (1982) The history of flora and vegetation during the Quaternary. Prog Bot 44:406–417

Frenzel B (1983) Die Vegetationsgeschichte Süddeutschlands im Eiszeitalter. In: Müller-Beck HJ (ed) Urgeschichte in Baden-Württemberg, 91–165, Theiß, Stuttgart

Frenzel B (2007) Quaternary palaeoecology: Isotopes as valuable aids in palaeoecological research. Prog Bot 68:336–355

Frenzel B, Pécsi M, Velichko AA (1992) Atlas of paleoclimates and paleoenvironments of the Northern Hemisphere; Late Pleistocene – Holocene. Geogr Res Inst, Hungar Acad Sci, Budapest, G Fischer, Stuttgart

Gehrels WR, Marshall WA, Gehrels MJ, Larsen G, Kirby JR, Eiríksson J, Heinemeier J, Shimmield T (2006) Rapid sea-level rise in the North Atlantic Ocean since the first half of the nineteenth century. Holocene 16:949–965

Gerth A, Becker-Haumann R (2007) Sedimentuntersuchungen an unterpleistozänen Schmelzwasserablagerungen und Periglazialschottern im Riß-Iller-Gebiet, deutsches Alpenvorland. Eiszeitalter Gegenwart 56:186–211

Giunta S, Negri A, Maffioli P, Sangiorgi F, Capotondi L, Morigi C, Principato MS, Corselli C (2006) Phytoplankton dynamics in the eastern Mediterranean Sea during marine isotope stage 5e. Palaeogeogr Palaeoclim Palaeoecol 235:28–47

Gliemeroth AK (1995) Paläoökologische Untersuchungen über die letzten 22'000 Jahre in Europa. Vegetation, Biomasse und Einwanderungsgeschichte der wichtigsten Waldbäume. Akademie der Wissenschaften und der Literatur Mainz, Paläoklimaforschung 18, Gustav Fischer, Stuttgart Jena New York.

Greenbaum N, Porat N, Rhodes E, Enzel Y (2006) Large floods during late Oxygen Isotope Stage 3, southern Negev desert Israel. Quatern Sci Rev 25:704–719

Grosval'd MG, Devirc AL, Dobkina ÈI (1961): K istorii golocena zemli Franca-Iosifa. Doklady Akademii Nauk SSSR 141:1175–1178

Guiot J, de Beaulieu J-L, Cheddadi R, David F, Ponel P, Reille M (1993) The climate in Western Europe during the last Glaciation/Interglacial cycle derived from pollen and insect remains. Palaeogeogr Palaeoclimatol Palaeoecol 103:73–93

Gusskov SA, Levtchuk LK (1999) Foraminiferal complexes and palaeooceanographic reconstructions of the Middle and Late Pleistocene interglacial basins in the north of Siberia. Sborník geologických věd –Antropozoikum 23:125–132

Gyllencreutz R, Backman J, Jakobsson M, Kissel C, Arnold E (2006) Postglacial palaeoceanography in the Skagerrak. Holocene 16:975–985

Habbe KA, Ellwanger D, Becker-Haümann R (2007) Stratigraphisdie Begriffe für das Quartär des süddeürfsdieu Alpenvarlaxdes. Eiszeitalfer Eegehwart 56:66–83

Hald M, Aspeli R (1997) Rapid climatic shifts of the northern Norwegian Sea during the last deglaciation and the Holocene. Boreas 26:15–28

Havlíček P, Smolikova L, Kovanda J, Břizová E (1994) Loess complex near Sedlec (southern Moravia). Sborník geologických věd – Antropozoikum 21:5–18

Hoffmann MH, Litt T, Jäger EJ (1998) Ecology and climate of the early Weichselian flora from Gröbern (Germany). Rev Palaeobot Palynol 102:259–276

Holtmeier F-K, Broll G (2006) Radiocarbon-dated peat and wood remains from the Finnish Subarctic: evidence of treeline and landscape history. Holocene 16:743–751

Hoof van TB, Bunnik FPM, Waucomont JGM, Kürschner WM, Visscher H (2006) Forest regrowth on medieval farmland after the Black Death pandemie. Implications for atmospheric CO_2 levels. Palaeogeogr Palaeoclimatol Palaeoecol 237:396–411

Houben C (2003) Die Wirbeltierfauna aus dem Letzten Interglazial von Lehringen (Niedersachsen, Deutschland). Eiszeitalter Gegenwart 52:25–39

Ingólfsson O, Norddahl H, Haflidason H (1995) Rapid isostatic rebound in southwestern Iceland at the end of the last glaciation. Boreas 24:245–259

Jensen JB, Bennike O, Witkowski A, Lemke W, Kuijpers A (1997) The Baltic Ice Lake in the southwestern Baltic: sequence-, chrono- and biostratigraphy. Boreas 26:217–236

Jiang H, Klingberg F (1996) The transition from the Younger *Dryas* to the Preboreal: a case study from the Kattegat, Scandinavia. Boreas 25:271–282

Joerin UE, Stocker TF, Schlüchter C (2006) Multicentury glacier fluctuations in the Swiss Alps during the Holocene. Holocene 16:697–704

Juvigné É, Kozarski S, Nowaczyk B (1995) The occurrence of Laacher See Tephra in Pomerania, NW Poland. Boreas 24:225–231

Kabaciński J, Bratlund B, Kubiak L, Makowiecki D, Schild R, Tobolski K (1999) The Hamburgian settlement of Mirkowice: recent results and research perspectives. Fol Quatern 70:211–238

Kaiser K, Schoknecht T, Janke W, Kloss K, Prehn B (2002): Geomorphologische, palynologische und archäologische Beiträge zur holozänen Landschaftsgeschichte im Müritzgebiet (Mecklenburg-Vorpommern). Eiszeitalter Gegenwart 51:15–32

Kerschner H, Hertz A, Gross G, Ivy-Ochs S, Kubik PW (2006): Surface exposure dating of moraines in the Kromer valley (Silvretta Mountains, Austria) – evidence for glacial response to the 8.2 ka event in the Eastern Alps? Holocene 16:7–15

Klimenko VV, Klimanov VA (2003) Kholodnyj klimat rannoj subatlantičeskoy ėpokhi v severnom polušarii. Doklady Akademii Nauk 391:393–397

Klotz S, Guiot J, Mosbrugger V (2003) Continental European Eemian and early Würmian climate evolution: comparing signals using different quantitative reconstruction approaches based on pollen. Glob Planet Change 36:277–294

Knudsen KL, Conradsen K, Heier-Nielsen S, Seidenkrantz MS (1996) Palaeoenvironments in the Skagerrak-Kattegat basin in the eastern North Sea during the last deglaciation. Boreas 25:65–77

Koenigswald, W von, Menger F (2002) Ein ungewöhnlich großer Schädel vom Auerochsen (*Bos primigenius*) aus dem letzten Interglazial von Groß-Rohrheim bei Darmstadt. Eiszeitalter Gegenwart 51:67–73

Kondratene O (1996) Stratigrafiya i paleogeografiya kvartera Litvy po paleobotaničeskim dannym. Academia Vilnyus, 213 p

Korhola A (1995) The Litorina transgression in the Helsinki region, southern Finland: new evidence from coastal mire deposits. Boreas 24:173–183

Kovanda J (2006) "*Fagotia*-Faunen" und quadriglazialistisches stratigraphisches System des Pleistozäns im nördlichen Alpenvorland im Vergleich zu einigen klassischen Fundorten im Bereich der nordischen Vereisung Deutschlands. Sborník geologických věd – Antropozoikum 26:5–37

Kovanda J, Smolíkovà L, Horáček I (1995a) New data on four classic loess sequences in Lower Austria. Sborník geologických věd – Antropozoikum 22:63–85

Kovanda J, Smolíkovà L, Ford DC, Kaminská L, Ložek V, Horáček I (1995b) The Skalka travertine mound at Hôrce-Ondrej near Poprad (Slovakia). Sborník geologických věd – Antropozoikum 22:113–140

Krąpiec M. Jędrysek MO, Skrezypek G, Kałużny A (1998) Carbon and hydrogene isotope ratios in cellulose from oak tree-rings as a record of palaeoclimatic conditions in southern Poland during the last millennium. Fol Quatern 69:135–150

Kremenetski C, Vaschalova T, Goriachkin S, Cherkinsky A, Súlerzhitska L (1997) Holocene pollen stratigraphy and bog development in the western part of the Kola Peninsula, Russia. Boreas 26:91–102

Krzyszkowski D, Böttger T, Junge FW, Kuszell T, Nawrocki J (1996) Ferdynandovian Interglacial climate reconstructions from pollen successions, stable-isotope composition and magnetic susceptibility. Boreas 25:283–296

Kühn P, Janetzko P, Schröder D (2002) Zur Mikromorphologie und Genese lessivierter Böden im Jungmoränengebiet Schleswig-Holsteins und Mecklenburg-Vormmerns. Eiszeitalter Gegenwart 51:74–92

Kultti S, Mikkola K, Virtanen T, Timonen M, Eronen M (2006) Past changes in the Scots pine forest line and climate in Finnish Lapland: a study based on megafossils, lake sediments, and GIS-based vegetation and climate data. Holocene 16:381–391

Lackschewitz KS, Baumann K-H, Gehrke B, Wallrabe-Adams H-J, Thiede J, Bonani G, Endler R, Erlenkeuser H, Heinemeier J (1998) North Atlantic ice sheet fluctuations 10000–7000 yr ago as inferred from deposits on the Reykjanes Ridge, Southeast of Greenland. Quatern Res 49:171–182

Lang G (1994) Quartäre Vegetationsgeschichte Europas. Methoden und Ergebnisse. Spektrum Verlag, Jena Stuttgart New York

van der Linden M, Geel B. van (2006) Late Holocene climate change and human impact recorded in a south Swedish ombrotrophic peat bog. Palaeogeogr Palaeoclimatol Palaeoecol 240:649–667

Link A, Preusser F (2005) Hinweise auf eine Vergletscherung des Kemptener Beckens (Südwest-Bayern) im Mittleren Würm. Eiszeitalter Gegenwart 55:64–87

Lippstreu L, Stackenbrandt W (2003) Jänschwalde und die Gliederung des Saale-Komplexes – ein Kommentar zum Beitrag von Werner Nowel. Eiszeitalter Gegenwart 52:84–90

Litt T, Behre K-E, Meyer K-D, Stephan H-J, Wansa S (2007) Stratigraphische Begriffe für das Quartär des norddeutschen Vereisungsgebietes. Quatern Sci J 56:7–65

Łopuszyńska M (2002) Differentiation of subfossil populations of snails from Vistulian loesses in southern Poland. Fol Quatern 73:101–189

Ložek V (2000) Palaeoecology of Quaternary Mollusca. Sborník geologických věd – Antropozoikum 24:35–59

Ložek V (2006a) Dolní Věstonice - Biostratigraphie der Sedimente einer Hangrinne (ein Beitrag zur Polygenese der holozänen Bodenbildungen). Sborník geologických věd – Antropozoikum 26:51–60

Ložek V (2006b) Last glacial paleoenvironments of the West Carpathians in the light of fossil malacofauna. Sborník geologických věd – Antropozoikum 26:73–84

Ložek V, Cílek V (1995) Late Weichselian – Holocene sediments and soils in mid-European calcareous areas. Sborník geologických věd – Antropozoikum 22:87–112

Ložek V, Horáček I (2006) Martina Cave (Bohemian Karst) – biostratigraphy of the entrance sediments. Sborník geologických věd – Antropozoikum 26:61–71

Lubinski DL, Korsun S, Polyak L, Forman SL, Lehman SJ, Herlihy FA, Miller GH (1996): The last deglaciation of the Franz Victoria Trough, northern Barents Sea. Boreas 25:89–100

Lyså A, Vorren TO (1997) Seismic facies and architecture of ice-contact submarine fans in high relief fjords, Troms, Northern Norway. Boreas 26:309–328

Madeyska T (1999) Palaeogeography of European lowland during the Late Vistulian. Fol Quatern 70:7–30

Magny M, Ruffaldi P (1995) Younger *Dryas* and early Holocene lake-level fluctuations in the Jura mountains, France. Boreas 24:155–172

Mahaney WC, Kalm V (1995) Scanning electron microscopy of Pleistocene tills in Estonia. Boreas 24:13–29

Mäkilä M (1997) Holocene lateral expansion, peat growth and carbon accumulation on Haukkasuo, a raised bog in southeastern Finland. Boreas 26:1–14

Margilewski W (2006) Records of the late glacial – Holocene palaeoenvironemental changes in landslide forms and deposits of the Beskid Makowski and Beskid Wyspowy Mts. area (Polish Outer Carpathians). Fol Quatern 76.149p

Margilewski W, Zernitskaya V (2003) Late Glacial – Holocene paleoenvironmental evidence recorded in the Hajduki peat bog (Beskid Średni Mts., outer western Carpathians). Fol Quatern 74:57–73

Margiewski W, Obidowicz A, Pelc S (2003) Late-Glacial – Holocene peat bog on Kotou Mt. and its significance for reconstruction of palaeoenvironment in the western outer Carpathians (Beskid Makowski Range, South Poland). Fol Quatern 74:35–56

Mathews A (2000) Palynologische Untersuchungen zur Vegetationsentwicklung im Mittelelbegebiet. Telma 30:9–42

Mayle FE, Lowe JJ, Sheldrick C (1997) The Late Devension Late Glacial palaeoenvironmental record from Whitrig Bog, SE Scotland. 1. Lithostratigraphy, geochemistry and palaeobotany. Boreas 26:279–295

McCarroll D, Ballantyne CK, Nesje A, Dahl S-O (1995) Nunataks of the last ice sheet in northwest Scotland. Boreas 24:305–323

Meyer KD (2005) Zur Stratigraphie des Saale-Glazials in Niedersachsen und zu Korrelationsversuchen mit Nachbargebieten. Eiszeitalter Gegenwart 55:25–42

Michaelis D, Skriewe S (2004) Die Lieper Posse in NO-Brandenburg – Braunmoostorfe, Akkumulationsraten und das Problem der Moorgenese. Telma 34:11–29

Moros M, Jensen KG, Kuijpers A (2006) Mid- to late-Holocene hydrological and climatic variability in Disko Bugt, central West Greenland. Holocene 16:357–367

Nita M (1999) Mazovian interglacial at Konieczki near Kłobuck (Silesian-Cracovian Upland). Acta Palaeobot 39:89–135

Nitychoruk J, Ber A, Hoefs J, Krzywicki T, Schneider J, Winter H (2000) Interglaziale Klimaschwankungen in Nordost-Polen – palynologische und isotopengeochemische Untersuchungen an organischen Seesedimenten. Eiszeitalter Gegenwart 50:86–94

Nowel W (2003a) Zur Korrelation der Glazialfolgen im Saale-Komplex Nord- und Mitteldeutschlands am Beispiel des Tagebaus Jänschwalde in Brandenburg. Eiszeitalter Gegenwart 52:47–83

Nowel W (2003b) Nochmals zur Altersstellung des Tranitzer Fluviatils (Anmerkungen zum "Kommentar" von Lippstreu und Stackebrandt). Eiszeitalter Gegenwart 53:114–123

Penck A, Brückner E (1909) Die Alpen im Eiszeitalter. I. Die Eiszeiten in den nördlichen Ostalpen. II. Die Eiszeiten in den nördlichen Westalpen. Tauchnitz, Leipzig

Peschke P, Hannss C, Klotz S (2000) Zur spätpleistozänen Vegetationsentwicklung der Banquette von Barraux (Grésivaudan, französische Nordalpen). Eiszeitalter Gegenwart 50:1–24

Ponel P (1995) Rissian, Eemian and Würmian Coleoptera assemblages from La Grande Pile (Vosges, France). Palaeogeogr Palaeoclimatol Palaeoecol 114:1–41

Ralska-Jasiewiczowa M, Latałowa M, Wasylikowa K, Tobolski K, Madeyska E, Wright Jr HE, Turner C (2004) Late Glacial and Holocene history of vegetation in Poland based on isopollen maps. W Szafer Institute of Botany, Polish Academy of Sciences, Kraków

Rehfuess K-E (2005) Zusammenfassung des Vortrages: Der Umgang mit dem Klimawandel: Anpassung und Vermeidung, von Hans von Storch. In: Rundgespräche der Kommission für Ökologie 28: Klimawandel im 20. und 21. Jahrhundert. Verlag Dr. Friedrich Pfeil, München, pp 123–124

Rioual P, Andrieu-Ponel V, Rietti-Shati M, Battarbee RW, Beaulieu de J-L, Cheddadi R, Reille M, Svobodova H, Shemesh A (2001) High resolution record of climate stability in France during the last interglacial period. Nature 413:293–296

Robertsson A-M (1997) Reinvestigation of the interglacial pollen flora at Leveäniemi, Swedish Lapland. Boreas 26:81–89

Robertsson A-M, Svedlund J-O, Andrén T, Sundh M (1997) Pleistocene stratigraphy in the Dellen region, central Sweden. Boreas 26:237–260

Rowinsky V, Strahl I (2004) Entwicklung von extrem tiefgründigen Kesselmooren im Plauer Stadtwald (Mecklenburg-Vorpommern). Telma 34:39–64

Ruiz F, Rodriguez-Ramirez A, Cáceres LM, Vidal JR, Carretero MI, Abad M, Olias M, Pozo M (2005) Evidence of high-energy events in the geological record: Mid-Holocene evolution of the southwestern Doña National Park (SW Spain). Palaeogeogr Palaeoclimatol Palaeoecol 229:212–229

Schaub M (2007) Lateglacial environmental conditions on the Swiss Plateau. A multi-proxy approach using tree rings and sediment-based proxies. Schriftenreihe Physische Geographie, Glaziologie und Geomorphodynamik 54, Geographisches Institut Zürich

Schaub M, Kaiser F, Kromer B (2004) Lateglacial tree-ring chronologies – A high resolution archive of the past. TRACE, Tree Rings in Archaeology, Climatology, and Ecology 3, Schriftenreihe des Forschungszentrums Jülich, Reihe Umwelt/Environment 53:96–101

Schaub M, Kaiser KF, Kromer B, Talamo S (2005) Extension of the Swiss lateglacial tree-ring chronologies. Dendrochronologia 23:11–18

Schaub M, Kaiser KF, Frank DC, Büntgen U, Kromher B, Talamo S (2007) Environmental change during the Allerød and Younger *Dryas* reconstructed from Swiss tree-ring data. Boreas 37:74–86

Schaub M, Büntgen U, Kaiser KF, Kromer B, Talamo S, Andersen KK, Rasmussen SO (2008) Lateglacial environmental variability from Swiss tree rings. Quatern Sci Rev, 27:23–41

Schirmer W (2000) Eine Klimakurve des Oberpleistozäns aus dem rheinischen Löß. Eiszeitalter Gegenwart 50:25–49

Schönwiese C-D (2005) Globale und regionale Klimaänderungen im Industriezeitalter – Beobachtungsindizien und Ursachen. Bayerische Akademie der Wissenschaften: Rundgespräche der Kommission für Ökologie 28:17–32

Seret G, Guiot J, Wansard G, Beaulieu de J-L, Reille M (1992) Tentative palaeoclimatic reconstruction linking pollen and sedimentology in La Grande Pile (Vosges, France). Quatern Sci Rev 11:425–430

Shalaboda VL (2001) Characteristic features of Muravian (Eemian) pollen succession from various regions of Belarus. Acta Palaeobot 41:27–41

Šibrava V (2006) Alpine foreland-problems of Quaternary stratigraphy, nomenclature and interpretation. Sborník geologických věd – Antropozoikum 26:39–49

Snyder JA, Forman SL, Mode WN, Tarasov GA (1997) Postglacial relative sea-level history: sediment and diatom records of emerged coastal lakes, north-central Kola Peninsula, Russia. Boreas 26:329–346

Stefanova I, Atanassova J, Delcheva M, Wright HE (2006) Chronological framework for the lateglacial pollen and macrofossil sequence in the Pirin Mountains, Bulgaria: Lake Besbog and Lake Kremensko-5. Holocene 16:877–892

Stephan S (2000) Bt-Horizonte als Interglazial-Zeiger in den humiden Mittelbreiten: Bildung, Mikromorphologie, Kriterien. Eiszeitalter Gegenwart 50:95–106

Straus LG, Otte M (1999) La Grotte du Bois Laiterie (Profondeville, Belgique): Halte de chasse Magdalénienne. Fol Quatern 70:101–113

Stuchlik L, Wojcik A (2001) Pollen analysis of Malopolanian Interglacial deposits at Łowisko (Kolbuszowa Upland, southern Poland). Acta Palaeobot 41:15–26

Svendsen JI, Elverhøi A, Mangerud J (1996) The retreat of the Barents Sea Ice Sheet on the western Svalbard margin. Boreas 25:244–256

Svoboda J, Horáček I, Ložek V, Svobodova H, Silar J (2000) The Pekárna Cave, Magdalenian stratigraphy, environment, and the termination of the loess formation in Moravian Karst. Sborník geologických věd – Antropozoikum 24:61–79

Terberger T, L De Klerk P, Helbig H, Kaiser K, Kühn P (2004) Late Weichselian landscape development and human settlement in Mecklenburg-Vorpommern (NE Germany). Eiszeitalter Gegenwart 54:138–175.

Valen V, Larsen E, Mangerud J (1995) High-resolution paleomagnetic correlation of Middle Weichselian ice-dammed lake sediments in two coastal caves, western Norway. Boreas 24:141–153

Vasiečuk YK, Vasilčuk AK, Suleržickij LD, Budanceva NA, Volkova EM, Čižova YuN (2003) Radiouglerodnaya khronologiya bugrov pučeniya Bolšezemelskoj Tundry. Doklady Akademii Nauk 393:101–105

Velichkevich FYu, Mamakowa K (1999) Taxonomic revision of the collections of plant macrofossils from some localities of Poland now referred to the Vistulian Glaciation. Acta Palaeobot 39:29–87

Walkling AP, Coope GR (1996) Climatic reconstructions from the Eemian/Early Weichselian transition in Central Europe based on the coleopteran record from Gröbern, Germany, Boreas 25:145–159

Wanner H, Casty C, Luterbacher J, Pauling A (2005) 500 Jahre Klimavariabilität im europäischen Alpenraum – raumzeitliche Strukturen und dynamische Interpretationen. Bayerische Akademie der Wissenschaften: Rundgespräche der Kommission für Ökologie 28:33–52.

Weymann H-J, Feldmann L, Bombien H (2005) Das Pleistozän des nördlichen Harzvorlandes–eine Zusammenfassung. Eiszeitalter Gegenwart 55:43–63

Yu G, Harrison SP (1995) Holocene changes in atmospheric circulation patterns as shown by lake status changes in northern Europe. Boreas 24:260–268

Water Relations in the Mycorrhizosphere

M.F. Allen

Contents

Abstract The role of mycorrhizae in plant water relations has been controversial since it was proposed over a century ago. Improvements in plant water throughput and associated carbon fixation by mycorrhizal plants have been demonstrated since the early 1970s, but the mechanisms remain poorly understood. Mechanism studies have concentrated in greenhouse pot studies. These studies are limited because roots readily become pot-bound, in contrast to the distinct spatial stratification in the field. Water transport studies also focused on symplastic water flow rates. Recent data from hyphal structure and using models suggest that apoplastic transport may be the focal mechanism under drought conditions. Hydraulic redistribution along mycorrhizal fungal hyphae indicates that apoplastic transport and water potential gradients, especially in arid lands, regulate water flow patterns along hyphae. New sensor technologies may open the possibility of greater study of water and nutrient dynamics in situ where these mechanisms can be studied in detail.

M.F. Allen
Center for Conservation Biology, Department of Plant Pathology,
University of California, Riverside, CA 92521-0334, USA
e-mail: michael.allen@ucr.edu

U. Lüttge et al. (eds.), *Progress in Botany 70*,
DOI:10.1007/978-3-540-68421-3,

1 Introduction

Mycorrhizae increase the carbon acquisition by host plants. In order to do that, and to allocate carbon to the fungus, increased water throughput is a necessity. In mesic to wet environments, that water flow is rarely crucial, and so is not often measured, even though plant production, carbon gain, and nutrient concentrations are directly related to water use. Most production ecosystems studied are concentrated in areas where water is not the limiting factor; N, P and pests generally regulate output. However, as the world's population continues to grow, increasingly semi-arid and arid ecosystems become critical agricultural and forest production lands depending upon seasonal precipitation or irrigation. But seasonal precipitation is often unpredictable, and irrigation expensive. For this reason, understanding the dynamics and mechanisms of interaction for biological mutualisms continues to become critical to management of productive lands. Mycorrhizae, in particular, appear to be crucial to supplying water to wildland plants in arid ecosystems, and practices that reduce activity can impose an additional water stress on these plants.

The interdependence of water and mycorrhizae has been studied since the work of Stahl (1900). Although he did not understand how mycorrhizae, water, and nutrient acquisition were interrelated, he postulated that water did play a critical role in the functioning of the symbiosis. The importance of hyphal transport of nutrients, particularly HPO_4^- and NH_4^+, that do not move by mass flow was demonstrated as early as the 1930s in ectomycorrhizae (EM) (Hatch 1937), and in arbuscular mycorrhizae (AM) by the 1960s (Gerdemann 1968). Importantly, NO_3^-, K, Ca and other nutrients that move by mass flow were also shown to increase, but not in all experiments, depending upon the experimental conditions (Mosse 1973; Marks and Kozlowski 1973). In the 1970s, a number of studies were initiated focusing on the importance of mycorrhizae on plant water relations (Safir et al. 1971). By this time, it was apparent that mycorrhizae could improve plant water relations. The problem is in understanding the mechanisms resulting in the improvements. Most studies to date have assumed that any mycorrhizal effect would be due to symplastic water flow. With this limitation, limiting steps such as flow within the cytoplasm, flow through semi-permeable membranes, and active transport of solutes, would limit the transport capacity. But water flows apoplastically as well as symplastically. This movement is simply dependent upon water potential gradients and has the potential to move water much faster, and over greater distances, than symplastic transport.

The difficulty with developing a mechanistic understanding of mycorrhizal responses and water are really 2-fold. First, water moves in response to gradients, and these are often difficult to control experimentally at spatial scales at which mycorrhizal fungi live. In particular, apoplastic flow is rarely considered in fungal dynamics, yet it may be the most critical mechanism and does not depend upon active transport. Second, glasshouse studies that form the experimental backdrop for mycorrhiza studies do not mimic field soil conditions; they have limited rooting volume with an enormous pot effect, and they do not mimic soil structure, crucial to understanding and measuring water flow patterns and rates. These conditions are especially limiting to understanding and measuring the dynamics of the fungal–soil–plant interfaces. While these experiments can define the range of mechanisms that potentially occur, they cannot eliminate alternative mechanisms, and cannot tell us which mechanisms are involved in the field at any particular time and under specific environmental conditions.

My goal with this review is to outline the range of mechanisms that may be involved in regulating mycorrhizal fungal–plant–soil water fluxes, to outline experiments undertaken in the field that illuminate some of these mechanisms, and to propose a type of experimentation that is becoming available to study the role of mycorrhizae in water-limited ecosystems. The vast majority of terrestrial plants form mycorrhizae. In this review, I will concentrate on EM and AM, as there is little or no relevant work on water relations and mycorrhizae in other types. That said, there are important studies to be undertaken regarding mycorrhizae, water relations and epiphytes (including orchids) that must acquire water from surfaces exposed to the atmosphere, including on ericoid mycorrhizae, especially those living in arid regions such as the sand plains of Western Australia, and myco-heterotrophs that depend upon a host plant for carbon, especially as the dry season approaches.

Mycorrhizal plants differ physiologically from nonmycorrhizal (NM) plants, often dramatically! Just as importantly, a plant with one fungal symbiont can differ from that with another, where the NM plant may express a growth rate and form intermediate between symbionts if one of the symbionts causes reduced plant response. Under most field conditions, a plant has multiple mycorrhizal fungal symbionts, and a mycorrhizal fungus is often attached to and drawing carbon from many hosts. In most cases, the NM plant is the unusual, making it very difficult to define what a control is. Creating a NM plant for experimental purposes requires an experimental condition that eliminates mycorrhizae, rather than a natural condition where mycorrhizae are present. By viewing a nonmycorrhiza as the experimental treatment, and by undertaking comparative studies among plants, fungi, and environmental conditions, maybe we can generate a better perspective on the actual role of mycorrhizae in field situations.

2 Mechanisms of Water Movement in Laboratory Studies

2.1 Plant Physiological Drivers

Safir et al. (1971) published one of the first studies on water relations improvement by mycorrhizae. This was a landmark paper because, previously, most researchers had envisioned that the action of mycorrhizae was through uptake of nutrients that did not move via mass flow. Safir and colleages demonstrated that AM could enhance even this resource. But they followed this study by postulating that the mechanism was due to improved P nutrition. Safir et al. (1972) noted that creating a NM plant using Benomyl® reduced P uptake, but not water uptake, and concluded that the increased P was the cause of the improved water transport. But, P uptake is an active transport process whereas water is passive. They neglected apoplastic flow as a potential mechanism for enhanced water transport. The hyphal structure could remain for several days following the addition of Benomyl®, and the increased concentrations of nutrients that preceded the death of the AM fungi would also still be present and driving water flow. Since that time, considerable evidence has accumulated that AM hyphae alter soil structure in a manner that affects water relations of plants (e.g., Augé et al. 2004; Rillig and Mummey 2006), and while EM reduce short root length, they also influence long-distance water transport (Duddridge et al. 1980) and total water throughput capacity (Hobbie and Colpaert 2004). Based on these studies, simple pot experimental data contrasting a plus/minus mycorrhiza is often equivocal.

NM plants tend to have lower nutrient concentrations than mycorrhizal plants. These lower concentrations include not only those elements that must be actively transported (P, NH_4^+), but also elements such as K, and Ca that move by mass flow (e.g., Allen and Cunningham 1983; Azcón-Aguilar and Barea 1997). These elements contribute to an increased osmotic concentration which allows a mycorrhizal plant to continue to take up water as soil water potentials decline (Allen and Boosalis 1983; Augé et al. 1986). In grapevines, drought tolerance and water use efficiency was higher in AM than NM plants corresponding to higher levels of proline (Valentine et al. 2006). In citrus, the improved drought stress tolerance appeared to be due to increased levels of K^+, Ca^{2+}, and Mg^{2+} (Wu and Xia 2006). NM plants also have lower levels of cytokinins, gibberellins, auxins and a higher level of abscisic acid (Allen 1980; Allen et al. 1980, 1982) under non-drought conditions that could make them less drought tolerant than mycorrhizal plants (Allen et al. 1981). In EM *Pinus sylvestris* seedlings, under low N concentrations, mycorrhizae increased water-use efficiency (WUE), but that pattern was not apparent under high N concentrations (Hobbie and Colpaert 2004). These results suggest a potential tie to proline, but that hypothesis has not been tested.

Most of these responses have been verified in a wide variety of other systems, plants, and plant-fungal combinations (e.g. Augé 2001). In NM plants, stomates close more rapidly with drought stress, and contain phytohormone levels that indicate greater drought stress. Osmotic potentials are higher in NM plants, again indicating a reduced capacity to tolerate drought stress.

One response often overlooked is really a relatively simple one. In order to increase carbon fixation resulting in increased plant production, mycorrhizae must either increase total water throughput, or increase the efficiency of CO_2 fixation per unit water loss, or both. Evidence for increased photosynthesis is widespread in glasshouse studies (e.g., Reid and Woods 1969; Kucey and Paul 1982; Allen et al. 1981, 1984). Evidence for enhanced WUE is mixed (Augé 2001). In estimating water use efficiency, Allen (1980) found no effect of mycorrhizae on the ratios. In a more direct study, Di and Allen (1991) measured $\delta^{13}C$ ratios and determined that the different varieties of the plant (*Agropyron cristatum*) showed different responses in water-use-efficiency to mycorrhizae. Different plant species also have varying WUE responses to AM (e.g. Davies et al. 2002). Mena-Violante et al. (2006) noted that fruit from AM plants resembled non-droughted fruit indicating less stress. In another case, Ruiz-Lozano et al. (2006) reported that the species of fungal symbiont determined whether or not water-use efficiency was affected by mycorrhizae. Thus, we can conclude that mycorrhizae increase water throughput, but that increasing water-use efficiency depends upon species, and even variety composition, and upon environmental conditions.

2.2 *Fungal Structural Drivers*

Fungi have a structure that makes their water relations extremely interesting. AM fungal hyphae are coenocytic, that is, they do not have regular cross-walls and water, organelles, and cytoplasm move linearly via Brownian movement and concentration gradients. Ascomycetes and Basidiomycetes have cross-walls (septa) that constrict, constraining symplastic flow, but actual membranes blocking flow are found at the hyphal tips and irregularly through the mycelium. Nuclei and other organelles can be observed squeezing through the septa, and water and other materials regularly flow, unless the pore is blocked by Woronin bodies or other materials (Alexopoulos et al. 1996). Membranes form the largest barrier to water movement (Nobel 1974) under saturated conditions.

Hardie and Leyton (1981) proposed that AM fungal hyphae may be responsible for increasing conduction of water from soil to plants. Allen (1982) measured water fluxes through glasshouse-potted plants, and contrasted AM and NM individuals. In that experiment, I measured the total seedling root surface area, the number of entry points, and the water uptake rates. The difference between an AM and a NM plant, if attributed to the hyphal entries, suggested that a 10-μm diameter hypha would have to transport 90 nl h^{-1} of water between the external network and the plant root. While high, this figure was within the range of transport for fungal hyphae. Hardie (1985) then performed an intricate experiment where she measured water fluxes, then carefully clipped the entry AM hyphal entry points. She re-measured water fluxes, and found that water uptake in AM plants with clipped hyphae was reduced to those of the NM plants. Thus, she concluded that the AM hyphae directly contributed the water difference to the host.

Recent experiments have substantiated the potential for hyphae to contribute to plant water balance. Allen (1996) observed water moving along hyphae from the tips toward the host in using a glass-plate observation system. Khalvati et al. (2005) used a mesh and evaluated water transport from a hyphal-only compartment. They found that about 4% of the water in the hyphal compartment was transferred to the plant compartment via hyphae. Although this was a small fraction (~1.5%) of the total plant transpiration, it does indicate that hyphae transport water. In a very interesting experiment, Augé et al. (2004) found that the presence of AM hyphae in soil could increase the hydraulic conductivity of that soil. The hyphae themselves served as a pathway for water flow in dry soils.

Duddridge et al. (1980) reported that many EM fungal rhizomorphs formed hollow tubes within the hyphae, resembling "vessels" capable of translocation large amounts of water long distances. This developmental pattern has the capacity to move both water and nutrients up to several meters. AM do not form the complex

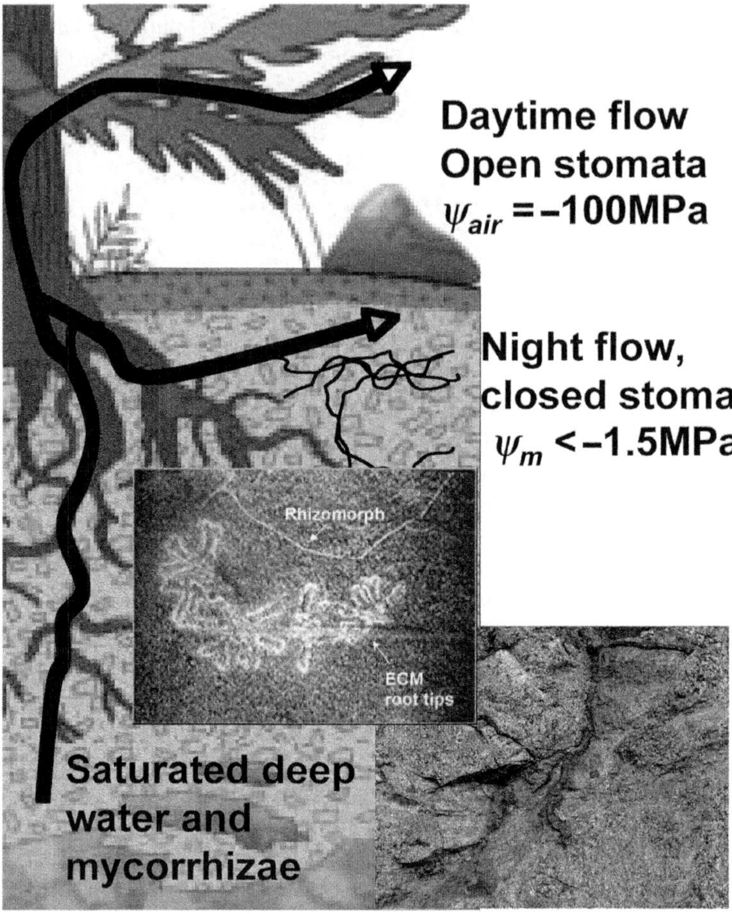

Fig. 1 Hydraulic redistribution and mycorrhizae. This occurs when surface soils are dry and there is deep soil moisture. During the day, water is taken up from deep water and from soil pores either by roots or hyphae. However, at night, when stomata close and surface ψ_m is well below field capacity, water can reverse flow direction and travel out fine roots and mycorrhizal fungal hyphae and out the hyphal tips (Querejeta et al. 2003b), or even into neighboring plants (Egerton-Warburton et al. 2008)

networks found in EM, but hyphae will wrap around themselves, forming a net that can translocate water apoplasmically (Allen 2007).

2.3 *Flow Directions and Environmental Conditions*

The importance of mycorrhizae has focused on flow from soil to plant via the fungus because of the emphasis on improving plant drought tolerance. One basic problem to studying flow is that water moves simply in response to gradients and barriers; there is not active transport. Organisms can adjust their solute composition, changing their osmotic potential (ψ_π) and even matric potential (ψ_m) within a membrane that restricts the solutes, but cannot redirect transport rates directly. By refocusing research on flow in any direction, recent studies have lead to interesting new insights into the functioning of mycorrhizae.

Recent studies have documented that hydraulic lift may play a critical role in sustaining mycorrhizal hyphae during drought (Querejeta et al. 2003b, 2007). The mechanism is relatively straightforward, mechanistically, but exceedingly complex in understanding the implications of such activity. Water flows in response to water potential gradients. In mesic ($\psi_m > -1.5$ MPa) soils, water flows from soil to plant as the plant maintains a ψ below that of the soil. As surface soil ψ_m drops below -1.5 MPa, larger pockets of water empty out and the ψ_s becomes lower than ψ_{roots}. Many hyphae can tolerate ψ well below -1.5 MPa (Mexal and Reid 1973). During the day, since the atmospheric concentration of water is comparatively extremely low (< -100 MPa), transpiration occurs if the roots reach deeper layers containing higher water potentials. But, during the night, the ψ_{soil} is below the re-hydrating ψ_{plant}. In this case, water then flows back out roots, and into mycorrhizal hyphae (Querejeta et al. 2003b). This water flow can move out of hyphal tips, across air gaps, and even into neighboring plants (Egerton-Warburton et al. 2007).

This pattern of water flow is diel; during the day, water flows from soil to fungus, to plant, and during night, from plant to fungus to soil (Fig. 1). The night-time water is exuded through the hyphal tips into the surrounding matrix (Querejeta et al. 2003b). Because the water is exuded during the night, and then taken back up during the day, that water can support a variety of bacteria mineralizing litter. The concentrated solutions with N, P and other resources is then re-absorbed by the hypha the next day and translocated back into the hypha and then to the plant (Egerton-Warburton et al. 2008). This diurnal hydraulic redistribution and nutrient uptake may play a critical role, especially for plants that have access to water from deep roots (Querejeta et al. 2007b) and from bedrock matrices (Egerton-Warburton et al. 2003; Bornyasz et al. 2005).

3 Field Experiments of Mycorrhizae and Water Flow

Studies in the laboratory and glasshouse delineate potential mechanisms that can result in changing water relations of mycorrhizae, but, they do not detail how mycorrhizae actually change plant physiological ecology in response to environmental stresses. For that, we must turn to actual field experiments. Fortunately, there are a number of published field experiments that provide valuable insights into mycorrhizal regulation of plant drought tolerance. The difficulty is that being mycorrhizal is the "normal" situation. Therefore, studying mycorrhizae in the field means eliminating the fungi, followed by re-inoculation. In most instances, this re-inoculation is experimental, while in a few cases naturally re-colonizing events can be studied. In many cases, it is difficult to get a complete elimination of mycorrhizae, so comparisons of very low infection or among mycorrhizal symbionts is undertaken.

Mount St. Helen's erupted in May 1980. The new pumice material provided a sterile substrate for plants to colonize. Most plants were NM, but those were interspersed with a few AM plants inoculated by soil moved by pocket gophers. In these cases, AM plants had increased stomatal conductance, increased N, and increased growth compared with NM plants (Allen et al. 1992). In the case of EM conifers, NM seedlings were observed colonizing the pumice annually, but all died during the August droughts. Beginning in 1986, the first seedlings were observed surviving the summer drought and establishing. All the surviving seedlings formed EM (Allen 1987).

Allen and Allen (1986) studied the role of AM in establishment of grasses on a strip mine in western Wyoming, USA. The overburden material comprising the site consisted of stockpiled B and C horizon material with virtually no biological activity. 0.5-m^2 plots were inoculated with 2 cm of "live" or steam-sterilized soil from the adjacent undisturbed area. Grasses receiving "live" fresh soil were mycorrhizal whereas those with no or steam-sterilized topsoil had no to low (1–2%) infection. These soils are high in available P and there was no evidence of an N deficiency. In this case, during the period when soils were either wet or very dry, AM had no measurable effect on plant water relations. But, during the dry-down period, AM plants were under significantly lower water stress than NM plants. The phenology also showed this reduced drought stress; under conditions that triggered early seed development in NM plants, AM plants were able to continue accumulating carbon, and thereby increase seed production and overall growth.

In a study examining the effects of inoculation on transplanted seedlings, Hickson (1993) found that different mycorrhizal fungi had markedly different effects on sagebrush seedling photosynthesis and water relations, both positive and negative. But a clear outcome was that transpiration was affected by the interaction of inoculum and stomatal conductance ($p = 0.08$). Photosynthesis responded to stomatal conductance and internal CO_2 concentration regardless of the inoculum type ($p < 0.001$), and inoculum type modified the stomatal conductance. Plants tended to respond more positively to their natural symbiont than to an exotic transplanted inoculum, but there was considerable variation. In this case, at the San Diego site, there were significant differences in water-use efficiency between two of the inocula. Teasing apart the mechanisms contributing to this change was tricky, however. In part, the mycorrhizal fungi changed the shape of the plant. The plants forming mycorrhizae with *Scutelospora calospora* grew short and wide, a pattern that would reflect adaptation to its origin in the central Great Basin, where winters are long and severe, and the only competition is short grasses. The plants forming mycorrhizae with *Acaulospora elegans* (from the warmer and more competitive south) grew taller and narrower, suggesting a need to get above competition from taller grasses and other shrubs. Plants with *S. calospora* had a greater proportion of sun leaves, and a greater need to fix C during the shorter growing season of the site of origin for the fungus. In this case, separating the impacts of direct mycorrhizal involvement in WUE from indirect growth pattern shifts proved indecipherable. Across the broader dataset at the other site, WUE showed no differences among inoculum types, although the total photosynthesis and transpiration varied greatly among the individual plants. Uninoculated plants had low–intermediate rates, but the various inocula varied greatly.

Isotopes of water can provide useful indicators of sources of soil water, and how that water might be transported. Querejeta et al. (2007b) were able to demonstrate that during the dry season, when ψ_{soil} of surface soils was below −4 MPa, water was redistributed from the groundwater by deep roots of live oak (*Quercus agrifolia*) and sustained both EM and AM fungi through the profile. In the interspaces with no oaks and only grasses, the surface water content evaporated and no mycorrhizal activity was observed during the late summer drought. Along hillsides, mature oaks were also observed to have water that had been extracted from granite matrix between the cracks where roots proliferated. This water may have moved along hyphae that permeated that granite matrix (Bornyasz et al. 2005; Allen 2006).

Natural abundances of isotopes can be used in the field, either via monitoring, or in experiments to illustrate actual flux patterns. As opposed to glasshouse experiments, these results demonstrate actual responses in the field, not just potential responses. Querejeta and his colleagues have undertaken a number of analyses of isotopic signatures in replicated field experiments showing a range in responses to mycorrhizae broadly, and to species and species combinations specifically. To try to tease these factors apart, Querejeta and colleagues added known (local vs exotic) inocula, in the form of single species or multiple species, to seedlings planted in experimental blocks across a severely disturbed site. In the first study, AM inoculation enhanced the WUE in *Olea europaea*, a long-lived slow-growing evergreen tree, but not in *Rhamnus lycioides*, a drought-deciduous shrub with a short lifespan (Querejeta et al. 2003a). These data indicate that the relative effect of AM on WUE and drought tolerance is species specific. In a second study, Querejeta et al. (2006) found that different AM fungal species combinations differentially affected *O. europaea* and *R. lycioides*. Foliar $\delta^{18}O$ indicated that native AM fungi enhanced stomatal conductance whereas an exotic *Glomus claroideum* did not. WUE was improved in *O. europaea* when inoculated with a mix of native *Glomus* species, in contrast to inoculation with *G. claroideum* or in the previous experiment with the exotic *G. intraradices*. Subsequently, in a third experiment involving *Pistacia lentiscus* and *Retama sphaerocarpa*, native AM enhanced shoot $\delta^{15}N$, shoot water content ($\delta^{18}O$), and P uptake. No differences in $\delta^{13}C$ were observed, indicating that the rates of increased transpiration corresponded directly to the enhanced photosynthesis (Querejeta et al. 2007a).

In summary, the species of both plant and fungus matter, as does the combination of species (Querejeta et al. 2003a, 2006, 2007a; Allen et al. 2003). More field work on the sources of water and on the impacts of total water flow and water-use efficiency are needed. Many researchers prefer to work in greenhouse pots but, because of the constraints of pots, these studies will not be as useful as field studies.

4 Models of Water Flow in Hyphae

Understanding the importance of mycorrhizae to plant water relations also requires characterizing the water dynamics of individual hyphae in the localized microenvironment. There are few technologies that allow for studying these processes in the field. For this reason, much of the understanding for these processes comes from models based on principles of biological and soil properties. An analysis of these models requires understanding two distinctive phenomena, changing matric potential with changing water content, and the structure and dynamics of fungal hyphae.

Matric water potential (ψ_m) in essence is the structure of the vapor–liquid water interface. A way to visualize this structure is as the curvature of the liquid surface. ψ_m is a function of surface tension (γ), and radius of the pore (r), where

$$\psi_m = 2\gamma{*}r^{-1} \tag{1}$$

(Stolp 1988). When all pores are full, roots are in contact, so r approaches 0. In wet soils, large macropores penetrable by roots have water. As macropores drain, roots grow to contact more pores. When water content remains in this capillary region, maximum plant growth can occur because O_2 tension allows respiration but adequate water is available for transport and internal cell processes. As the soil dries below field capacity, the curvature of the liquid film increases, with films pulling back along the soil particles, and into ever-smaller pores. As the soil continues to dry, larger soil pores become drained and water contracts back into ever-smaller pores. Roots cannot reach into pores smaller than macropores and the tips are no longer in contact with these films. At $\psi_s < -1.5$ MPa ("permanent" wilting point), the size of the pores restricts root growth, water potential reaches the adsorption region, and roots become less able to access water. Films become inaccessible, r increases and ψ_m declines. Eventually, roots can no longer reach into pores and, in micropores <5 μm, $r \rightarrow \infty$. On the other hand, hyphae can readily penetrate larger ultramicropores (down to 1 or 2 μm). As a root depletes the soil water that it can access, the fungal hyphae can continue to extract water (Fig. 2). Importantly, in situations where roots occupy the entire soil volume, such as a glasshouse pot, roots effectively access all pores well before drought is imposed.

Fungal ultrastructure is organized such that water relations are difficult to measure experimentally. At this point, we are restricted to constructing flow patterns based on physical models of adhesion and water potential gradients. Membranes generally do not separate individual "cells" of hyphae. Those generally occur at the tips. The pattern of flow may be modeled using AM symbioses, but recognize that the principles are physical and should be widely applicable. AM fungal hyphae branch dichotomously, with each branch becoming smaller, probably depending upon either taxonomic or energy-flow to regulate that size characteristic. Friese and Allen (1991) described a single network per entry point, while Bago et al. (1998) described a network of networks. Both would have the same transport characteristics, although the rates would vary depending on the distances between water source in soil, and water sinks in the plant root. I will initially use the Friese and Allen (1991) model for this discussion, as it is discrete. The Bago et al. (1998) model adds another level of complexity that requires additional modeling work.

In the Friese and Allen model, a hypha penetrates the root forming an infection unit. That hypha is approximately 10 μm in diameter and extends approximately

Fig. 2 Shifting importance of mycorrhizal fungal hyphae as soil dries, draining macropores, and then successively smaller pores. As the soil dries $\psi = -\gamma r^{-1}$. Where r is the pore diameter. In dry soils, for a plant, $r = \infty$, for a fungus, $r = n, <\infty$. This reduces the gradient required for water extraction

0.5 cm into the soil. AM hyphae branch dichotomously forming two slightly smaller hyphal lengths (8 μm in diameter). Those two hyphae again branch forming four slightly smaller hyphal lengths (6 μm). This process continues until a limit, potentially resulting from internal size limits to expansion. This appears to be somewhat less than 2 μm, at about eight branching orders. In a model system, each segment length is about 0.5 cm in length. Thus, the hyphal numbers can be modeled using a geometric series, where hyphal length (h)

$$h = a(1) + a(2) \ldots + a(128). \quad (2)$$

Because this is a geometric series, we can model the length based on the branch ratio (=2), and the number of branches (1–8). Thus,

$$h = \sum_{1}^{n} [ar^{n} - 1)] / (r - 1), \quad (3)$$

and assuming that the number of branches in limited to 8 by C flow, h = 122 cm per entry point (Allen et al. 2003).

As hyphae grow from one pore to the next, they cross air-filled pores. Older hyphal wall layers are hydrophobic (Allen 2006). What this means is that the water picked up in a micropore by a hydrophilic tip can be transported apoplastically

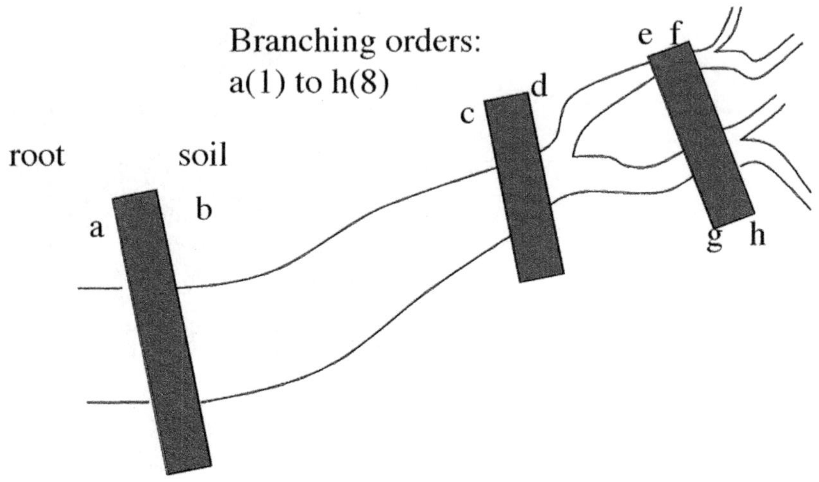

Fig. 3 Hyphal pipe of AM fungi. Water flux from *b* to *a* must equal from *d* to *c*, and the sum of *f* to *e* plus *h* to *g*. This extends to eight branching orders (see text for details)

between the membrane and the walls, or symplastically through a linear hypha without crossing a membrane. This structure dramatically reduces the tortuosity (Γ) of water flow from a relatively hydrated region to the root surface, where water is depleted (Allen 2007). Several of these individual hyphae can even wrap around each other, forming a lattice of hyphae or, in the case of EM, rhizomorphs, where additional water can flow apoplastically (Duddridge et al. 1980; Allen 2007).

Water flow can be modeled where the flow into the root must equal the flows throughout the network. That is, the fine tips must provide as much water as is transferred to the host plant. This can be visualized as a pipe (Fig. 3), where the flow from *b* to *a* must equal *d* to *c*, and must equal the combined amount of *h* to *g*, plus *f* to *e*. In this case, the flux (J)

$$J_r == \int_a^b \psi A ds = \int_c^d \psi A ds = \int_g^h \psi A ds + \int_e^f \psi A ds \quad (4)$$

If the tips are equally hydrated, then:

$$\mathrm{Jr} = \int_{a1}^{a2} \psi A_1 ds = 128 * \int_{h1}^{h2} \psi A_h ds, \quad (5)$$

and if they are different, then

$$J_r = \int_{a1}^{a2} \psi A_1 ds = \sum_1^{128} \int_{h1}^{h2} \psi A_h ds \quad (6)$$

The argument has been made that tips cannot provide an adequate amount of water to support the flow of water into the plant entry point. But, there are many hyphal tips. And, even as the flux rate per tip declines precipitously (as a single hypha gets smaller to the power of 4), the numbers are increasing dramatically (geometrically increasing with each branching). If the hyphal tips are in water, then the flux at the collective tips can equal that of the root surface (Fig. 4), based on the model above.

It is important to understand how fungi grow through the soil. Multiple hyphae parallel each other, often wrapped, through the soil matrix. This is important in that the extraction of water from a pore has been modeled as flow between parallel plates, across a filled duct or pore (Warrick 2003). In this case, the flux:

$$J = \alpha - L^2/\eta^{*dp}/dx, \tag{7}$$

where L is the length, η is viscosity, p is the pressure across a distance (x). The proportionality is regulated by the shape of the pore and interpore. As water is extracted, the conductivity increases in complexity based on the shapes of the water droplets and air pockets. Water diminishes to a thick film, a thin film, and finally contracts into the corners of soil pores. Water flow in unsaturated soil can then be modeled as a suite of bundles or viscous strings from a saturated soil pore to a plant root (Jury et al. 1991). When hyphae are present, instead of these water "bundles" bending and turning around each soil particle, they can coalesce around the hypha growing straight across the pore (Allen 2007). This may well be the mechanism

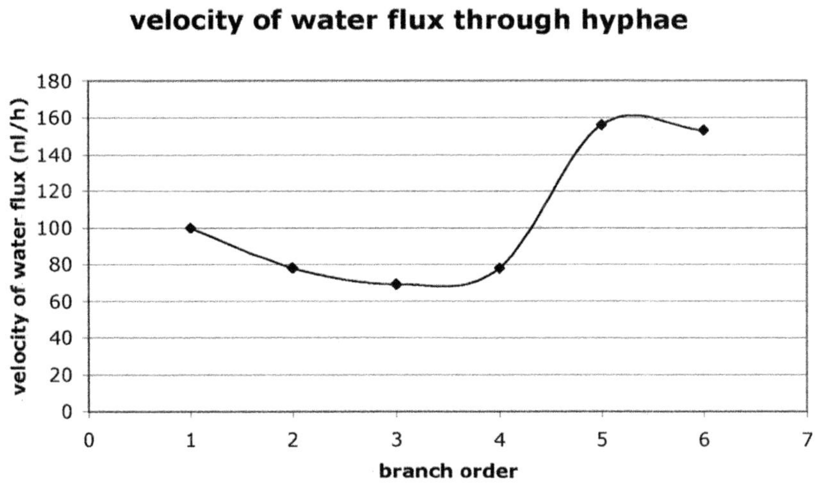

Fig. 4 Water flux through cumulative branching orders of the hyphal network. The assumptions are that the coenocytic hyphal network acts as a pipe from the sum of the 64 hyphal tips in the six branches, to the single hypha (branch order 1) entering the root. The model is based on Friese and Allen (1991) refined by Allen et al. (2003)

whereby plants access water within granite matrices below or between rooting zones (Allen 2006) in soils with low ψ_m.

The actual fluxes become more complex because a single infection point does not always remain a single branching network. Bago et al. (1998) illustrated a much more complex branching pattern consisting of networks of hyphae where there is a major arterial hypha and multiple branching networks emerged from the arterial hypha. In EM there is a range in variation. In some fungi, such as *Cenococcum* spp., the fungus forms a dense, but localized, hyphal network. In other fungi, such as *Pisolithus* spp., rhizomorphs form complex chords that extend for up to several m to find water patches, and even connect many individual plants (Allen 2007). The shear complexity of these observations clearly provides a rich modeling environment for future study.

5 Field Sensors and Tracking Dynamics

The current limitation to measuring mycorrhizal water transport and modeling hyphal flow is measuring soil properties, water, and mycorrhizal activity in the field at scales at which mycorrhizae actually function. This means observing hyphal growth and respiration in situ simultaneously with the environmental conditions at the same scale. We have begun to develop technologies and place them in the field observing mycorrhizal activity and measuring respiration and environmental (water, temperature, N) conditions. Some minirhizotrons enable observations of large hyphae in situ. Surprisingly, we were able to observe AM hyphae expanding rapidly at ψ_{soil} <-5 MPa, levels that should be below the growth capacity of the hyphae (Allen 2007). Across a 0.2-cm^2 window, the AM hyphal network was observed to expand from a root across the window in 2–days. This growth is well within the estimated expansion rate for a hyphal network postulated by Friese and Allen (1991) and Allen et al. (2003). We were also able to observe rhizomorphs expanding across a 2-cm^2 image across a 2-week period when ψ_{soil} <-5.0 MPa (Allen 2007). We postulated that at least part of this expansion was using hydraulically-lifted water, as the soil water potential was likely too low for active growth of these fungi (Querejeta et al. 2007b). Just as importantly, root, hyphal and rhizomorph growth initiated predominantly when the soils were wet, and soil temperatures were low.

We monitored different regions along a ridge to compare moisture and the potential for hydraulic lift in different vegetation types. This includes areas under a pine forest canopy and areas in meadows dominated by grasses and annual forbs. During one 2-month period, we observed that, during the summer drought, nearly under the ectomycorrhizal pines (which had roots reaching to the groundwater) surface water always accumulated by 1–2% soil moisture in the afternoons after stomata of the trees had closed. Deep water showed no diurnal change. After a monsoonal rainfall, the moisture accumulation was no longer observed. In the meadow, no diurnal pattern was observed, despite the fact that there was more deep water than under the pines. The increasing surface moisture under the pines could result from two possible mechanisms: moisture rising with heating and hydraulic redistribution. We do not have the isotopic evidence yet to distinguish among these mechanisms. However, additional data point to hydraulic redistribution, regulated by the roots and fungal hyphae. The relative numbers of roots, EM root tips, rhizomorphs, and coarse hyphae were all significantly greater under the pine canopy than in the meadow ($p < 0.001$), despite the dramatically lower soil moisture. Soil respiration remained high under the pine canopy, again, despite the low soil moisture. Finally, if soil temperature were the driver, I would have expected greater moisture movement from deeper soil layers in the meadow, which had a greater temperature gradient than the pine canopy. In other studies under an oak canopy, we found that oaks redistributed groundwater through the profile, whereas grassland soil had a much larger deviation from meteoric water indicating that it had been depleted by earlier evapo-transpiration (Querejeta et al. 2007b).

Our preliminary studies show interesting dynamics (Fig. 5). For example, the relationship between hyphal counts and soil moisture is not statistically significant, when monitored weekly through the growing season. However, there are distinct time periods where relationships appear. During the time when soil is cooling, but moisture increasing with melting snow, the hyphae show a curve-linear relationship with soil moisture, with a distinct lag. This lag appears to be caused by the cool temperatures, snow melt preceding the growth of the hyphae. Then, as the soil warms and dries out, the hyphae remain active until there is a precipitous drop as soil moisture drops below 9%, ($\psi_s < -2$ MPa). This is a classical hysteresis effect.

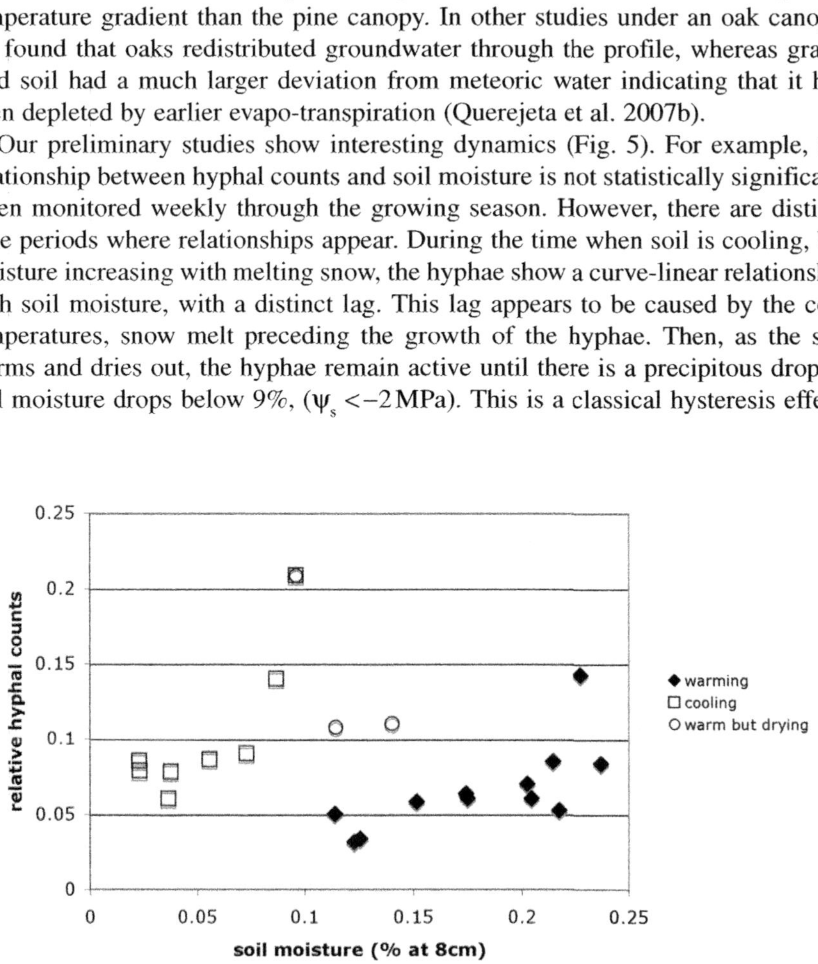

Fig. 5 Lags in relative counts of coarse hyphae (>8 μm). The *closed symbols* represent the hyphal counts as the soil warms, showing a lag in new hyphal production as θ increases, where hyphal counts $= 0.083 - 0.774\theta + 3.586\theta^2$, $r^2 = 0.522$, $p = 0.034$. The *open symbols* indicate the warming and drying period, again showing a lag with extensive hyphae remaining as the soil dries, where hyphal counts $= 0.06 + 0.816\theta - 2.088\theta^2$, $r^2 = 0.24$, $p = 0.03$. Note that the slopes between the two curves are opposite, indicating a hysteresis effect

Although in these images, we cannot determine saprobic from mycorrhizal hyphae, the size and shape of the hyphae coupled with our molecular analyses suggest that a large fraction are mycorrhizal (unpublished data).

Roots and mycorrhizal fungi do not respond linearly to changing conditions. Assessing root and microbial activity in the field is difficult at best. We have been using minirhizotron tubes as a window into field dynamics by coupling sensors to minirhizotron readings (Allen et al. 2007). Importantly, the growth and turnover dynamics of hyphae occur at daily time frames, and there are numerous root and hyphal fragments to be followed. Automating root and fungal growth in the images is imperative if we are to characterize dynamics. Further, using arrays of sensors, we can measure processes at the scales at which they act. Each organism and process responds to events at independent time scale. There are lags, hysteresis responses, and daily dynamics that may be captured by using arrayed sensor technologies in the field (Vargas et al. 2008).

6 Synthesis and Conclusions

Mechanisms of action are often based on laboratory studies. Studies of mycorrhizal action are no exception. But, the actions may result from the experimental conditions imposed. Some of these are intended, such as temperature, watering regimes, nutrient additions. Others result from unanticipated responses. Factors such as limited rooting volume, vapor pressure deficits, and solar radiation can have major influences on the direction and change in water dynamics. These are often not accounted for in mechanistic studies. Even laboratory or glasshouse "experiments" at best only describe a range in potential behavior. One common outcome is no effect of inoculation; which can be from the experimental conditions imposed rather than the mechanism of interaction between host and fungus. Such a response is also clearly evident when roots become pot-bound.

Water does not move only symplastically through hyphae! Mycorrhizal impacts on plant growth are usually associated with increasing uptake of nutrients that do not move by mass flow caused by the increased hyphal surface area. This mechanism explains most of the response in moist soils where roots are in contact with soil moisture pockets. However, in dry soil, apoplastic flow results in a large fraction of the water movement. This mechanism only appears when the entire soil–plant–atmosphere continuum is intact and surface soils are below field capacity. That flow can occur through root vessels, through rhizomorphs, or along individual or groups of hyphae traversing soil pores. The flow also occurs in the direction dictated by water potential gradients.

An additional issue in studying mycorrhizae is that "mycorrhizal" responses are often extrapolated from a single plant–fungus–soil system. But, there are hundreds (AM) to thousands (EM) of fungal associates and thousands of plant partners, each with slightly different morphological structure and physiological capacity. In some cases, fungi are connected to multiple plants of the same or different species. In the field, a plant almost always supports multiple species of fungi. There do appear to be limits (Allen et al. 2002), but these associations are still diverse enough to make generalization about physiological interactions difficult. Combinations often act differently than individual species, and act differently when inoculated at different times (Allen et al. 2003).

While some interesting mechanisms can still be found in glasshouse or laboratory studies, I believe that the most important studies involve field experiments, modeling, field observations, and more modeling. Field experimental inoculations coupled with isotopic studies of water sources and use are essential to determine which interactions actually occur, as opposed to are possible. Observational studies provide interesting insights into system behavior and the role of mycorrhizae and are critical to understanding responses to events. We have worked to develop an arrayed network of observational platforms that are coupled with sensors (Hamilton et al. 2007; Allen et al. 2007). We can observe mycorrhizal roots and fungi in the field as they grow, respire, and die and decompose. Not surprisingly, there are distinct time periods when temperatures and moisture facilitate growth, and also turnover. They may or may not be coupled to nutrient release or water transport, factors that must be teased apart, then observed as a system. There are lags resulting in classical hysteresis behavior. Our ability to utilize these systems is only beginning.

Understanding the dynamics of water in mycorrhizal relationships, and in turn, mycorrhizal symbioses on plant, fungal and soil water is crucial to understand, predict, and manage agricultural and wildland resources. The simple glasshouse experiments have largely been done and explored. We have reached a point where we need to go beyond potential mechanisms to describing which organisms are involved, when the range of potential mechanisms are operational, and how all the various components interact to result in improved (or reduced) plant growth. The newer technologies of natural abundance isotope measurements, molecular assessments of participants and gene regulation, and sensor arrays describing detailed environmental dynamics are just beginning to allow us to take this integrative step.

Acknowledgements This work was funded by the US National Science Foundation numbers EF 0410408, CRR-0120778 and DEB 9981548, and the University of California Agricultural Experiment Station. I thank Edith Allen for helpful comments on the manuscript.

References

Alexopoulos CJ, Mims CW, Blackwell M (1996) Introductory mycology, 4th edition. Wiley, New York

Allen MF (1980) Physiological alterations associated with vesicular-arbuscular mycorrhizal infection of *Bouteloua gracilis*. Ph.D. Dissertation, University of Wyoming, Laramie

Allen MF (1982) Influence of vesicular-arbuscular mycorrhizae on water movement through *Bouteloua gracilis*. New Phytol 91: 191–196

Allen MF (1987) Reestablishment of mycorrhizae on Mount St. Helens: migration vectors. Trans Br Mycol Soc 88: 413–417

Allen MF (1996). The ecology of arbuscular mycorrhizae: A look back into the 20th century and a peek into the 21st centenary review article, British mycological society. Mycol Res 100: 769–782

Allen MF (2006). Water dynamics of mycorrhizas in arid soils. In: Gadd GM (ed) Fungi in biogeochemical cycles. Cambridge University Press, New York, pp 74–97

Allen MF (2007) Mycorrhizal fungi: highways for water and nutrients in arid soils. Vadose Zone 6: 291–297

Allen EB, Allen MF (1986) Water relations of xeric grasses in the field: interactions of mycorrhizas and competition. New Phytol 104: 559–571

Allen MF, Boosalis MG (1983) Effects of two species of vesicular-arbuscular mycorrhizal fungi on drought tolerance of winter wheat. New Phytol 93: 67–76

Allen EB, Cunningham GL (1983) Effects of vesicular-arbuscular mycorrhizae on *Distichlis spicata* under three salinity levels. New Phytol 93: 227–236

Allen MF, Moore TS Jr, Christensen M (1980) Phytohormone changes in *Bouteloua gracilis* infected by vesicular-arbuscular mycorrhizae: I. Cytokinin increases in the host plant. Can J Bot 58: 371–374

Allen MF, Smith WK, Moore TS Jr, Christensen M (1981) Comparative water relations and photosynthesis of mycorrhizal and nonmycorrhizal *Bouteloua gracilis* H.B.K. Lag ex Steud. New Phytol 88: 683–693

Allen MF, Moore TS Jr, Christensen M (1982) Phytohormone changes in *Bouteloua gracilis* infected by vesicular-arbuscular mycorrhizae: II. Altered levels of gibberellin-like substances and abscisic acid in the host plant. Can J Bot 60: 468–471

Allen MF, Allen EB, Stahl PD (1984) Differential niche response of *Bouteloua gracilis* and *Pascopyrum smithii* to VA mycorrhizae. Bull Torrey Bot Club 111: 361–365

Allen MF, Crisafulli C, Friese CF, Jeakins SL (1992) Re-formation of mycorrhizal symbioses on Mount St. Helens, 1980–1990: Interactions of rodents and mycorrhizal fungi. Mycol Res 98: 447–453

Allen MF, Lansing J, Allen EB (2002) The role of mycorrhizal fungi in composition and dynamics of plant communities: A scaling issue. Prog Bot 63: 344–367

Allen MF, Swenson W, Querejeta JL, Egerton-Warburton LM, Treseder KK (2003). Ecology of mycorrhizae: A conceptual framework for complex interactions among plants and fungi. Annu Rev Phytopath 41: 271–303

Allen MF, Vargas R, Graham EA, Swenson W, Hamilton M, Taggart M, Harmon TC, Tatko A, Rundel P, Fulkerson B, Estrin D (2007) Soil sensor technology: Life within a pixel. BioScience 57: 859–867

Augé RM (2001) Water relations, drought and vesicular-arbuscular mycorrhizal symbiosis. Mycorrhiza 11: 3–42

Augé RM, Schekel KA, Wample RL (1986) Greater leaf conductance of well watered VA mycorrhizal plants is not related to phosphorus nutrition. New Phytol 103: 107–116

Augé RM, Sylvia DM, Park S, Buttery BR, Saxton AM, Moore JL, Cho K (2004) Partitioning mycorrhizal influence on water relations of *Phaseolus vulgaris* into soil and plant components. Can J Bot 82: 503–514

Azcón-Aguilar C, Barea JM (1997) Mycorrhizal dependency of a representative plant species in Mediterranean shrublands (*Lavandula spica* L.) as a key factor to its use for revegetation strategies in desertification-threatened areas. Appl Soil Ecol 7: 83–92

Bago B, Azcon-Aguilar C, Piche Y (1998) Architecture and developmental dynamics of the external mycelium of the arbuscular mycorrhizal fungus *Glomus intraradices* grown under monoxenic conditions. Mycologia 90: 52–62

Bornyasz MA, Graham RC, Allen MF (2005). Ectomycorrhizae in a soil-weathered granitic bedrock regolith: linking matrix resources to plants. Geoderma 126: 141–160

Davies FT, Olalde-Portugal V, Aguilera-Gomez L, Alvarado MJ, Ferrera-Cerrato RC, Boutton TW (2002) Alleviation of drought stress of Chile ancho pepper (*Capsicum annuum* L. cv. San Luis) with arbuscular mycorrhiza indigenous to Mexico. Scientia Horticulturae 92: 347–359

Di JJ, Allen EB (1991) Physiological responses of six wheatgrass cultivars to mycorrhizae. J Range Manage 44: 336–341

Duddridge JA, Malibari, A, Read DJ (1980) Structure and function of mycorrhizal rhizomorphs with special reference to their role in water transport. Nature 287: 834–836
Egerton-Warburton LM, Graham RC, Hubbert KR (2003). Spatial variability in mycorrhizal hyphae and nutrient and water availability in a soil-weathered bedrock profile. Plant Soil 249: 331–342
Egerton-Warburton LM, Querejeta JI, Allen, MF (2007). Common mycorrhizal networks provide a potential pathway for the transfer of hydraulically lifted water between plants. J Exp Bot 58: 1473–1483
Egerton-Warburton LM, Querejeta JI, Allen MF (2008) Efflux of hydraulically lifted water from mycorrhizal fungal hyphae during imposed drought. Plant Signal Behav 3: 68–71
Friese CF, Allen MF (1991) The spread of VA mycorrhizal fungal hyphae in the soil: inoculum types and external hyphal architecture. Mycologia 83: 409–418
Gerdemann JW (1968) Vesicular-arbuscular mycorrhiza and plant growth. Annu Rev Phytopath 6: 397–418
Hamilton MP, Graham EA, Rundel PW, Allen MF, Kaiser W, Hansen MH, Estrin DL (2007) New approaches in embedded networked sensing for terrestrial ecological observatories. Environ Eng Sci 24: 192–204
Hardie K (1985) The effect of removal of extraradical hyphae on water uptake by vesicular-arbuscular mycorrhizal plants. New Phytol 101: 677–684
Hardie K, Leyton L (1981) The influence of vesicular-arbuscular mycorrhiza on growth and water relations of red-clover. 1. In phosphate deficient soil. New Phytol 89: 599–608
Hatch AB (1937) The physical basis of mycotrophy in *Pinus*. Black Rock Forest Bull 6: 1–168
Hickson LD (1993) The effects of vesicular-arbuscular mycorrhizal fungi on the light harvesting, gas exchange, and architecture of sagebrush (*Artemisia tridentata.* ssp. *tridentata*). M.S. Thesis, San Diego State University, San Diego, CA
Hobbie EA, Colpaert JV (2004) Nitrogen availability and mycorrhizal colonization influence water use efficiency and carbon isotope patterns in *Pinus sylvestris*. New Phytol 164: 515–525
Jury WA, Gardner WR, Gardner WH (1991) Soil physics, 5th edition. Wiley, New York
Khalvati MA, Hu Y, Mozafar A, Schmidhalter U (2005) Quantification of water uptake by arbuscular-mycorrhizal hyphae and its significance for leaf growth, water relations, and gas exchange of barley subjected to drought stress. Plant Biol 7: 706–712
Kucey RMN, Paul EA (1982) Carbon flow, photosynthesis, and N2 fixation in mycorrhizal and nodulated faba beans (*Vicia faba* L.). Soil Biol Biochem 14: 407–412
Marks GC, Kozlowski TT (1973) Ectomycorrhizae; their structure and function. Academic, New York
Mena-Violante HG, Ocampo-Jimenez O, Dendooven L, Martinez-Soto G, Gonzalez-Castaneda J, Davis FT, Olalde-Portugal V (2006) Arbuscular mycorrhizal fungi enhance fruit growth and quality of chile ancho (*Capsicum annuum* L. cv San Luis) plants exposed to drought. Mycorrhiza 16: 261–267
Mexal J, Reid CPP (1973) The growth of selected mycorrhizal fungi in response to induced water stress. Can J Bot 51: 1579–1588
Mosse B (1973) Advances in study of arbuscular-arbuscular mycorrhiza. Annu Rev Phytopath 11: 171–196
Nobel PS (1974) Biophysical plant physiology. Freeman Press, New York
Querejeta JI, Barea JM, Allen MF, Caravaca F, Roldan A (2003a) Differential response of $\delta^{13}C$ and water use efficiency to arbuscular mycorrhizal infection in two aridland woody plant species. Oecologia 135: 510–515
Querejeta JI, Egerton-Warburton LM, Allen MF (2003b) Direct nocturnal water transfer from oaks to their mycorrhizal symbionts during severe soil drying. Oecologia 134: 55–64
Querejeta JI, Allen MF, Caravaca F, Roldan A (2006). Differential modulation of host plant $\delta^{13}C$ and $\delta^{18}O$ by native and nonnative arbuscular mycorrhizal fungi in a semiarid environment. New Phytol 169: 379–387
Querejeta JI, Allen MF, Alguacil MM, Roldan A (2007a) Plant isotopic composition provides insight into mechanisms underlying growth stimulation by AM fungi in a semiarid environment. Funct Plant Biol 34: 683–691

Querejeta JI, Egerton-Warburton LM, Allen MF (2007b) Hydraulic lift may buffer rhizosphere hyphae against the negative effects of severe soil drying in a California oak savanna. Soil Biol Biochem 39: 409–417

Reid CPP, Woods FW (1969) Translocation of C14-labeled compounds in mycorrhizae and its implications in interplant nutrient cycling. Ecology 50: 179–181

Rillig MC, Mummey DL (2006) Mycorrhizas and soil structure. New Phytol 171: 41–53

Ruiz-Lozano JM, Porcel R, Aroca R (2006) Does the enhanced tolerance of arbuscular mycorrhizal plants to water deficit involve modulation of drought-induced genes? New Phytol 171: 693–698

Safir GR, Boyer JS, Gerdemann JW (1971) Mycorrhizal enhancement of water transport in soybeans. Science 172: 581–583

Safir GR, Boyer JS, Gerdemann JW (1972) Nutrient status and mycorrhizal enhancement of water transport in soybeans. Plant Physiol 49: 700–703

Stahl E (1900). Der Sinn der Mycorrhizenbildung. Jahrbücher für wissenschaftliche Botanik 34: 539–668

Stolp H (1988) Microbial ecology: Organisms, habitats, activities. Cambridge University Press, New York

Valentine AJ, Mortimer PE, Lintnaar A, Borgo R (2006) Drought responses of arbuscular mycorrhizal grapevines. Symbiosis 41: 127–133

Vargas R, Allen MF, Allen EB (2008) Biomass and carbon accumulation in a fire chronosequence of a seasonally dry tropical forest. Global Change Biol 14: 109–124

Warrick AW (2003) Soil water dynamics. Oxford University Press, New York

Wu QS, Xia RX (2006) Arbuscular mycorrhizal fungi influence growth, osmotic adjustment and photosynthesis of citrus under well-watered and water stress conditions. J Plant Physiol 163: 417–425

Index

GPSR Compliance

The European Union's (EU) General Product Safety Regulation (GPSR) is a set of rules that requires consumer products to be safe and our obligations to ensure this.

If you have any concerns about our products, you can contact us on ProductSafety@springernature.com

In case Publisher is established outside the EU, the EU authorized representative is:

Springer Nature Customer Service Center GmbH
Europaplatz 3
69115 Heidelberg, Germany

Batch number: 10755053

Printed by Printforce, the Netherlands